AF805301

Optical Spectroscopic and Microscopic Techniques

Harekrushna Sahoo
Editor

Optical Spectroscopic and Microscopic Techniques

Analysis of Biological Molecules

Springer

Editor
Harekrushna Sahoo
Biophysical and Protein Chemistry
Laboratory, Department of Chemistry
National Institute of Technology (NIT)
Rourkela
Rourkela, Odisha, India

Center of Nanomaterials
National Institute of Technology (NIT)
Rourkela
Rourkela, Odisha, India

ISBN 978-981-16-4549-5 ISBN 978-981-16-4550-1 (eBook)
https://doi.org/10.1007/978-981-16-4550-1

© The Editor(s) (if applicable) and The Author(s), under exclusive license to Springer Nature Singapore Pte Ltd. 2022
This work is subject to copyright. All rights are solely and exclusively licensed by the Publisher, whether the whole or part of the material is concerned, specifically the rights of translation, reprinting, reuse of illustrations, recitation, broadcasting, reproduction on microfilms or in any other physical way, and transmission or information storage and retrieval, electronic adaptation, computer software, or by similar or dissimilar methodology now known or hereafter developed.
The use of general descriptive names, registered names, trademarks, service marks, etc. in this publication does not imply, even in the absence of a specific statement, that such names are exempt from the relevant protective laws and regulations and therefore free for general use.
The publisher, the authors, and the editors are safe to assume that the advice and information in this book are believed to be true and accurate at the date of publication. Neither the publisher nor the authors or the editors give a warranty, expressed or implied, with respect to the material contained herein or for any errors or omissions that may have been made. The publisher remains neutral with regard to jurisdictional claims in published maps and institutional affiliations.

This Springer imprint is published by the registered company Springer Nature Singapore Pte Ltd.
The registered company address is: 152 Beach Road, #21-01/04 Gateway East, Singapore 189721, Singapore

Contents

1 **Absorption Spectroscopy: What Can We Learn About Conformational Changes of Biomolecules?** 1
Manali Basu and Padmaja Prasad Mishra

2 **Circular Dichroism Spectroscopy: Principle and Application** 19
Suchismita Subadini, Pratyush Ranjan Hota, Devi Prasanna Behera, and Harekrushna Sahoo

3 **Steady-State Fluorescence Spectroscopy as a Tool to Monitor Protein/Ligand Interactions** 35
Roopa Kenoth, Balamurali M. M., and Ravi Kanth Kamlekar

4 **Fluorescence Anisotropy: Probing Rotational Dynamics of Biomolecules** 55
Gourab Prasad Pattnaik and Hirak Chakraborty

5 **Fluorescence Lifetime: A Multifaceted Tool for Exploring Biological Systems** 77
Subhrajit Mohanty and Usharani Subuddhi

6 **From Ensemble FRET to Single-Molecule Imaging: Monitoring Individual Cellular Machinery in Action** 113
Farhana Islam, Manali Basu, and Padmaja Prasad Mishra

7 **Nanosecond Time-Resolved Fluorescence Assays** 143
Yan-Cen Liu and Andreas Hennig

8 **Fluorescence Correlation Spectroscopy: A Highly Sensitive Tool for Probing Intracellular Molecular Dynamics and Disease Diagnosis** 177
Bikash Chandra Swain, Anand Kant Das, Janmejaya Rout, Shrutidhara Biswas, and Umakanta Tripathy

9 **Principles and Applications of Fluorescence Microscopy** 197
Bibhu Ranjan Sarangi

10 Analysis of Biomolecular Dynamics Under Fourier Transform Infrared Spectroscopy 215
Sanjeev Kumar Paikra and Monalisa Mishra

11 Raman Spectroscopy in Biology: Perspectives and Emerging Frontiers 243
Pradeep K. Sengupta

About the Editor

Harekrushna Sahoo is currently working as an Associate Professor of Chemistry at the NIT (National Institute of Technology) in Rourkela, India. He previously served as a guest scientist at the Max-Bergmann Center (Dresden, Germany). He completed his Ph.D. at Jacobs University Bremen (Bremen, Germany) in 2006, prior to engaging in postdoctoral research at the University of Massachusetts (Amherst, USA) and Technical University Dresden (Dresden, Germany). His research chiefly focuses on using biophysical chemistry to understand the kinetics and dynamics of various types of proteins, along with the impacts of environmental stress. He has served as a reviewer for a number of international peer-reviewed and reputed journals, including *Journal of Physical Chemistry B*, *ACS Omega*, *ACS Sustainable Chemistry and Engineering*, *Journal of Photochemistry and Photobiology*, and *International Journal of Biological Macromolecules*. He has more than 9 years of teaching experience in General Chemistry, Physical Chemistry, Supramolecular Chemistry, Structure and Function of Biomolecules, Biophysical Chemistry, and Optical Spectroscopy.

He has also published more than 50 research articles in peer-reviewed international journals and authored or coauthored several book chapters. He is a member of many international scientific societies and organizations, e.g., American Chemical Society, Biophysical Society, Indian Photobiology Society, and Orissa Chemical Society.

Absorption Spectroscopy: What Can We Learn About Conformational Changes of Biomolecules?

1

Manali Basu and Padmaja Prasad Mishra

Abstract

The quantification of interaction between electromagnetic radiation and matter serves as an effective tool in the characterization of materials in order to identify and quantify specific substances. Absorption spectroscopy is based on the phenomenon of wavelength-dependent absorption, that is, the attenuation of radiation intensity when passed through solution containing sample in turn helps to quantify the concentration as well as the nature of substances present within the sample.

1.1 Introduction

Since the advent of researches on atoms and molecules, there has been an enormous curiosity among chemists, physicists, and other scientists about the structure of molecules; nonetheless, the infinitesimal dimension of molecules has posed to be a major setback in visualizing their structure. However, the initial comprehension of molecular structure could be indirectly derived from the technique known as **spectroscopy**. Spectroscopy deals with the precise measurement of the molecular interactions with electromagnetic waves leading to the transitions between energy levels upon absorption of suitable radiations dictated by the quantum mechanical selection rules [1]. The energy (frequency to be precise) of electromagnetic radiation is used to promote electrons to an excited state from the ground state when the energy of photons is comparable to the difference in energy between the two molecular energy levels. A spectrum thus obtained is measured as a function of its

M. Basu · P. P. Mishra (✉)
Chemical Sciences Division, Saha Institute of Nuclear Physics, Kolkata, West Bengal, India

HBNI, Mumbai, India
e-mail: padmaja.mishra@saha.ac.in

© The Author(s), under exclusive license to Springer Nature Singapore Pte Ltd. 2022

H. Sahoo (ed.), *Optical Spectroscopic and Microscopic Techniques*,
https://doi.org/10.1007/978-981-16-4550-1_1

wavelength or frequency and is termed as absorption spectrum. The origin of spectral lines in molecular spectroscopy not only arises due to absorption of photons (e.g. UV-Vis, IR, NMR), emission (luminescence and fluorescence), and scattering of photons (Raman spectroscopy) but also actively contributed due to changes in the energy level of molecules. However, in this chapter, we will be focusing on absorption, namely UV-Vis spectroscopy in particular.

1.2 Origin of UV/Vis Spectrum

Visible light forms a small part of electromagnetic radiation (Fig. 1.1) comprising of oscillating electric and magnetic fields that are mutually perpendicular to each other, whose energy "*E*" is given by:

$$E = h\vartheta = hc/\lambda \tag{1.1}$$

where h, c, λ, and ϑ are Planck's constant, speed, wavelength, and frequency of the light. If the energy of a light photon is the same as the difference between the energy of the ground and the excited state, then the molecule absorbs the photon. The redundant energy may be released in the form of a photon with lower energy (fluorescence or phosphorescence) or as radiation of heat. Occasionally, it ends up inducing chemical changes in the absorbing molecule. The symmetry of molecules and the selection rule tunes the probability of the absorption of a photon of a given energy. This further depends on the difference between the energy configuration of the molecular orbitals and the next highest energetically allowed electronic configuration. The commonly observed low-energy electronic transitions of molecules are $n \rightarrow \pi^*$ and $\pi \rightarrow \pi^*$ type (where * represents an excited state), although $\sigma \rightarrow \sigma *$ and $n \rightarrow \sigma *$ transition occur as higher-energy transitions falling in the lower wavelength region of spectra (Fig. 1.2.).

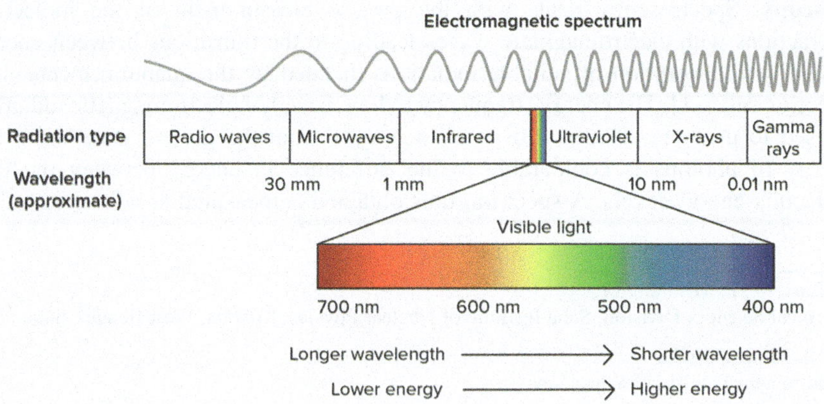

Fig. 1.1 Comparison of energy, frequency, and wavelength of the electromagnetic spectrum

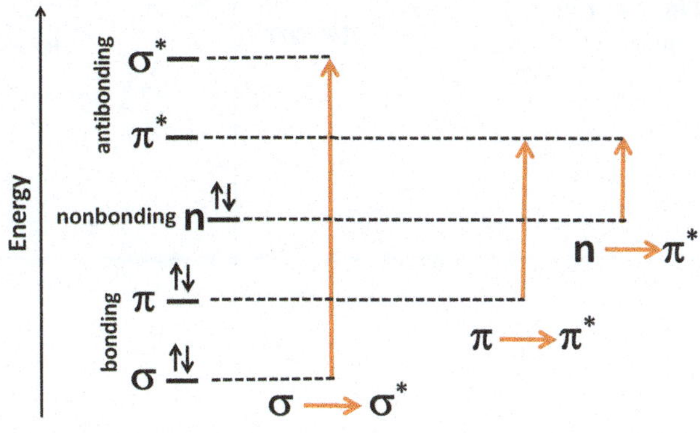

Fig. 1.2 Schematic representation of all the possible electronic transition of any molecules

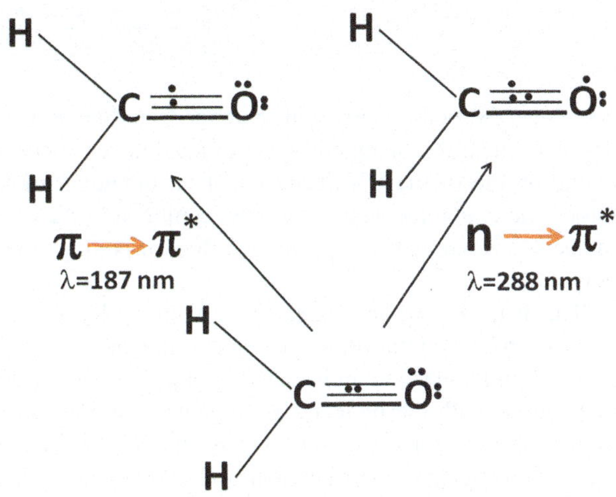

Fig. 1.3 Possible electronic transitions in case formaldehyde

In general, aromatic systems with delocalized electrons absorb light in the visible (400–800 nm) or the near-UV (150–400 nm) region. If the electronic configurations of molecules contain bonding and nonbonding electrons, they are readily excitable at optical wavelengths. For instance, heteroatomic molecular systems like carbonyls have both π bonding electrons (C=O) and nonbonding (n) electrons at the O atom (Fig. 1.3.). Although both of these types of electrons can be excited, they certainly absorb at different wavelengths that could be used to distinguish the nature of transitions.

However, the extent of conjugation is one of the pivotal factors affecting the wavelength of absorption by a molecule. Molecules having extended conjugation are defined as a system having connected p orbital with delocalized π electrons, i.e., containing alternating double and single bonds, which are stable compared to their nonconjugated counterparts. The bonding orbitals of organic molecules are almost always filled (HOMO), and the antibonding orbitals are usually always empty (LUMO). Greater is the extent of conjugation of molecules, less energy is required to transfer the electrons, and thus the energy difference between HOMO and LUMO

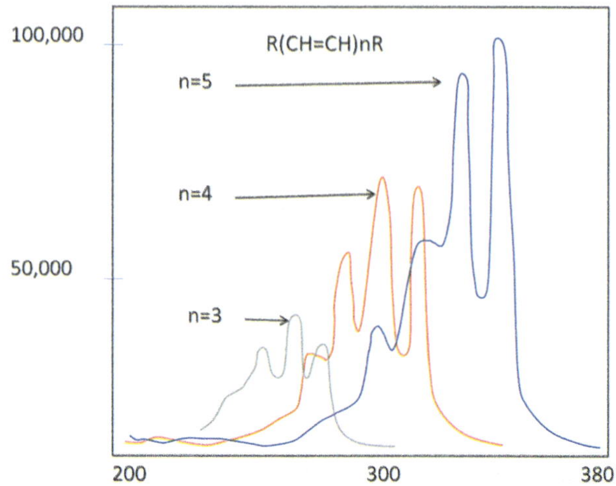

Fig. 1.4 Effect of conjugation on λ_{max}

decreases. So compounds with extended conjugation usually absorb in the visible region of the electromagnetic spectrum and hence appear colored. Thus, investigation of the profile of the UV spectrum of an unknown molecule hints at the structural information and the nature of any conjugated π electron systems present in the molecule. Figure 1.4 represents the effect of conjugation on the electronic spectra of poly-alkene.

The distinct band that appears in the ultraviolet spectra displays the discrete nature of these electronic transitions. However, the energy associated with different orbitals varies slightly due to vibrations and rotations about covalent bonds and also due to the interactions with the molecule's immediate environment. This arises due to several vibrational and rotational states corresponding to both the ground and the excited states. Nevertheless, the transition not only occurs from a particular energy level to another. Rather, a combination of probable transitions between various vibrational and rotational states ends up in the broadening of the absorption bands. The output of the distribution of bond energy appears wide with a Gaussian shape for the majority of the absorption bands of molecules in the solution. The dissemination of these bands increases if the populations of similar types of chromophores experience different microenvironmental effects. Thus, in the solution phase (especially in polar solvents), absorption spectra of molecules are broad. The substructure in the band is usually caused due to distinct vibrational transitions or unique microenvironments.

1.3 Lambert–Beer's Law

Ideally, when a bunch of light rays having intensity I_0 pass through a transparent solution containing chromophore molecules, it experiences a decrease in intensity (I). This decrease in the intensity is directly proportional to the concentration of the chromophore and (c) and the width of the sample layer (dl)

$$-\mathrm{d}(I/I_0) = kc\,\mathrm{d}l \tag{1.2}$$

K being a constant.

Integration of above results,

$$\ln{(I_0/I)} = kcl \tag{1.3}$$

Thus, the absorbance (*A*) is defined as:

$$A = \log{(I_0/I)} = \epsilon cl \tag{1.4}$$

where ϵ is the molar extinction coefficient, *c* is molar concentration, and l is path length in cm. The above relation is known as the Beer–Lambert law. log (I_0/I) is referred to as optical density (OD) if the processes involve additional phenomena like light scattering, etc., other than pure absorption.

Divergence from this behavior often occurs under the following conditions:

(a) For a sample having high absorbance (i.e. >1.5).
(b) If the sizes of the molecules and the wavelength of the incident light in solution are equivalent, a significant portion of the incident light beam is scattered. However, for larger particles, the chromophores masks begin to optically obscure one another, giving rise to the phenomenon known as absorption flattening.
(c) For highly fluorescent samples.
(d) For chemically and photochemically unstable samples.

Ultraviolet spectrophotometry has been used as a popular detection technique for studying the effect of the microenvironment and structural alternation due to interaction with other molecules during protein folding and structural integrity due to its fast, accurate, quantitative, and nondestructive nature. Additionally, it provides some unique features that have resulted in a moderate resurgence in its use in certain applications.

1.4 Instrumentation of UV Spectroscopy

The principle of measurement of the absorption spectrum using a UV spectrophotometer is relatively a straightforward method. The spectrometer measures the transmittance or absorbance of the incident light by a sample as a function of the wavelength. The five basic components of all spectrophotometers are (Fig. 1.5):

(a) Source of radiation
(b) Monochromator
(c) Sample chamber

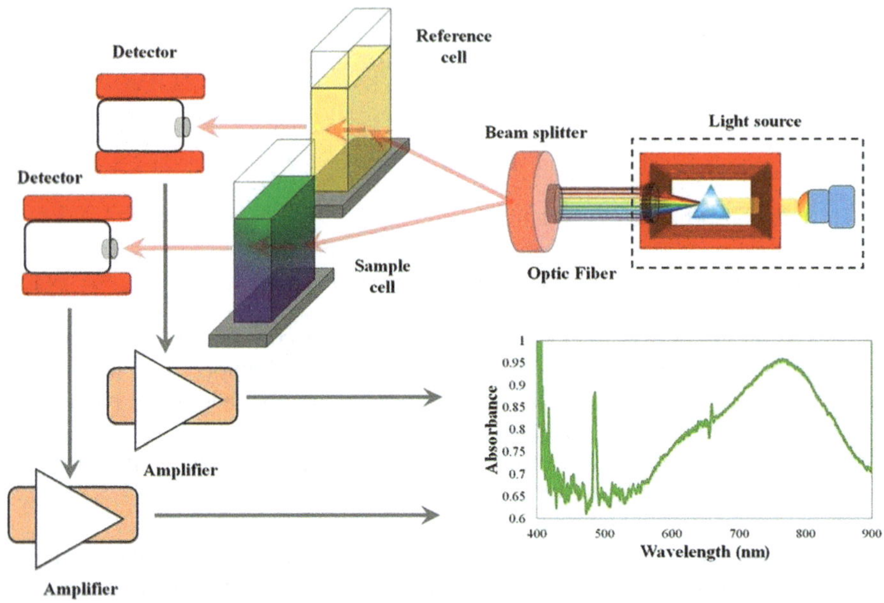

Fig. 1.5 Experimental setup in a spectrophotometer. (Image source: https://doi.org/10.1149/2.0601805jes)

(d) Detector
(e) Data analysis system

(a) **Source of radiation:** The most widely employed light sources are tungsten filament lamps and Hydrogen-Deuterium lamps, which fall in the UV region. The intensity of tungsten filament lamps falls in the red region of radiations; they emit radiations of 375 nm, while the intensity of Hydrogen-Deuterium lamps falls below 375 nm.
(b) **Monochromator:** A monochromator is an optomechanical component for measurement of the light spectrum. Composed of prisms and slits, they generally act as a double-beam spectrophotometer. The main components of monochromator are as follows:
   - An entrance slit
   - A collimating lens
   - A dispensing device (usually a prism or a grating)
   - A focusing lens
   - An exit slit

   The incident light beam is usually focused on the input slit and subsequently diffracted through a grating, thus separating light waves according to their wavelengths. In this way, only one color is transmitted through the output slit at a given time. Spectra are then selected wavelength by wavelength, changing the angle of the grating so that a series of continuously increasing wavelengths pass through the slits for recording purposes. In a double-beam spectrophotometer, the output beam is further divided into two beams with the help of another prism and sent to the sample side and the reference side simultaneously.

(c) **Sample and reference cells:** One out of the two divided beams is passed through the cuvette containing sample solution, and the second beam is passed through the cuvette with the reference solution only. The cuvettes are usually made from silica or quartz materials.

(d) **Detector and amplifier:** The photomultiplier tube (PMT), consisting of the photoemissive cathode (a cathode that emits electrons when struck by photons of radiation), several dynodes (emitting several electrons for each electron striking them), and an anode, is the commonly used detector in UV-Vis spectrophotometer. The instrument has two PMTs, one of which receives the beam from the sample cell, while the second one receives the beam from the reference. The radiation from the reference cell is stronger than that of the sample cell, resulting in the generation of pulsating or alternating currents in the PMTs. This alternating current is then transferred to the amplifier, connected with a small servometer. As a low-intensity current is generated in the PMTs, the primary purpose of the amplifier is to amplify the signals many times so we can get clear and recordable signals.

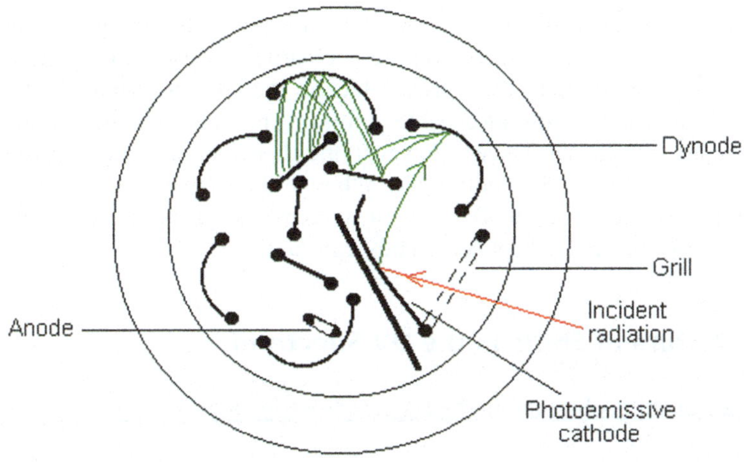

Cross-section of a PMT

(e) **Display systems:** The amplifier is coupled to a pen recorder of a computer or a visual unit of the computer that displays the spectrum. The computer stores all the generated data, and the spectrum of the desired compound is produced.

1.5 Chromophores and Auxochromes

Chromophore: A molecule with a covalently bonded group and absorbs light in the UV-Vis region (200–800 nm) is known as a chromophore. Chromophores are primarily divided into two categories:

- Those containing p electrons and capable of undergoing $\pi \rightarrow \pi*$ transitions, for example, ethylenes and acetylenes.
- Those containing both p and nonbonding electrons thus capable of undergoing two types of transitions; π to π^* and nonbonding (n) to π^*. For example, carbonyl, nitriles, azo compounds, nitro compounds, etc. A table of some characteristic UV absorptions for various chromophores is given in Table 1.1.

Auxochromes: An auxochrome is any group associated with the chromophore, which does not itself act as a chromophore but induces a shift of the absorption band toward the longer wavelength of the spectrum. –OH, –OR, –NH_2, –NHR, –SH, etc. are examples of auxochromic groups.

1.6 Factors Affecting Absorption Spectra

The position, as well as the intensity of UV/VIS bands in spectra, is governed by a variety of factors. Nature of the solvent (e.g. polarity, viscosity, pH, etc.), temperature, and concentration are the causes affecting the spectral behavior [2, 3]. For example, water and alcohols are capable of forming hydrogen bonds, resulting in the shift of bands of polar substances. However, the change in the band intensity due to change in temperature is less noticeable, although the simple thermal expansion of the solution may be sufficient. To render this effect negligible, the sample chambers of spectrophotometers are kept sufficiently constant.

1.7 Nature of Shifts in the UV Spectrum

Four types of shifts observed in the UV spectroscopy are:

Table 1.1 List of some common chromophores and their light absorption characteristics

Chromophore	Example	Excitation	λ_{max}, nm	ε	Solvent
C=C	Ethene	$\pi \rightarrow \pi^*$	171	15,000	Hexane
C≡C	1-Hexyne	$\pi \rightarrow \pi^*$	180	10,000	Hexane
C=O	Ethanol	$n \rightarrow \pi^*$	290	15	Hexane
		$\pi \rightarrow \pi^*$	180	10,000	Hexane
N=O	Nitromethane	$n \rightarrow \pi^*$	275	17	Ethanol
		$\pi \rightarrow \pi^*$	200	5000	Ethanol
C–X, X = Br, I	Methyl bromide	$n \rightarrow \sigma^*$	200	200	Hexane
	Methyl iodide	$n \rightarrow \sigma^*$	255	360	Hexane

Fig. 1.6 Schematic representation of the shifts in wavelength and absorbance in UV-Vis spectrophotometry

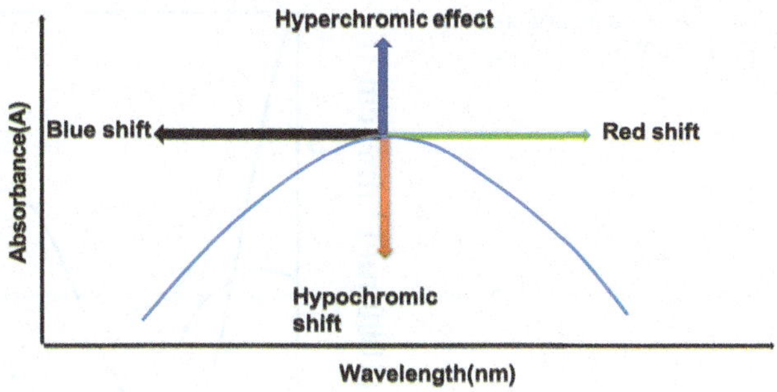

(a) *Bathochromic effect*: Also known as the redshift in wavelength, the bathochromic shift is an effect under which the absorption maximum is shifted toward the longer wavelength as a result of the presence of an auxochrome or change in solvents (Fig. 1.6). For example, the n → π* transition of carbonyl compounds.
(b) *Hypsochromic effect*: The hypsochromic shift refers to the shift in absorption maximum toward the shorter wavelength (Fig. 1.6). It is commonly referred to as the blue shift. This happens if the conjugation is removed or the polarity of the solvents is altered due to some reason.
(c) *Hyperchromic effect*: It refers to the increase in the intensity of absorption maximum, generally due to the introduction of an auxochrome in the compound (Fig. 1.6).
(d) *Hypochromic effect*: If the geometry of the molecule is distorted due to the introduction of a new group, thus resulting in a decrease in intensity of absorption maximum, this is called as Hypochromic effect (Fig. 1.6).

Scientific Term	Nature of shift
Bathochromic shift(redshift)	Toward longer wavelength
Hypochromic shift(blue shift)	Toward shorter wavelength
Hyperchromic effect	Toward longer absorbance
Hypochromic effect	Toward shorter absorbance

1.8 Fundamental Absorption Characteristics of Biomolecules

1.8.1 Characteristic UV-Vis Spectra of Nucleic Acid Bases

Since all the nucleotide bases contain aromatic conjugated systems, they possess their characteristic electronic spectrum. The spectrum toward the easily accessible region of the spectrum (down to 180 nm) is due to π → π* transitions. They are all polarized in the plane of the bases and so approximately perpendicular to the helix axis in B-DNA [4]. The absorbance spectra of guanine (G), adenine (A), thymine

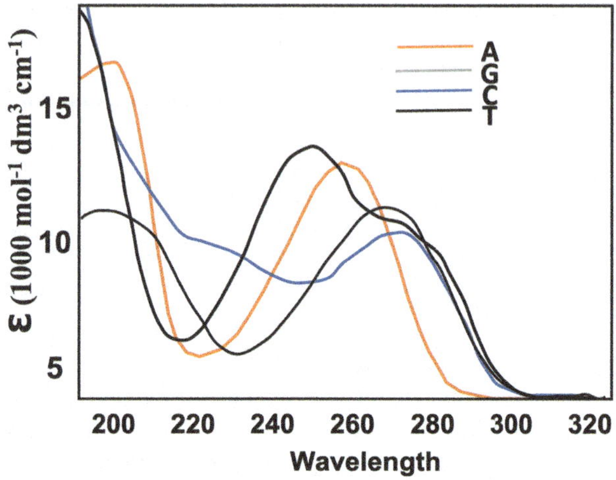

Fig. 1.7 Absorption spectra of four nucleic acid bases

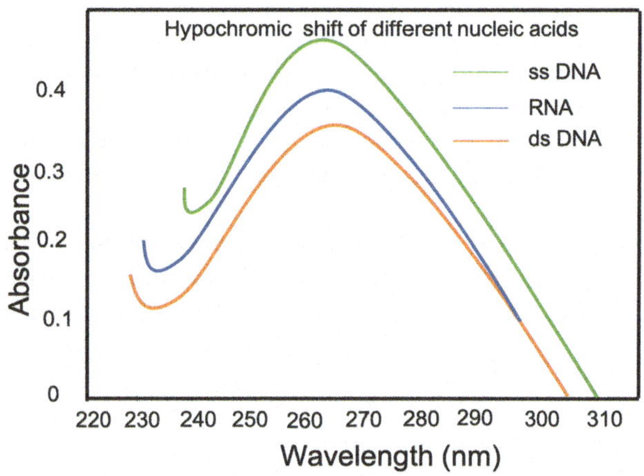

Fig. 1.8 Absorption spectra of DNA and RNA

(T), and cytosine (C) and the current best estimates of their transition polarizations are represented in Fig. 1.7.

1.8.2 Electronic Spectroscopy of DNA and RNA

Due to exclusive transitions of the planar purine and pyrimidine bases in DNA and RNA, both possess a strong absorbance in the range between 200 and 300 nm, predominately due to $\pi \rightarrow \pi^*$ transitions of the bases (Fig. 1.8) [5, 6]. Nitrogen purging is required to access the spectrum below 200 nm, as the absorption oxygen in this region interferes with the spectrum. Though the absorbance spectra of the bases (Fig. 1.7) possess two bands, each of the bands is a combination of multiple transitions. This makes the exhaustive analysis of the absorption spectrum bit complicated. However, the nature of the spectrum is very sensitive to any perturbation in the structure or microenvironment, making this tool a useful qualitative or empirical probe. The π–π stacking interactions significantly perturb the base transitions, and hence, the nucleotide sequence as well as the structure of the nucleic acid affects the wavelength maxima and transition intensities.

1.8.3 Nucleic Acid Denaturation

Hydrophobic and van der Waals interactions between the aromatic rings act as the driving forces to hold both of the strands of a DNA or a double-strand RNA [7]. Heat is one of the factors to break these weak interactions and denature the duplex structure, known as "melting". This is reflected by an increase in absorbance near 260 nm. Plotting absorbance as a function of temperature gives a sigmoidal "melting" curve, and the temperature corresponding to the 50% of the initial absorbance is known as the "melting temperature" of the double-helical nucleic acid, which is well studied using this tool. The melting temperature depends on the nature of the solvent as well as the ratio of the nucleotide pairs comprising the helical structure [8]. The repulsive force between the negatively charged phosphate backbone lowers the stability of the duplex and is progressively screened when counter ions (typically in the form of NaCl) are added in increasing concentrations. The stability of the double helix increases with the increase in salt concentration. Furthermore, the melting temperature of a duplex with more G–C content is higher than that with more A–T content. This is simply due to the existence of three H bonds in G–C base pairs, whereas A–T pairs have only two hydrogen bonds. This factor governs the melting temperature, too. The G–C content of DNA can be determined from its melting temperature. However, duplexes with multiple regions of G–C content have a complex melting temperature profile encompassing several melting points.

1.8.4 Electronic Spectroscopy of Proteins

The observed UV/visible spectrum in the case of peptides and proteins is determined by the spectroscopy of the amide bonds, the side chains, and any prosthetic groups (such as hemes) [9]. The UV absorption spectra of proteins are categorized into two groups, namely the "far" and "near" UV regions. The far UV (<250) region of the spectra is primarily by transitions of the peptide backbone within the molecule, although transitions from some side chains also contribute to this region [9]. However, the near UV regions, also described as the aromatic region, fall within 250–300 nm and originate mainly due to the transitions of disulfide bonds (cysteines) present within the biomolecules.

Tryptophan, tyrosine, and phenylalanine show absorbance in the near-UV region of the spectrum. The tryptophan has an absorption band in 240–290 nm region that tributes to three distinct transitions of the indole side chain, with a maximum extinction coefficient $\epsilon_{279\ \mathrm{nm}}$~5000 $\mathrm{mol}^{-1}\ \mathrm{dm}^3\ \mathrm{cm}^1$. Tyrosine and phenylalanine

Fig. 1.9 Absorption spectra of amino acids at pH = 6 [10]

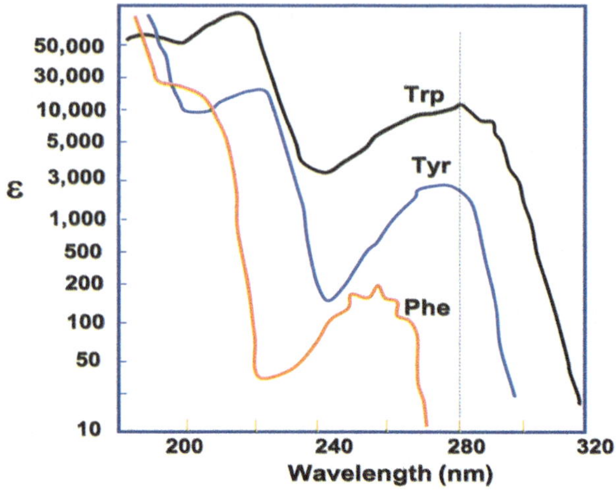

have one transition with $\epsilon_{274\ \mathrm{nm}}$ ~1400 $\mathrm{mol^{-1}\ dm^{3}\ cm^{1}}$ and $\epsilon_{258\ \mathrm{nm}}$ 190 $\mathrm{mol^{-1}\ dm^{3}\ cm^{-1}}$, respectively (Fig. 1.9). The disulfide bond in cysteine shows an absorption band between 250 and 270 nm, with an extinction coefficient ϵ~ 300 $\mathrm{mol^{-1}\ dm^{3}\ cm^{-1}}$. Although tryptophans have by far the most intense transitions in the aromatic region, proteins with few tryptophans compared with the other aromatic groups show absorbance. Similarly, aromatic side chains, disulfide cysteines, arginine, asparagine, aspartic acid, glutamine, glutamic acid, and histidine show far UV transitions. The σ-bonding electrons and nonbonding electrons on the oxygen and the nitrogen atoms, and *p*-electrons that are delocalized to some extent over the carbon, oxygen, and nitrogen atoms of peptide chromophore show absorption bands in the far-UV region. In the case of peptide chromophore, $n \rightarrow \pi^*$ transition is the lowest energy transition followed by the $\pi \rightarrow \pi^*$ transition.

Several methods have been developed for the quantitative determination of the phenylalanine, tyrosine, and tryptophan content in proteins, considering their characteristic absorbance in the near-UV spectral regions. Depending on the content of aromatic amino acids and the microenvironment surrounding those amino acids, each protein possesses a characteristic spectral signature in UV derivative spectroscopy. Levine and Federici (II) have demonstrated that second derivative spectroscopy can be effectively used to measure accurately the aromatic amino acid content in guanidine-HCl denatured protein samples [11]. The first derivative of the absorbance in the Beer–Lambert (Eq. 1.4) is:

$$\begin{aligned} \mathrm{d}A(\lambda)/\mathrm{d}\lambda &= \mathrm{d}[\log I_0(\lambda) - \log I(\lambda)]/\mathrm{d}\lambda \\ &= 0.43\ \mathrm{d}I_0(\lambda)/I_0(\lambda)\mathrm{d}\lambda - 0.43\ \mathrm{d}I(\lambda)/I(\lambda)\mathrm{d}h \\ &= cl(\mathrm{d}\epsilon(\lambda)/\mathrm{d}\lambda) \end{aligned} \tag{1.5}$$

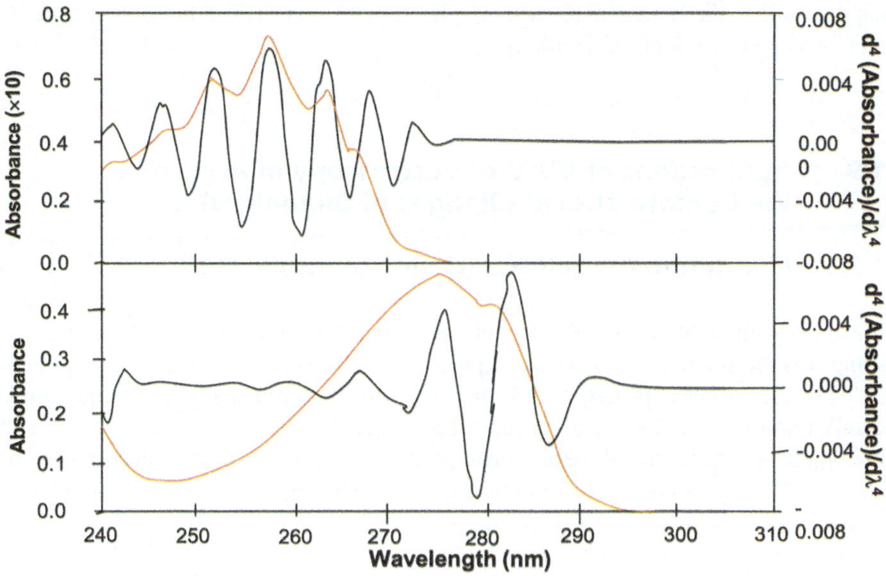

Fig. 1.10 The characteristic fourth-order derivative peak of tyrosine (lower panel) and phenylalanine (top panel)

The above equation holds effective for all higher-order derivatives, second and fourth being the most frequently used. Generally, due to the uniqueness of the second-derivative spectrum, the proteins and DNA allow sensitive detection of the presence of both of these components. Being potentially a unique signature, the derivative spectrum for each protein in a mixture of several proteins can be analyzed by multiple linear regression (Fig. 1.10). This helps in estimating the concentrations of each protein present without resorting to the actual physical separation of the components.

1.8.5 Folding and Unfolding of Protein

The native structure of a protein is distorted and unfolded due to the presence of denaturants s.a. urea or guanidinium chloride (GdmCl) or heat. This unfolding of the protein structures is reflected as a change in their absorbance at 287–292 nm, as a function of the denaturing factor. The temperature reduces the refractive index of the solution due to the thermal expansion, thus altering the difference in the absorbances of native and unfolded proteins. Similarly, the concentration of protein and the ionization of dissociable groups also affect the spectral intensity and position. These effects influence protein absorbance to a negligible extent, and therefore, the change in absorbance only due to temperature is usually small in the absence of structural transitions. However, the disappearance of absorbances of tyrosine and tryptophan at 287 nm and 291 nm, respectively, with the addition of denaturants happens due to the change in refractive index (i.e. the polarity) of the solvent with the increased concentration of denaturants.

1.9 Applications of UV-Vis Spectroscopy in Accessing the Conformational Changes of Biomolecules

1.9.1 Monitoring the Self-Association of Insulin [12]

Insulin is a small protein with characteristic association nature that can be monitored using this simple tool from the absorption spectrum. The monomer of the protein (5.7 kDa) associates strongly to a dimer and further polymerizes to give predominantly tetramers and hexamers at increased concentrations. At 25 °C, pH ~ 2, and low ionic strength, the dimeric form predominates even at a concentration of 10 mg ml^{-1}. However, the formation of tetramer and hexamer is favored with increasing the ionic strength or raising the pH to 4. The concentration-dependent absorbance change for insulin (Fig. 1.11) has been delineated through the association of monomer to dimer and is independent of higher-order polymerization. It is observed that more than 80% of the change in absorbance occurs between 0 and 2 mg ml^{-1} of the protein at 25 °C. The change in the spectral profile reflects the formation of the monomer to the dimer. From the concentration dependence of the difference, spectrum equilibrium constant for the monomer–dimer reaction under the various conditions could be determined.

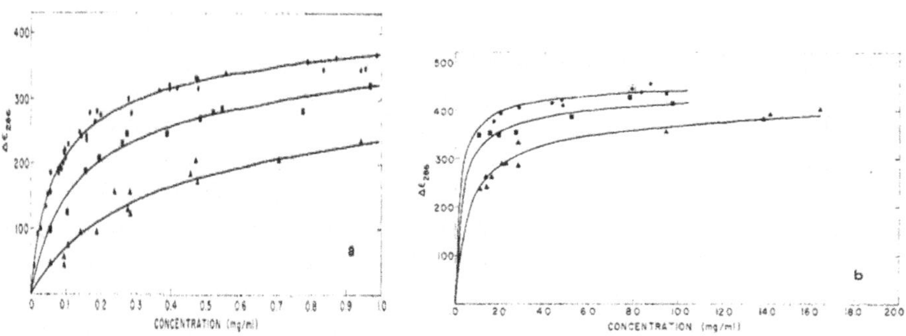

Fig. 1.11 Absorbance changes as a function of zinc insulin concentration. Change in extinction coefficient with zinc insulin concentration at pH = 2 and $\mu = 0.1$ and 25(•) and 35(▪) and 44° (Δ). (**a**) Expanded 0–10 mg ml^{-1} concentration range, (**b**) 0–20 mg ml^{-1} concentration range. (Image Courtesy: https://doi.org/10.1021/bi00746a014)

1.9.2 Determination of Protein and Nucleic Acid Content of the Virus [13]

The nucleic acid content of virus-like particles assembled in vitro has been of great interest to genetic information in virus-based transfection systems, especially in the context of gene therapy. Although the absorbance spectra of viruses and other large nucleoprotein complexes are difficult to interpret due to several associated factors, the UV absorption of protein and nucleic acid overlaps, light scattering systematically distorts the spectra, and there is the possibility that structural change of the nucleic acid may alter its hyperchromicity. Porterfield et al. in 2010 have employed light scattering corrected UV absorption spectroscopy for accurate determination of the RNA and protein concentration of virus capsids. They have examined the RNA content of Brome mosaic virus (BMV) and Hepatitis B virus (HBV) capsids containing their native genome and an ill-defined mixture of RNA derived from the *E. coli* expression system respectively. The results were verified using size exclusion multiangle laser light scattering (SEC-MALS) and validated the methodology for estimating RNA and protein concentration from UV spectra, as shown in Fig. 1.12, that reflects the RNA absorbance is not greatly affected due to encapsidation. Combining absorbance data with a known T number, they confirmed that BMV packages 3100 nucleotides/capsid, consistent with its genome, while

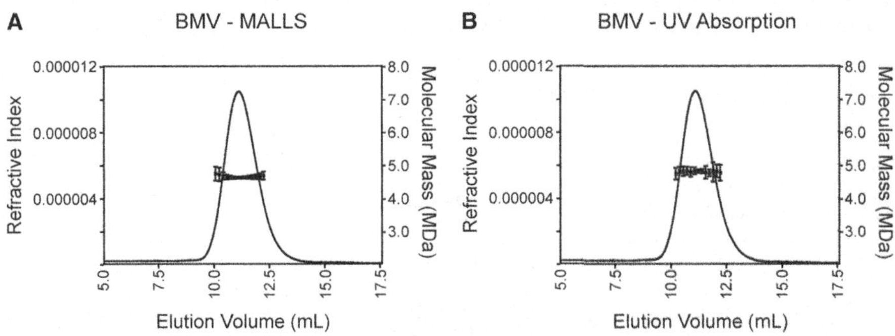

C

Method For MW Estimation	Molecular Mass	Standard Deviation	RNA Length
Theoretical	4.63 MDa		3045 nt
MALLS - Mw	4.64 MDa	0.06 MDa	3050 nt
UV - Absorption	4.78 MDa	0.16 MDa	3480 nt

Fig. 1.12 Light scattering corrected UV absorbance of BMV capsids in accordance with a virus with uniform RNA content. (**a**) SEC-MALLS—Elution of BMV capsids was observed by differential refractive index (**b**) UV absorption spectrophotometry to evaluate the absorbance for the concentration of RNA and BMV capsid protein. (**c**) Tabulated data showing good agreement between the average theoretical molecular mass of RNA-filled BMV capsids, the average molecular mass calculated using SEC-MALLS, and UV absorption. (Image courtesy: https://doi.org/10.1016/j.virol.2010.08.015)

HBV packages HBV capsid protein (Cp183, +15 charge) and a less basic mutant (Cp183-EEE, +12 charge) host RNA based on capsid charge and volume.

This is proven to be a sensitive method to differences in nucleic acid content and has been explored as a useful measure of the RNA content of virus capsids [13]. It also applies to other viruses and nucleoprotein complexes and is useful for characterizing artificial virus-like particles, too.

1.9.3 Location of Abnormal Tyrosines in Actin [14]

Mihashi and Ooi [14] with the help of spectrophotometric titration concluded and identified the helical segments that are located near the actin surface (Fig. 1.13) and the location of Tyr at these segments that are represented as one kind of buried residues. They have observed the existence of abnormal Tyr residues at 6 M urea content, though the helical segments of the protein are likely to be completely unfolded at this concentration of denaturant. The structural rigidity is mainly maintained due to the interaction of the embedded Tyr residues within the actin surface with other side-chain groups, thus the actin monomer core is formed. Using spectrophotometric studies, researchers concluded that even though 6 M urea efficiently disrupts the helical structure of the molecule, the interaction of deeply buried Tyr with other groups in the core remains less affected. Contrary to this, guanidine-HCl efficiently disrupts both the helical region as well as the region where abnormal Tyr residues are embedded. Hence, UV-Vis spectroscopic studies could effectively be used to gain an understanding of the structural integrity of molecules without using any sophisticated techniques.

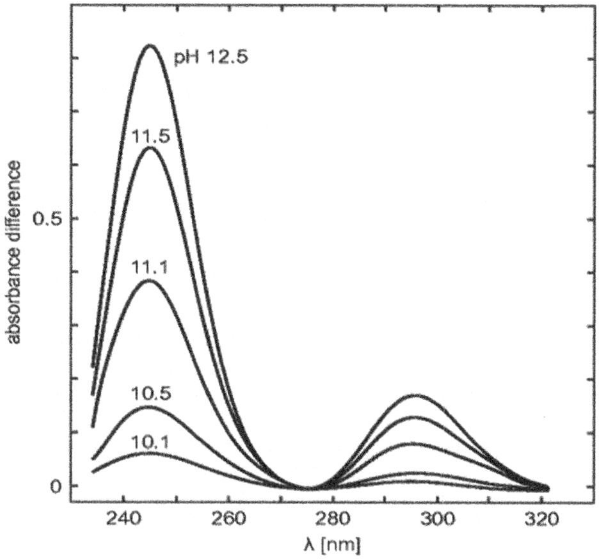

Fig. 1.13 The variation in the spectra of actin with change in the solvent pH. The concentration of the Protein is 0.17 mg ml^{-1} [14]

1.10 Conclusion

Although with rapid developments in technology, there are a plethora of techniques that could be precisely used to study protein stability and folding, one would be skeptical about the efficacy of low-resolution instruments, like UV spectroscopy. However, recent developments in UV spectroscopy in terms of both instrumentation and data handling capability make it a suitable tool for monitoring the conformational alternations during the unfolding and folding of protein under certain conditions. The several potential advantages and exclusive capabilities of this tool include:

1. The accessibility and convenience of modern UV-Vis absorption spectrophotometers, with increased precision and accuracy, allow the absorption intensity and peak position acutely well defined. Also, the enhanced time sensitivity of the measurement due to the advancement of the detector technology enables recording of the spectrum profile in a substantially short period, thus making this approach efficient in studying the fast kinetics during protein folding and unfolding pathway.
2. The development of monochromators and detectors with increased efficiency makes it possible to record high-resolution spectra, and incorporation of this technique is useful to study of protein folding–unfolding mechanism.
3. The UV spectrum profile of proteins in the 190–350 nm range gives information about secondary conformation (190–210 nm) and tertiary conformation (250–300 nm). Similarly, the aggregation phenomena can be well studied by monitoring the light scattering between 300 and 350 nm. Thus, the spectrum profile of a protein monitored with increased resolution and various path lengths in the UV region effectively provides extensive information regarding structural integrity and stability.
4. Higher-order derivative spectral profile of protein recorded with limited light scattering or the presence of solutes with strongly absorbing nature could provide essential information during folding studies.
5. Protein molecules having tyrosine and phenylalanine residues that are capable of intrinsic fluorescence are found to be insensitive to changes in protein microenvironments. In such a case, the information regarding the comprehensive kinetic details of residual populations can be obtained from the multiple peaks profile that arises due to multiple residues of the aromatic side chain of the biomolecule.

Considering the linear dependency of absorbance with its concentration of a solute, absorption spectroscopy is widely employed for purpose of quantitative measurements. Also, the sensitivity of absorption range and absorbance intensity depends not only on the chemical structure of the chromophores but also on the immediate microenvironment. Hence, UV-Vis spectrophotometer is proven to be an ideal technique to monitor the interaction of proteins/DNA with small external molecules, enzymatic activity, and conformational alternations in nucleic acids and proteins.

References

1. Ball DW (2001) The basics of spectroscopy, vol 49. Spie Press, Bellingham
2. Baruah M et al (2006) Solvent and pH-dependent fluorescent properties of a dimethylaminostyryl borondipyrromethene dye in solution. Chem A Eur J 110(18):5998–6009
3. Kirby EP, Steiner RF (1970) Influence of solvent and temperature upon the fluorescence of indole derivatives. J Phys Chem 74(26):4480–4490
4. Bravaya KB et al (2010) Electronic structure and spectroscopy of nucleic acid bases: ionization energies, ionization-induced structural changes, and photoelectron spectra. Chem A Eur J 114(46):12305–12317
5. Ullrich S et al (2004) Electronic relaxation dynamics in DNA and RNA bases studied by time-resolved photoelectron spectroscopy. Phys Chem Chem Phys 6(10):2796–2801
6. Šponer J, Leszczynski J, Hobza P (2001) Electronic properties, hydrogen bonding, stacking, and cation binding of DNA and RNA bases. Biopolymers 61(1):3–31
7. Hannon MJ (2007) Supramolecular DNA recognition. Chem Soc Rev 36(2):280–295
8. Mandel M, Marmur J (1968) [109] Use of ultraviolet absorbance-temperature profile for determining the guanine plus cytosine content of DNA. In: Methods in enzymology. Elsevier, Amsterdam, pp 195–206
9. Mach H et al (1995) Ultraviolet absorption spectroscopy. In: Protein stability and folding. Springer, pp 91–114
10. Wetlaufer D (1963) Ultraviolet spectra of proteins and amino acids. In: Advances in protein chemistry. Elsevier, pp 303–390
11. Levine RL, Federici MM (1982) Quantitation of aromatic residues in proteins: model compounds for second-derivative spectroscopy. Biochemistry 21(11):2600–2606
12. Lord R, Gubensek F, Rupley J (1973) Insulin self-association. Spectrum changes and thermodynamics. Biochemistry 12(22):4385–4392
13. Porterfield JZ, Zlotnick A (2010) A simple and general method for determining the protein and nucleic acid content of viruses by UV absorbance. Virology 407(2):281–288
14. Mihashi K, Ooi T (1965) Location of abnormal tyrosines in actin. Biochemistry 4(5):805–813

Circular Dichroism Spectroscopy: Principle and Application

2

Suchismita Subadini, Pratyush Ranjan Hota, Devi Prasanna Behera, and Harekrushna Sahoo

Abstract

Circular dichroism spectroscopy is generally used to determine the secondary and tertiary structure of proteins. Its principle is based on differential absorption of left- and right-handed polarized light. To be detected by CD, the molecule should be asymmetrical; hence, the molecule should have one or more chiral chromophore. As optically active molecules absorb left- and right-handed circular polarized light differently, it becomes the basis of determining the biomolecule. There are different types of CD techniques depending upon the scanning spectrum range. Protein secondary structure is determined by UV CD technique. Similarly, vibrational CD and IR CD are used to study the structure of small organic molecules, proteins, and DNA, respectively.

Suchismita Subadini, Pratyush Ranjan Hota and Devi Prasanna Behera have contributed equally to this work.

S. Subadini · P. R. Hota · D. P. Behera
Biophysical and Protein Chemistry Laboratory, Department of Chemistry, National Institute of Technology (NIT) Rourkela, Rourkela, Odisha, India

H. Sahoo (✉)
Biophysical and Protein Chemistry Laboratory, Department of Chemistry, National Institute of Technology (NIT) Rourkela, Rourkela, Odisha, India

Center of Nanomaterials, National Institute of Technology (NIT) Rourkela, Rourkela, Odisha, India
e-mail: sahooh@nitrkl.ac.in

© The Author(s), under exclusive license to Springer Nature Singapore Pte Ltd. 2022

H. Sahoo (ed.), *Optical Spectroscopic and Microscopic Techniques*,
https://doi.org/10.1007/978-981-16-4550-1_2

2.1 Introduction

Circular dichroism spectroscopy is generally used to determine the secondary and tertiary structure of proteins. Its principle is based on differential absorption of left- and right-handed polarized light. To be detected by CD, the molecule should be asymmetrical; hence, the molecule should have one or more chiral chromophore. As optically active molecules absorb left- and right-handed circular polarized light differently, it becomes the basis of determining the biomolecule. There are different types of CD techniques depending upon the scanning spectrum range. Protein secondary structure is determined by UV CD technique. Similarly, vibrational CD and IR CD are used to study the structure of small organic molecules, proteins, and DNA, respectively.

The circularly polarized light is comprised of two plane waves having equal amplitude by differing 90° in phase and when two plane waves having different amplitudes are related in phase by 90°, or when the relative phase is other than 90°, then the light is known as elliptically polarized (Fig. 2.1).

Circularly polarized light can be absorbed by a medium. As optically active chiral compound absorbs, the two circularly polarized lights (left & right) differently,

$$\Delta A = A_{\mathrm{L}} - A_{\mathrm{R}}$$

Beer–Lambert law says,

$$A = (\varepsilon_{\mathrm{L}} - \varepsilon_{\mathrm{R}})\mathrm{cl}$$

The molar circular dichroism is the difference in the molar absorptivity

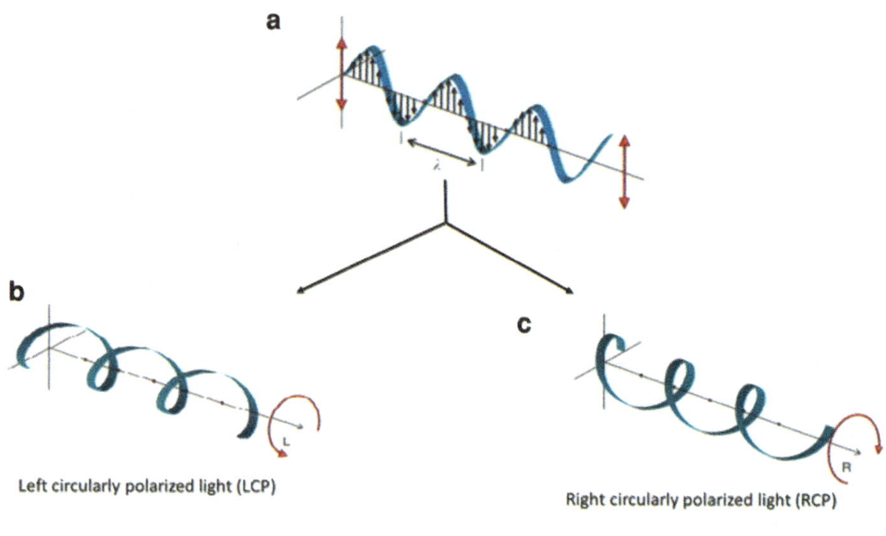

Fig. 2.1 Representation of circularly polarized light where the helix represents the path followed by the component electric field of a light beam. (**a**) Left polarized light. (**b**) Right polarized light

$$\Delta\varepsilon = \varepsilon_L - \varepsilon_R$$

When a circular polarized light falls on the asymmetric sample or medium, the amplitude of the incident radiation changes. As the intensity is directly proportional to the square of the amplitude of the radiation, it also changes the intensity of the light. In the chiral medium both left circularly polarized and right circularly polarized absorb differently resulting in different molar absorptivity and due to this difference the superimposed light becomes elliptically polarized light.

The CD spectrum is often represented by degrees of ellipticity (θ) that measures the ellipticity of the polarization given by

$$\tan\theta = \frac{E_R - E_L}{E_R + E_L}$$

where E is the magnitude of the electric field vector and the signals measured in radians ($\theta = 2.303/4(A_L - A_R)$) and θ can be converted to degrees by multiplying by $180/\pi$ that gives

$$\theta = 32.98\,\Delta A$$

Widely used unit for CD experiments is molar ellipticity, $[\theta]$, which removes the dependence on concentration and path length

$$[\theta] = 3298\,\Delta\varepsilon$$

2.2 Instrumentation

There are three methods of the determination of CD data. (a) Modulation: In this method, incident radiation is switched continuously between left and right circular polarized light. (b) Direct subtraction method: In this method, absorbance of two circular polarized lights are recorded separately and subtracted from each other. (c) Ellipsometric method: In this method change in ellipticity of the transmitted light is taken for consideration. The modulation method is widely used method but the direct subtraction and ellipsometric methods have advantages in time-resolved circular dichroism technique.

2.2.1 Modulation Method

The difference in the absorptivity of the both left and right polarized light by a chiral sample is very small (usually less than one part in 10^3). Hence, it needs sophisticated instruments to calculate the difference in absorption. Difference in absorbance can be related with the intensities by the following equation:

$$\Delta A = \log(I_0/I_L) - \log(I_0/I_R) = \log(I_R/I_L)$$

From the above expression, we need only to consider the intensity of transmitted light. The intensity of the transmitted light is quite nearer to the average value of both left and right polarized light. When monochromatic light passes through the polarizer, the two linearly polarized lights are produced having equal intensities. The polarizer is placed at 45° along the principal axis of the quarter-wave retarder. AC voltage can change the refractive index of the retarder in different directions and the width of the retarder is so accurate that while passing through the retarder, the two linearly polarized light emerge having a phase difference of Π/2. Hence it makes the light right circularly polarized. By rotating the retarder by 90°, the left polarized light can be obtained. Thus, it is possible to create a modulated quarter-wave retarder which alternately reverses the refractive indices in different directions of retarder. The magnitude of the ac voltage determines the wavelength for which the device is a quarter-wave retarder, and can be changed as the wavelength is scanned. Then the two circularly polarized lights are made to pass through the sample chamber and they get differentially absorbed by the sample and rest detected by the detector. The DC signal obtained is proportional to the intensity ratio, while the AC signal is proportional to the intensity difference of two circularly polarized lights after passing through the sample chamber. The AC signal is faint in comparison to the DC signal but as the AC signal is the required signal, lock-in amplifier is used to enhance it (Fig. 2.2).

The CD signal of the sample is proportional to the AC signal divided by the DC signal. Most of the equipment used in CD operation use the DC voltage as constant and during calibration it becomes the part of the proportionality constant and the output signal is recorded.

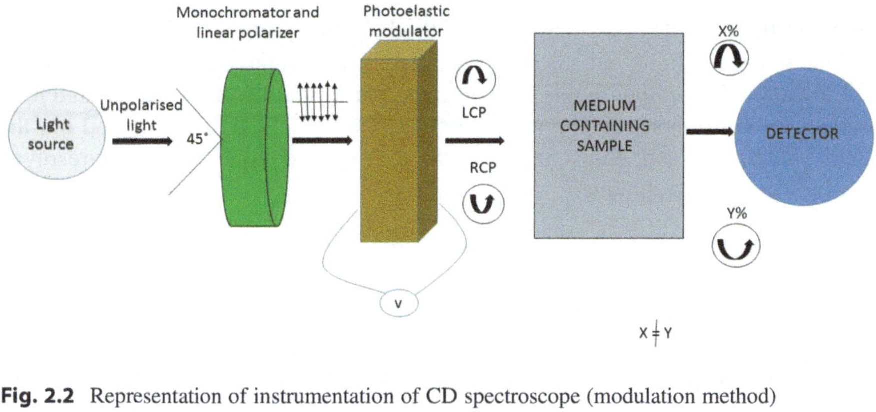

Fig. 2.2 Representation of instrumentation of CD spectroscope (modulation method)

2.2.2 Direct Subtraction Method

In this method both left and right polarized light are made to pass through the sample separately and the absorption is calculated in later stage subtracted from each other. The light passes through the conventional polarizer that separates the two orthogonal parts of the wave. Then both the waves are made to pass through the half wave retarder forming left and right circularly polarized light. After being circularly polarized, both these waves enter the sample chamber and pass through it. Two separate detectors are used to detect the signals separately which are subtracted through computer software. In this method, the instrument uses the standard modulated quarter-wave retarder to make a flat baseline.

2.2.3 Ellipsometric Method (Fig. 2.3)

In this method, light from the source is made to pass through a vertical linear polarizer. Then highly elliptically polarized lights are produced in alternating rotations when it passes through a strain plate at 45° to linearly polarized light. The chiral sample changes this ellipticity, and a linear polarizer, which placed horizontally, transmits the minor axis of the elliptically polarized light. Then the detector detects this alternating intensity of minor axes for the two rotations of elliptically polarized light and converted to CD data.

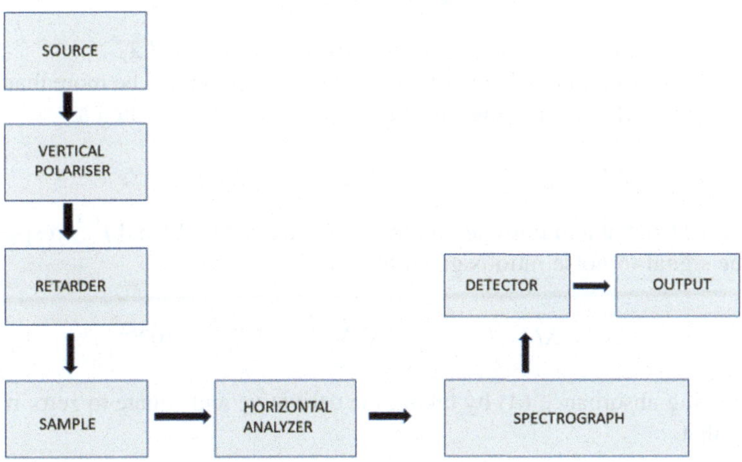

Fig. 2.3 Flow chart representation of ellipsometric method of CD spectroscope instrumentation. (Adapted from Johnson WC (1996) Circular dichroism instrumentation, in Circular Dichroism and the Conformational Analysis of Biomolecules. Springer: 635–652)

2.3 Sample Preparation and Electronic CD Measurement

(a) Nitrogen Purging:
CD instrument is generally flushed with nitrogen prior to turning on the source irrespective of which range is to be scanned. It is done to remove oxygen from the instrumental duct. Oxygen generally absorbs below 200 nm and intense light source also converts oxygen to ozone which can act as strong oxidizing agent, ultimately degrading the instrument optics. Hence, nitrogen is used prior to turning on the source and throughout any measurement.

(b) Monochromator handling:
To get perfect result, the first mirror should be placed just after the source. The prism should be kept clean, the aluminized side should not be touched, and the other side should also be cleaned with ethanol and soft cotton. If enough intensity of radiation is lacked, then after sample chamber a second mirror should be placed in such a manner that it reflects maximum amount of intensity.

To the signal to be detected by the detector, there should be enough intensity near the detector. The important fact for CD measurement is that, the solvent should not absorb in the range which is scanned for the sample. By decreasing the cell length, the concentration of the solvent can be extended to some extent. Generally, 0.1–0.05 mm cells are used for the measurement of CD spectra. To ensure the path length, the cells must be placed in UV and the interference caused by reflections from two closely spaced windows. The path length can be calculated by

$$l = (\Delta n \lambda_1 \lambda_2 / 2) / (\lambda_1 - \lambda_2)$$

where Δn is the number of maxima or minima between λ_1 and λ_2.

In the case of solution (solvent + protein), the absorption will be more than that of the pure solvent. By considering this, the Beer's law can be written as

$$\Delta I = I_{\mathrm{L}} - I_{\mathrm{R}} = -2.303 I \Delta A = -2.303 I \, \Delta\varepsilon A / \varepsilon$$

Again, CD signal and noise are directly proportional to ΔI and $I^{1/2}$, respectively, hence the signal-to-noise ratio is given by

$$S/N \propto \Delta I / \sqrt{I} = -(2.303 \, \Delta\varepsilon A / \varepsilon) \left(I_0^{1/2} \times 10^{A/2} \right)$$

Optimizing absorbance (A) by taking the derivative and equate to zero, it can be obtained that

$$2.303A = 2$$

Absorbance (A) comes as 0.869. When a graph is plotted between the ratio and A, we can see that the ratio for the sample solution is optimum at 0.3. Hence to keep the A below 1, the signal-to-noise ratio should be below 0.3 or can be 0.3. For larger value of the ratio which is greater than 0.3, A becomes bigger. This ultimately

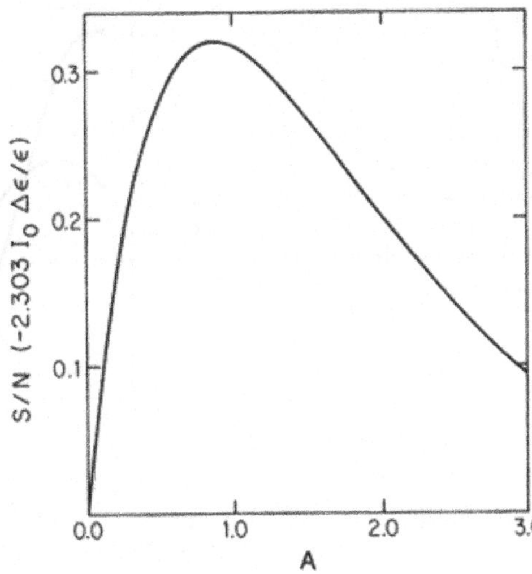

Fig. 2.4 Signal-to-noise ratio as a function of absorbance. (Adapted from Johnson WC (1996) Circular dichroism instrumentation, in Circular Dichroism and the Conformational Analysis of Biomolecules. Springer: 635–652)

decreases the signal strength to reach the detector. But in some cases the ratio can be higher when the solvent is transparent [1] (Fig. 2.4).

2.4 Application to Protein Structure

Circular dichroism (CD) spectroscopy is a useful tool used in identifying structural aspects of proteins, carbohydrates, nucleotides, peptides, DNA, etc. Because of asymmetric peptide bonds present in protein, "-NH_2-" units of "-NH_2-CO-" moiety influence the CD spectra in Far-UV region. The major secondary structure of protein constitutes α-helix, β-sheet structure. Two negative bands at 222 nm ($n \rightarrow \pi^*$), 208 nm ($\pi \rightarrow \pi^*$) and a positive band at 190 nm ($\pi \rightarrow \pi^*$) are the characteristics of α-helix, whereas a positive band at 198 nm and a broad negative band around 220 nm is of the β-sheet. In addition, at 200 nm a band can be seen due to presence of a random coil, if any.

A protein coil constitutes 20 different amino acids out of which only aromatic residue containing chains such as phenyl alanine (Phe), tyrosine (Tyr), and tryptophan (Trp) can show CD spectrum due to $\pi \rightarrow \pi^*$ transition around 250–300 nm range, i.e., near-UV zone. Here disulfide bond also contributes to spectrum around 260 nm due to $n \rightarrow \sigma^*$ transition [2]. Phe exhibits a weak and sharp band in the range of 255–270 nm and Tyr in the range of 275–282 nm. However, at longer wavelength region of Tyr, the band due to Trp overlaps with the former, which in turn shows a band near 290 nm. Vibrionic transition between different vibrational levels of respected residues causes the appearance of fine structure. Near-UV CD study is mostly used to monitor the tertiary structure of protein. There are so many factors on which shape and size of near-UV spectra depends that include number of aromatic amino acid, H-bonding between them, polar groups present and the polarizability [3] (Fig. 2.5).

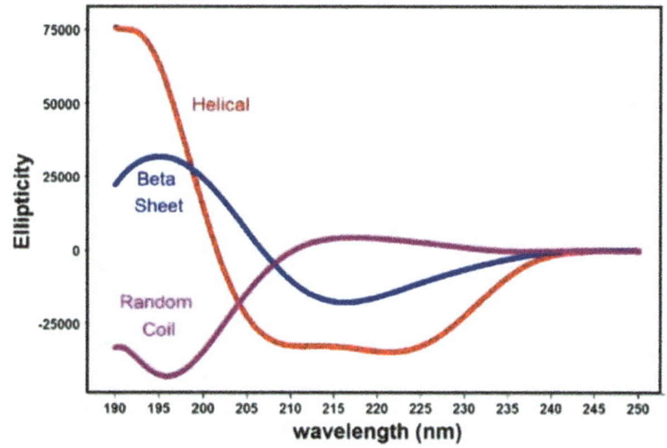

Fig. 2.5 Far-UV CD spectra of proteins with α-helix (red) β-sheet (blue) and random (pink) structure in predominance. (Adapted from CD spectra by Christopher Berndsen, source: https://ceberndsen.github.io/CEBerndsen.github.io/CD.html#circular_dichroism)

It can also be useful in the study of non-protein chromophores such as haem (CD bands at 410 nm approximately piculiarly known as Soret band), flavin, flavocytochromes, photosynthetic pigments [4].

2.4.1 Protein–Lipid Interaction

Apocytochrome C, in aqueous solution, gets unfolded due to interaction with lipid bilayers of membrane and acquires helical structure. Similarly, when it bound with dioleoyl phosphatidylserine (with single lamella), its α-helical structure gets enhanced while its β-sheet counterpart remains the same [5]. In case of MBP or myelin basic protein, α-helix constitutes 20% of its total structure when it bound to phosphatidyl seine vesicles [6], whereas in aqueous solution it lacks helical structure that is confirmed from CD study. Similarly, melittin upon interacting with any lipid exhibits α-helix conformation, which can be studied by CD [7].

2.4.2 Protein–Ligand Interactions

A number of studies have been done for protein–ligand interactions. CD spectrum of HSA protein shows that α-helix structure reduced from 53.31% to 48.4% due to binding of 5-FU in hydrophobic part of protein, i.e. 5-FU destabilizes the protein (from Far-UV CD study) [8]. However, in case of Lyz, the drug acts as stabilizing ligand as it enhances its helical structure upon binding. In case of Lyz, a positive triplet signals at 283, 289, and 295 nm can be seen due to transition of Trp residues present in active site. Here 5-FU acts as a stabilizing ligand and induces the signal [9]. Addition of tetracycline antibiotic (TC) to serum albumins (BSA and HSA) in an equimolar proportion causes destabilizing, whereas by increasing its concentration its effect can be altered, i.e. it will act as stabilizing ligand (as in Fig. 2.6) [10]. Upon interaction with triethyloctyl ammonium bromide ionic liquid, BSA experiences a conformational change around disulfide bond due to which its negative ellipticity

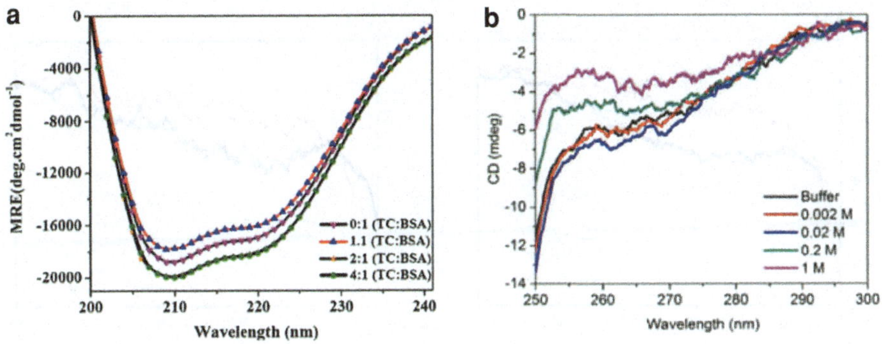

Fig. 2.6 (**a**) Far-UV CD spectra of BSA in the presence of TC in different proportions. (Adapted from Journal of Physical chemistry B (2011) 115, 6312–6320) [10] (**b**) Near-UV CD spectra of BSA in presence of IL of different proportions [11]

value decreases suggesting constructive role of IL (ionic liquid) upon tertiary structure of BSA protein, confirmed from near UV-CD study, though Far-UV study could not reveal anything due to IL interference [11].

2.4.3 Denaturation Study of Proteins

Along with conformational changes, CD spectroscopy also monitors thermal and chemical stability, i.e., on varying the temperature or on adding any chemical denaturants or by changing the pH of the medium. If denaturation continues, protein starts to unfold attaining a random coil structure rather a compact structure. As the spectra for a random coil and a folded structure is different, denaturation study can be done after choosing a suitable wavelength.

2.4.3.1 Thermal Denaturation

Lakkoji Satish and group studied thermal folding and refolding studies in ionic liquid medium by heating followed by cooling of BSA protein. To avoid protein aggregation, usually seen in case of the lower concentrated solution of ionic liquid, they use higher concentration (0.2 M, 1 M). Protein starts getting unfolded when it is heated up to 90 °C as we can see from Fig. 2.7. The negative band due to disulfide bonds, Phe residues at 261 and 268 nm get increased. The CD spectra behaves normally and shows a similar kind of spectra as they get before heating, i.e. in normal condition suggesting refolding process. For refolding to be happened, desorption of hydrophobic cations of ionic liquid are driven by refolding energy [11].

2.4.3.2 Chemical Denaturation

It includes the chaotropes like urea, GdHCl, surfactant molecules that usually help in unfolding of biomolecules like protein by disrupting non-covalent, hydrophobic interaction causing destabilizing and these molecules get unfolded ultimately. Figure 2.8b shows the study of BSA protein in presence of GdHCl denaturant. We

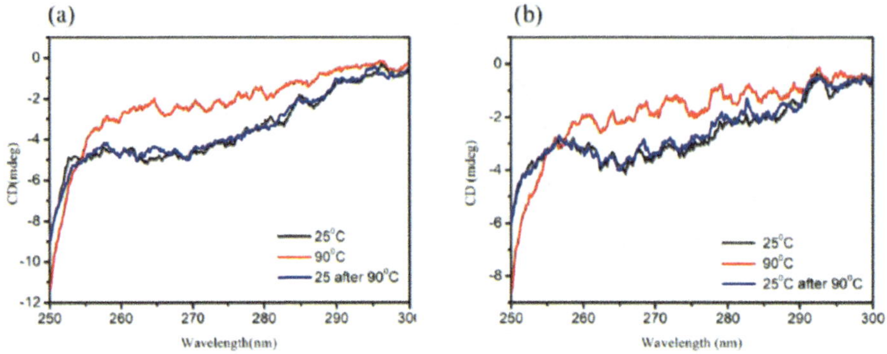

Fig. 2.7 Thermal folding and refolding of BSA protein in presence of ionic liquid at (**a**) 0.2 M (**b**) 1 M concentration [11]

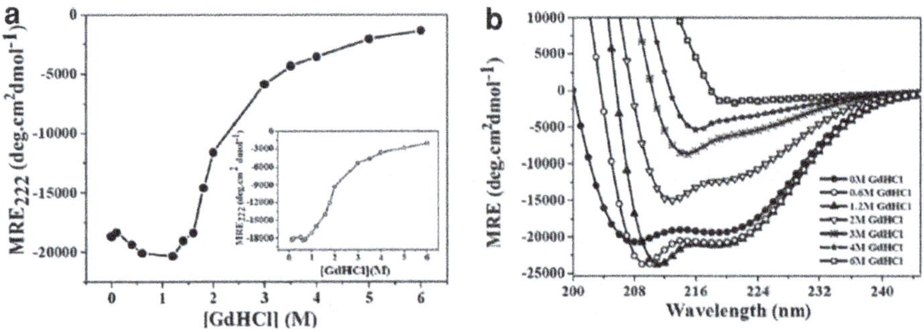

Fig. 2.8 (**a**) Mean residue ellipticity value at 222 nm vs various concentrations of GdHCl (**b**) CD spectra of BSA in presence and absence of GdHCl. (Adapted from Physical Chemistry Chemical Physics (2011) 13, 20418–20426) [12]

can observe that at low concentration, the denaturant behaves as stabilizing agent and helps in increasing the helicity but up to a threshold value. After crossing these concentrations, GdHCl drastically changes the MRE value. BSA protein exhibits two characteristic peaks of helix at 208, 222 nm from which the effect is studied at 222 nm by Anand and co-workers. In its native state, BSA constitutes approximately 66% of α-helix which got enhanced up to 71% at 0.6 M additive and the trend continued to 1.2 M and at 1.6 M GdHCl, helicity reduced to 41% and became 5% at 6 M concentration. So, we can conclude that, at high concentration of denaturant, protein starts to unfold while at lower concentration of additive, protein shows refolding phenomena [12].

2.4.4 Change in pH Induced Transition of Proteins

Hemoglobin (Hb) in tris-buffer is very much sensitive to any change in its pH value. Comparing the spectra at pH 3.5 and 1.9, a major structural change in the secondary structure of Hb can be observed. At pH 1.9, the helicity dips down to 14%, whereas till 3.5 from the physiological pH α-helix constitutes 61% [13]. At pH 6.35, *β*-lactoglobulin exhibits its characteristic band at 218 nm. From pH 3 to 4, we can observe a drastic change, however, in acidic range there is no significant change can be noticed. But in case of alkaline pH, it goes through a significant change where at pH 12.35, spectrum looks like an unfolded polypeptide instead of a sheet. At pH 6 and 13, author marked two sigmoidal transitions at 208 nm. From near-UV CD study, it can be observed that over the range of 0.9–5.7 pH, there is a slight change in tertiary structure, whereas in alkaline medium, it begins to lose its tertiary structure rapidly and at pH 12.5, (approximately) it loses its tertiary structure. At 293 nm, a similar transition can be noticed like in 208 nm at pH 6 and 13 [14].

2.4.5 Thermal Stability

Melting point (T_m) is the temperature at which protein remains in both folded and unfolded states in approximately equal proportion or we can say both are in equilibrium with each other. BMP-2 has a melting point of 70.66 °C. Upon adding nanoparticles, it was clearly seen that they act as stabilizing ligand, thus enhancing the melting points to a different extent depends upon types of nanoparticles added (from Fig. 2.9a) [15]. Sulfate salts do not cause any drastic changes to BMP-2 rather sulfates behave quite like unfolded native protein. Unlike sulfates, phosphates harshly affect thermal stability of BMP-2 as one can see from Fig. 2.9b. Na_2PO_4, $CaPO_4$ behave as stabilizing ligand, whereas $MgPO_4$ changes T_m and destabilizes the protein [16] (Fig. 2.10).

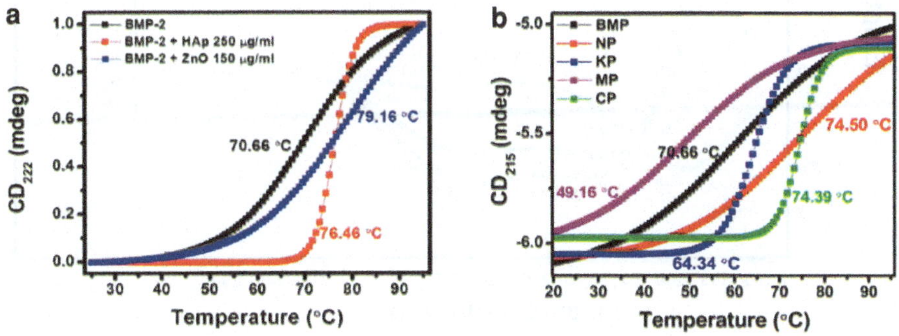

Fig. 2.9 Thermal stability of BMP-2 in presence and absence of (**a**) HAp, ZnO nanoparticles [15], (**b**) Na_2PO_4, KPO_4, $MgPO_4$, $CaPO_4$ [16]

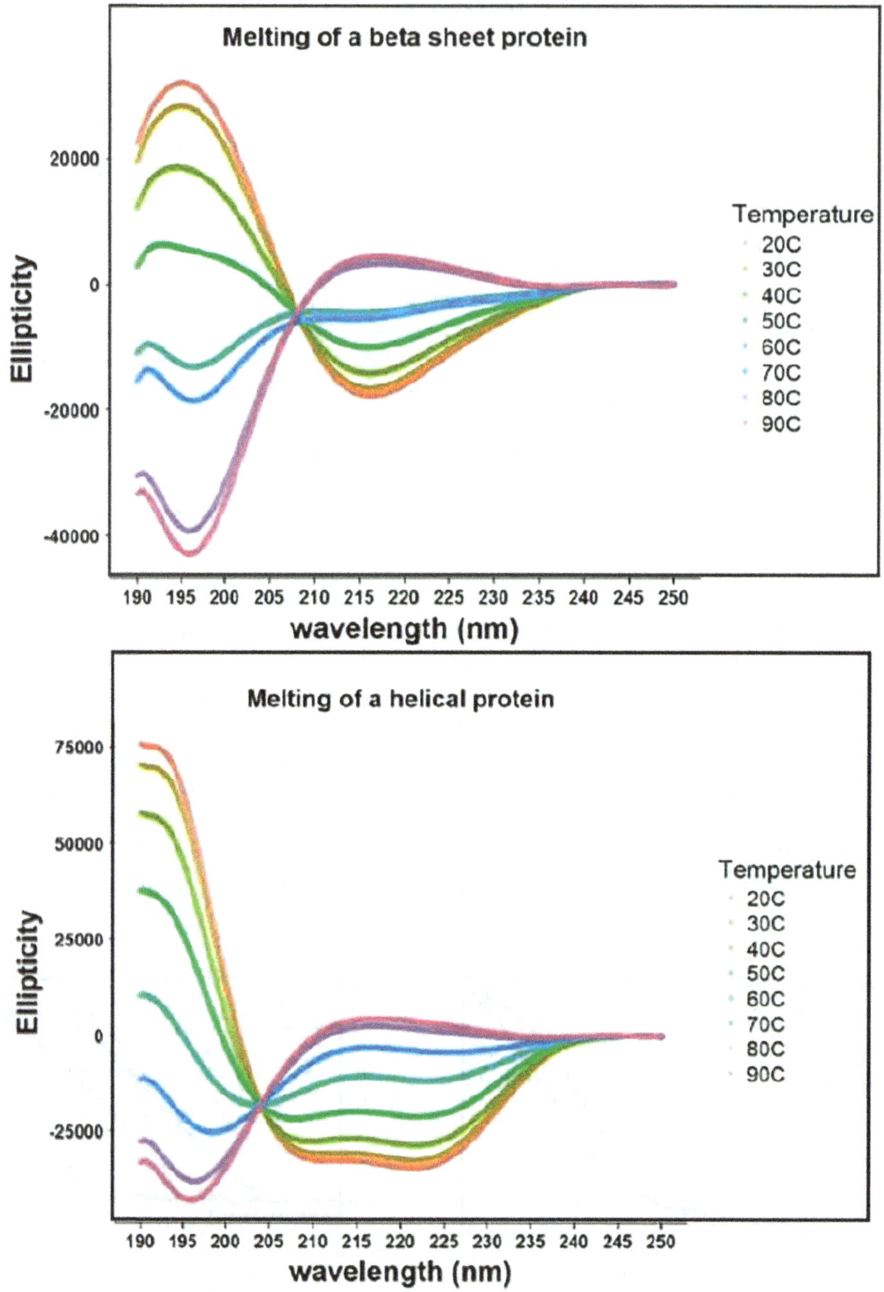

Fig. 2.10 CD spectra of melting of proteins with α-helix (below) β-sheet (above) structure in predominance. (Adapted from CD spectra by Christopher Berndsen, source: https://ceberndsen.github.io/CEBerndsen.github.io/CD.html#circular_dichroism)

2.5 CD Study of Nucleic Acids

CD study is also used for conformational analysis of nucleotides. Nucleotides exhibits spectra in the range of 200–320 nm due to nucleotide bases present in them (Below 200 nm buffers and RNA both absorbs UV light). DNA exhibits several forms out of which B-form is mostly observed. It contains right-handed helical structure with base pairs perpendicular to helices. Generally, it shows a positive band in the range of 260–280 nm ($\pi \rightarrow \pi^*$ transition) and a negative band at 245 nm. On increasing Adenine and Thymine bases, the negative bands deepen and conformation may vary. A-form is also right handed, but 25% shorter than B-form and exhibits positive band at 260 nm and negative band at 210 nm and spectra usually depends on A-form of corresponding RNA or its hybrid. Z-form is left-handed and just opposite to above two forms. It shows a negative band at 290 nm, positive at 260 nm, and extremely deep negative band at 205 nm [17]. RNA can also be found as Z- and A-form as left and right-handed coil, respectively. However, nature of spectra is quite opposite to those of DNA. A-form, like Z-DNA gives a negative minimum at 295 nm with a second positive band, whereas Z-form like A-DNA shows a broad maximum at 260 nm [18].

2.5.1 Denaturation Study of DNA

From denaturation study of B-DNA, a moderate change can be seen above 210 nm. T7 phage DNA constitutes around 52%of adenine and thiamine base pairs. After thermal denaturation above 230 nm, no significant changes can be found whereas, below 230 nm there is an appreciable change in magnitude of CD band which includes an increase in magnitude at 220 nm. Denaturation shifts the crossover to 261 nm from 258 nm. As there is no change at longer wavelengths, it suggests the CD band is due to base-pair [19] (Fig. 2.11).

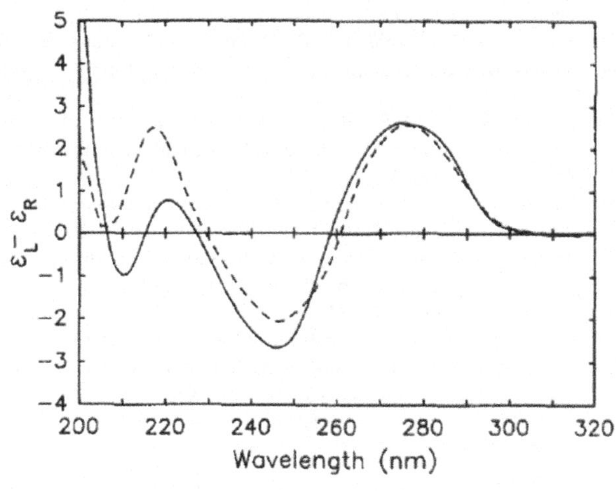

Fig. 2.11 CD spectra of T7 phage DNA before (solid line) and after (dash line) denaturation. (Adapted from Methods in Enzymology (1992) 211, 389–406) [19]

2.6 Conclusion

Nowadays, CD spectroscopy is widely used and preferred when it comes to investigate the interactions of biomolecules or to determine their structures. So, "why it is most preferable?" and the answer is simple, viz. preparation of the samples understudy is economical and study of interactions is not time consuming, in addition the technique is noninvasive. Though there is a limitation in the study of carbohydrates, it has been an important tool for studying many biological molecules in 3D aspects, for protein and peptide designing.

References

1. Johnson WC (1996) Circular dichroism instrumentation, in circular dichroism and the conformational analysis of biomolecules. Springer, Berlin, pp 635–652
2. Corrêa DH, Ramos CH (2009) The use of circular dichroism spectroscopy to study protein folding, form and function. Afr J Biochem Res 3(5):164–173
3. Kelly SM, Jess TJ, Price NC (2005) How to study proteins by circular dichroism. Biochim Biophys Acta Prot Proteom 1751(2):119–139
4. Kelly SM, Price NC (2000) The use of circular dichroism in the investigation of protein structure and function. Curr Protein Pept Sci 1(4):349–384
5. de Jongh HH, de Kruijff B (1990) The conformational changes of apocytochrome c upon binding to phospholipid vesicles and micelles of phospholipid based detergents: a circular dichroism study. Biochim Biophys Acta Biomembr 1029(1):105–112
6. Keniry MA, Smith R (1979) Circular dichroic analysis of the secondary structure of myelin basic protein and derived peptides bound to detergents and to lipid vesicles. Biochim Biophys Acta Prot Struct 578(2):381–391
7. Chen YH, Yang JT, Martinez HM (1972) Determination of the secondary structures of proteins by circular dichroism and optical rotatory dispersion. Biochemistry 11(22):4120–4131
8. Chinnathambi S, Karthikeyen S, Kesherwani M, Velmurugan D, Hanagata N (2016) Underlying the mechanism of 5-fluorouracil and human serum albumin interaction: a biophysical study. J Phys Chem Biophys 6:214–222
9. Millan S, Satish L, Bera K, Konar M, Sahoo H (2018) Exploring the effect of 5-fluorouracil on conformation, stability and activity of lysozyme by combined approach of spectroscopic and theoretical studies. J Photochem Photobiol B Biol 179:23–31
10. Anand U, Jash C, Boddepalli RK, Shrivastava A, Mukherjee S (2011) Exploring the mechanism of fluorescence quenching in proteins induced by tetracycline. J Phys Chem B 115(19):6312–6320
11. Satish L, Sabera M, Sasidharan VV, Sahoo H (2018) Molecular level insight into the effect of triethyloctylammonium bromide on the structure, thermal stability, and activity of bovine serum albumin. Int J Biol Macromol 107:186–193
12. Anand U, Jash C, Mukherjee C (2011) Protein unfolding and subsequent refolding: a spectroscopic investigation. Phys Chem Chem Phys 13(45):20418–20426
13. De S, Girigoswami A (2006) A fluorimetric and circular dichroism study of hemoglobin—effect of pH and anionic amphiphiles. J Colloid Interface Sci 296(1):324–331
14. Taulier N, Chalikian TV (2001) Characterization of pH-induced transitions of β-lactoglobulin: ultrasonic, densimetric, and spectroscopic studies. J Mol Biol 314(4):873–889
15. Konar M, Sahoo JK, Sahoo H (2019) Impact of bone extracellular matrix mineral based nanoparticles on structure and stability of purified bone morphogenetic protein 2 (BMP-2). J Photochem Photobiol B Biol 198:111563–111570

16. Konar M, Sahoo H (2019) Phosphate and sulphate-mediated structure and stability of bone morphogenetic protein-2 (BMP-2): a spectroscopy enabled investigation. Int J Biol Macromol 135:1123–1133
17. Kypr J, Kejnovska I, Renciuk D, Vorlickova D (2009) Circular dichroism and conformational polymorphism of DNA. Nucleic Acids Res 37(6):1713–1725
18. Miyahara T, Nakatsuji H, Sugiyama H (2016) Similarities and differences between RNA and DNA double-helical structures in circular dichroism spectroscopy: a SAC–CI study. J Phys Chem A 120(45):9008–9018
19. Gray DM, Ratliff RL, Vaughan MR (1992) Circular dichroism spectroscopy of DNA. Methods Enzymol 211:389–406

Steady-State Fluorescence Spectroscopy as a Tool to Monitor Protein/Ligand Interactions

3

Roopa Kenoth, Balamurali M. M., and Ravi Kanth Kamlekar

Abstract

Fluorescence spectroscopy is an ideal and powerful methodology for the potent and reliable study of protein–ligand interactions. Enhanced susceptibility accompanied with comparative easiness forms the prominent key factor for the application of fluorescence techniques in these studies. In fluorescence technique, protein–ligand interactions are often studied at very low concentrations compared to other optical methods with a thousandfold higher sensitivity. In this method, we measure the variation in quantum yield upon ligand binding, by observing variations in ligand fluorescence, intrinsic protein fluorescence, or fluorescence of covalently or noncovalently bound fluorescent probes that are sensitive to ligand binding. Sensitivity of a fluorescent ligand to the environment, energy transfer from protein to ligand producing reduction in protein fluorescence or amplification of ligand fluorescence, a conformational change in protein/ ligand on binding to the protein, etc., contribute to the changes in quantum yield upon ligand binding [1–3]. This chapter will outline a sketch of steady-state fluorescence techniques for defining molecular interactions and calculation of binding constants.

R. Kenoth · R. K. Kamlekar (✉)
Department of Chemistry, School of Advanced Sciences, VIT, Vellore, Tamilnadu, India
e-mail: ravikanth.k@vit.ac.in

B. M. M.
Division of Chemistry, School of Advanced Sciences, VIT, Chennai, Tamilnadu, India

© The Author(s), under exclusive license to Springer Nature Singapore Pte Ltd. 2022

H. Sahoo (ed.), *Optical Spectroscopic and Microscopic Techniques*,
https://doi.org/10.1007/978-981-16-4550-1_3

3.1 Introduction

Fluorescence spectroscopy is an ideal and powerful methodology for the potent and reliable study of protein–ligand interactions. Enhanced susceptibility accompanied with comparative easiness forms the prominent key factor for the application of fluorescence techniques in these studies. In fluorescence technique, protein–ligand interactions are often studied at very low concentrations compared to other optical methods with a thousandfold higher sensitivity. In this method, we measure the variation in quantum yield upon ligand binding, by observing variations in ligand fluorescence, intrinsic protein fluorescence, or fluorescence of covalently or noncovalently bound fluorescent probes that are sensitive to ligand binding. Sensitivity of a fluorescent ligand to the environment, energy transfer from protein to ligand producing reduction in protein fluorescence or amplification of ligand fluorescence, a conformational change in protein/ ligand on binding to the protein, etc., contribute to the changes in quantum yield upon ligand binding [1–3]. This chapter will outline a sketch of steady-state fluorescence techniques for defining molecular interactions and calculation of binding constants.

3.1.1 Basic Concepts

Emission and excitation spectra along with quantum yield and anisotropy are the distinguishing features of protein fluorescence. These specifications get altered during the association of a protein with another protein or ligand based on the immediate surroundings of the fluorophore. Therefore, these parameters are regularly used to measure the magnitude of complex formation associated with a particular interaction.

3.1.2 Intrinsic Protein Fluorescence

Intrinsic protein fluorescence is employed to gain knowledge about ligand binding. Intrinsic fluorescence of proteins is contributed largely by aromatic amino acid residues like tryptophan (Trp), tyrosine (Tyr), and phenylalanine (Phe) [4]. The quantum yield and molar extinction coefficient of Trp residues are high in accordant with stronger fluorescence compared to the Tyr and Phe. Energy transfer from Tyr and Phe to Trp is also possible. Protein–ligand interaction often alters the fluorescence of proteins. Information on conformational and structural changes during the interaction with various ligands is derived from the sensitivity of tryptophan residues to the polarity of the immediate environment [4, 5]. Trp residues are mostly found in the binding sites or at the subunit interfaces as they possess a greater tendency to interact through hydrophobic interactions as well as hydrogen bonds. Solvent accessibility and local mobility are often limited by ligand binding. The spectrum will shift to blue with a change in intensity and polarization during this process. Fluorescence intensity decreases if ligand binding leads to conformational change [which brings quenching group in the vicinity of the fluorophore or if the ligand is close enough to tryptophan(s)]. For most practical applications, Trp residues are excited at 295 nm to avoid contributions from other residues especially Tyr. In a nonpolar environment, Trp residues fluoresce at 320 nm while emission occurs at significantly longer wavelengths in polar surroundings [6].

3.1.3 Extrinsic Fluorescent Probes

Extrinsic fluorescent probes are employed when there is a poor response in intrinsic fluorescence upon protein–ligand interaction [7]. By conjugating these probes to proteins, nucleic acids, or lipids, the sensitivity and quality of the information provided by fluorescence spectrometry can be improved. Hydrophobic fluorescent ligands such as ANS are strongly fluorescent only when bound to the protein. Its derivatives like Prodan and Nile Red are also suitable means to explore protein–ligand interactions. The second category of solvent-sensitive fluorescent probe is Ketocyanines. Competition experiments can be carried out in which other ligands compete with this fluorophore. Thus fluorescence of both intrinsic and extrinsic may therefore be used for estimating dissociation constant (K_d) values of ligand binding with proteins. Fluorescent compounds, such as fluorescein, and rhodamine derivatives in addition to cyanine compounds may be combined with the ligand and employed for studying protein–ligand association. Fluorescence properties of cofactors such as NAD, FAD, and porphyrins are also helpful for in vitro applications [7–9].

3.2 Steady-State Fluorescence Analysis

Steady-state fluorescence is a general method used to measure the binding interactions of proteins. High sensitivity and capacity to execute measurements even in dilute protein solutions are the major advantages of fluorescence studies. The high sensitivity of tryptophan to its surroundings assists to detect protein association reactions by fluorescence. Fluorescence emission maximum or quantum yield of tryptophan residues are altered subjected to a change in solvent or a change in the vicinity of quenching groups [10–12].

The tertiary structure of the protein is often altered during protein–ligand binding. This change could affect the environment of intrinsic or extrinsic fluorophore of the protein, and as a consequence, the fluorescence spectrum gets altered and the changes observed at certain wavelengths may help to calculate the dissociation constant (K_d) of the protein–ligand complex. K_d is an estimate of the attraction of the protein for a particular ligand and is evaluated as shown below in Eq. (3.1).

$$\begin{gathered} P + L \rightleftarrows PL \\ K_d = [P]\,[L]/[PL] \end{gathered} \tag{3.1}$$

3.2.1 Fluorescence Analysis of Protein–Ligand/Drug Interactions

Fluorescence titrations are the most commonly used method for studying the protein–ligand interactions provided one can obtain an apparent variation in fluorescence as a result of binding to the ligand (Fig. 3.1). The ratio of protein to the ligand is changed gradually by changing their concentrations generally by sequential dilution during a fluorescence titration. Binding sites get saturated with an increase in the concentration of ligand (Fig. 3.2).

Half the concentration of the ligand at half saturation point is considered as the dissociation constant [13–17]. The fluorescence change observed when all ligand-binding sites are occupied (ΔF_{max}) is obtained at saturating concentrations of the ligand. When it is not possible to saturate acceptor with ligand under the conditions of assay owing to factors such as a large inner filter effect at high ligand concentration, ΔF_{max} may be obtained by extrapolation procedures. ΔF_{max} can also be obtained by titrating t solution of fixed total ligand concentration with an increasing concentration of protein.

In systems where the total concentration of ligand ($[L_t]$) is much bigger than the total protein concentration, ($[P_t]$), the free ligand concentration, ($[L_f]$), can be compared with the total ligand concentration. Under such circumstances, F_{max} can be obtained from a conventional double reciprocal plot of $1/\Delta F$ versus $1/[L_t]$ by extrapolating to infinite concentrations of ligand. If the bound ligand is significantly

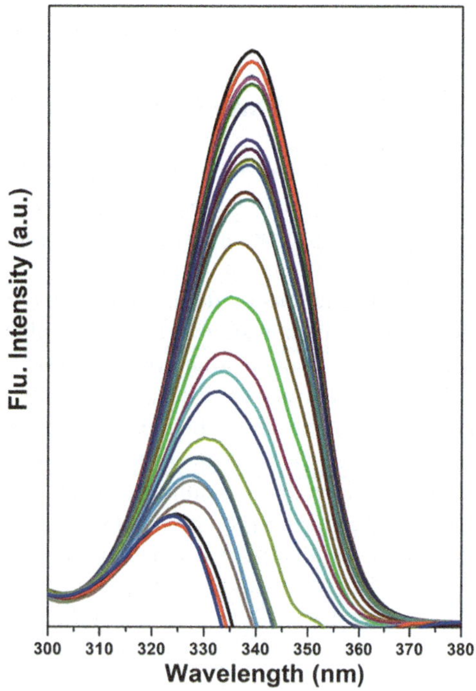

Fig. 3.1 Change in fluorescence intensity of protein upon ligand binding

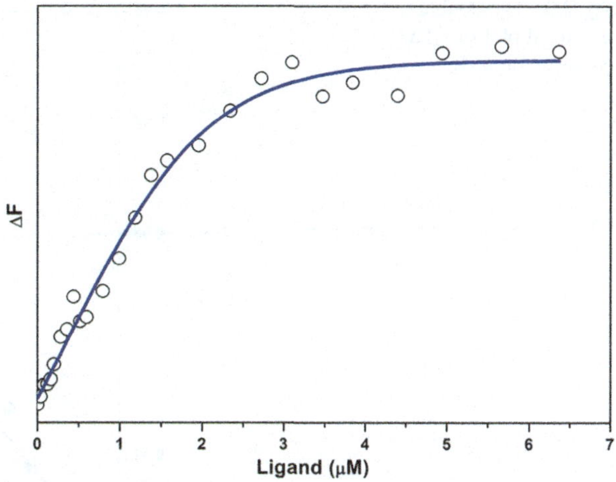

Fig. 3.2 A Plot of ligand concentration versus the change in fluorescence depicts the binding interactions between the protein and ligand

comparable to the total concentration of the ligand, this method may result in the wrong estimation of $\Delta F_{max,}$ as a linear relationship will not exist between $1/\Delta F$ and $1/[L_t]$ under such conditions. The ratio $\Delta F/\Delta F_{max}$ matches the fractional possession of total protein-binding sites (α) by ligand. The association constants related to the interaction can be obtained from conventional double reciprocal plots (Chipman Analysis) [18] or Scatchard plots [19].

3.2.1.1 Chipman Analysis

Double reciprocal plot or the plot of $1/\Delta F$ versus $1/[L_t]$, where ΔF is change in fluorescence intensity of the protein at a given ligand concentration during the titration, yields a straight line (Fig. 3.3) [18, 20]. From the ordinate of the plot, $F_{\alpha,}$ the fluorescence intensity when all the ligand molecules are bound to the protein was calculated. The binding constant, K_a, can be obtained from the abscissa of the plot of log $\{\Delta F/\ (F_c - F_\alpha)\}$ versus log $[L_f]$.

According to Chipman et al. (1967):

$$\text{Log}\ \{\Delta F/(F_c - F_\alpha)\} = \log K_a + \log\ [L_f] \tag{3.2}$$

where $[L_t]$ is the total ligand concentration in solution, $[L_f]$ is the concentration of free ligand (without bound protein), and $[P_t]$ is the total concentration of protein (Fig. 3.4).

$$[L_f] = [L_t] - P_t[\Delta F/\Delta F_\alpha] \tag{3.3}$$

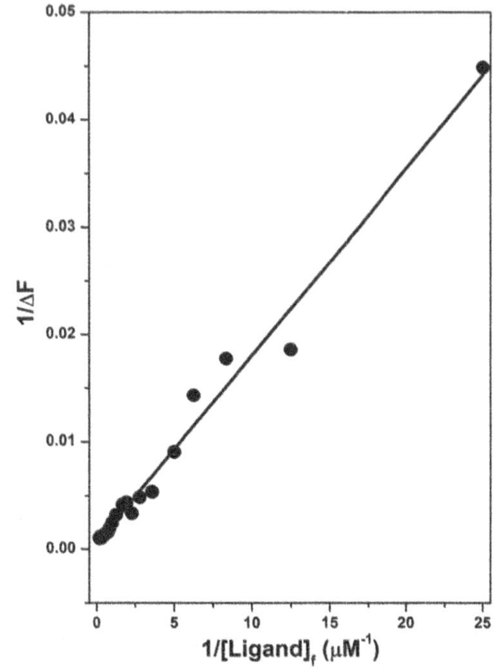

Fig. 3.3 The double reciprocal plot of $1/[\Delta F]$ versus 1/[Ligand]

3.2.1.2 Scatchard Plot Analysis

For the Scatchard plot analysis of fluorescence titrations, protein quantity is maintained at very low concentrations such that $[L]_{\text{bound}} << [L]_{\text{total}}$, and for further investigation of titration data, $[L]_{\text{total}}$ could be used as a good estimate for free ligand concentration. Plot of $\{F_r/[L]\}$ versus F_r, where F_r is ratio of corrected fluorescence intensity (F_{corr}) [$F_{\text{corr}} = F_{\text{obs}}$ antilog [$(\text{OD}_{\text{ex}} + \text{OD}_{\text{em}})/2$] and maximum fluorescence intensity (F_{max}) of the binding data, is used to analyze the binding (Fig. 3.2b). Association constant K_a and the number of binding sites are obtained respectively from the slope and x intercept of the Scatchard plot (Fig. 3.5) [20]. A typically reversible binding for the protein "P" and ligand "L" can be expressed as

$$P + nL \leftrightarrow PLn,$$

where n represents the total number of binding sites available on the protein for the ligand of interest, and PL_n denotes the protein–ligand macromolecular complex. The average number of ligand molecules "r" that can bind to the protein molecule at any given concentration of $[P]$ and $[L]$ is given as:

$$r = [PL_n]/[P] \tag{3.4}$$

While the Scatchard equation is expressed as:

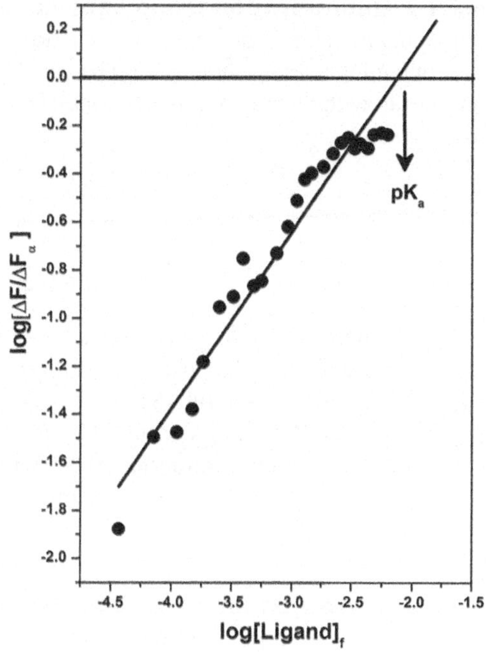

Fig. 3.4 A linear plot representing the Chipman analysis to follow the binding of ligand to the protein and estimate the associate constant (K_a)

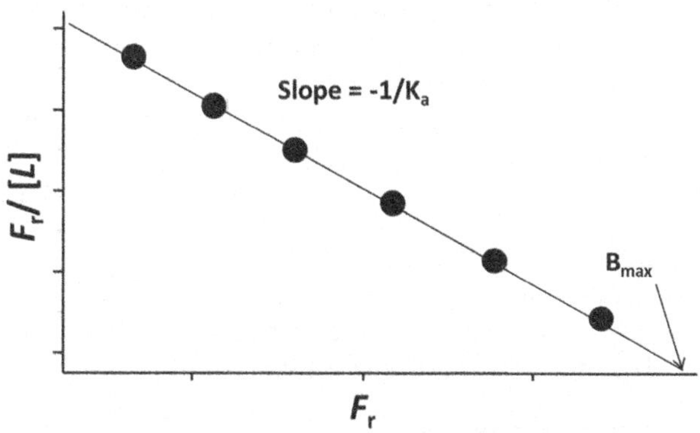

Fig. 3.5 Representation of a Scatchard plot of $F_r/[L]$versus F_r

$$r/[L] = nK_a - rK_a \tag{3.5}$$

A plot of $r/[L]$ versus r provides a slope of $-K_a$, and K_a represents the association (affinity) constant of the protein for the specific ligand. A deviation from linearity is observed if the interaction involves multiple binding sites on the protein (positive or negative cooperativity).

3.2.1.3 Quantification Using Labeled Ligand

For quantification of binding reactions using fluorescent ligand, ligand (L) containing a fluorophore should be capable of forming a complex (ASL) without interfering with the binding interactions of protein A with compound S. The reaction may be shown as

$$A + S \overset{K}{\rightleftarrows} AS$$

$$AS + L \overset{K_L}{\rightleftarrows} ASL$$

While interacting with the complex AS, the spectroscopic properties of the ligand L alter substantially. Supposing L is present in large excess comparative to the complex, $[ASL] = K_L[L][AS] \approx K_L[L]_{tot}[AS] = K_L[L]_{tot}K[A][S] = K_{obs}[A][S]$. The observed binding constant (K_{obs}) can be obtained by tracking the amount of ASL at several concentrations of S (with the concentration of A fixed). The binding constant ($K_d = 1/K$) can be estimated by knowing the values of K_L and $[L]_{tot}$ [21–23].

3.2.1.4 Correction for Inner Filter Effect

Because of the absorption of the incident and emitted light passing through the cell, the recorded fluorescence intensity of the sample for which the emission is detected usually varies disproportionally to sample concentration, and this process is called the inner filter effect. In other words, the inner filter effect is caused by the reabsorption of emitted light by the quencher.

Inner filter effect can be calculated by the equation:

$$F_{corr} = F_{obs}\text{Antilog}\left[(\text{OD}_{ex} + \text{OD}_{em})/2\right] \quad (3.6)$$

where F_{corr} and F_{obs} are the corresponding corrected and observed fluorescence intensities, and OD_{ex} and OD_{em} are the respective absorption intensities of the sample at the excitation wavelength and emission wavelength [22].

To reduce the inner filter effect,

- reduce the concentration of the sample. To get good results, optical density is kept preferably <0.1.
- excitation at 10–50 nm below the absorption maximum is recommended to avoid secondary inner filter effects.
- Decrease the chosen path length. To reduce the path length and consecutively the absorbance of the sample, choose a small-capacity cuvette or a triangular cuvette or move the cuvette holder such that the cuvette area measured is near the corner.

3.2.1.5 Applications

Steady-state fluorescence spectroscopy is useful for studying protein–substrate interactions, protein–nucleic acid interactions, protein–nucleotide interactions, protein–protein interactions, etc. The binding of drugs with carrier proteins is essential for the transport and metabolism of drugs. The microenvironment and architecture experienced by the ligand is determined largely by the folding/unfolding and/or transformation of protein structure during the interaction. In drug development for example, information about the binding site is very crucial, and this information is provided by spectroscopic properties of ligand housed in the microenvironment of proteins like serum proteins, which in turn will reflect the immediate surroundings of the molecule inside the target proteins [24–28]. Changes in fluorescence properties associated with protein–nucleotide interactions may be used as sensors for enzymatic activities or as probes for screening inhibitors/natural substrates [29–31]. A 2.5-fold increase in fluorescence is observed upon binding of Mant-ATP, a nucleotide analog to the cyclin-dependent kinase (Cdk2) that allows one to determine the dissociation constant [32, 33]. Fluorescently labeled oligonucleotides were used to investigate the mechanism of several DNA or RNA polymerases as well as to monitor the binding of regulatory proteins [34].

A combination of steady-state transient kinetic, and FRET single-molecule assay using dyes attached to the enzyme or its different partners helped in unraveling the structure and mechanism of the reverse transcriptase (RT) of HIV (human immunodeficiency virus), a key enzyme in the virus replication cycle that catalyzes a chain of reactions to convert the single-stranded HIV RNA genome into a double-stranded DNA for further integration in the host cell genome. The large change in dye fluorescence upon binding of primer/template oligonucleotide to RT was used to follow this interaction. These investigations have shown that catalysis of RT is dependent on the binding orientation of the substrate, which adopts opposite conformations for DNA and RNA duplexes [35–38].

Protein/protein interaction studies provide better a understanding of the different parameters that rule the association of protein complexes. Analyses of these parameters enable the development of shorter peptides with similar interaction properties as well as in the design of novel inhibitors of specific protein/protein interactions as well as biosensors of protein complexes [39–42]. Information obtained using fluorescence spectroscopy regarding the structure, organization, and dynamics of membrane proteins could be crucial for a better understanding of their role in health and diseases, as large a number of cellular functions are mediated through membrane proteins, which also play a vital role in pathogenicity [43].

3.2.2 Fluorescence Analysis of Protein–Lipid Interaction

Tryptophan (Trp) fluorescence measurements are one of the important methods used to investigate membrane interaction of human Glycolipid transfer protein (GLTP) (Fig. 3.6a). GLTP is excited at 295 nm, and emission spectra are measured from 310 to 420 nm. To avoid inner filter effects, protein concentration should be maintained at $A_{295} < 0.1$ [44]. While investigating the membrane interactions, the variation in emission signals from the Trp of GLTP (1 μM) is monitored with the addition of small aliquots of the phosphatidylcholine (PC) vesicles with/without

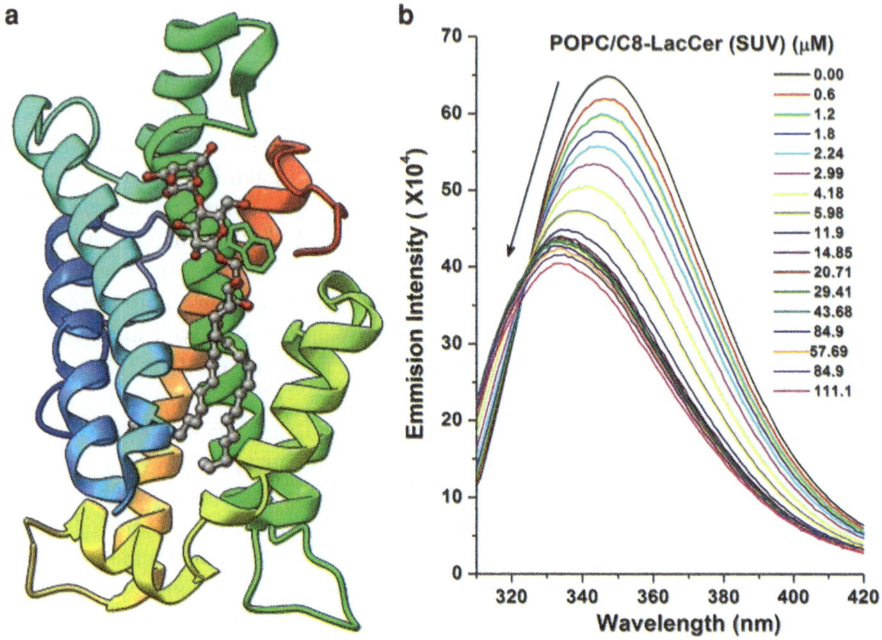

Fig. 3.6 (**a**) Model depicting the binding of GLTP with glycolipid, lactosylceramide (LacCer), and the probe residue (Trp96) is shown in stick model, and ligand is represented as ball and stick. (**b**) Decrease in Trp fluorescence of GLTP due to addition of POPC lipids vesicles having LacCer aliquots

glycolipid (20 mol%) (Fig. 3.6b). To assess the direct binding of glycolipid to GLTP, micro addition of glycolipids dissolved in ethanol can be followed [44, 45]. Measurements were accomplished with constant stirring along with the addition of 0.1 mM GSL to 1 μM of protein.

3.2.2.1 Binding (Partitioning) Coefficient Analysis

The total number of binding sites (α) occupied by glycolipids can be determined by the following Eq. (3.7):

$$\alpha = (F - F_o)/F_{\max} \tag{3.7}$$

where F_o and F are the emission intensities of the Trp residues in GLTP (in the absence and presence of glycolipid, respectively). The emission intensity of the completely saturated GLTP (at excess glycolipid) is $F_{\max}$ [46, 47]. By plotting a graph $1/(F - F_o)$ versus $1/L$, $F_{\max}$ can be determined (extrapolating $1/L = 0$, where L = total glycolipid concentration). The maximum change in the protein fluorescence (ΔF_m) is observed at saturating concentrations of glycolipid, and it is estimated by plotting $1/L$ against $1/\Delta F$. The following relationship was used to calculate the concentration of bound glycolipid to the protein:

$$[\text{Bound Lipid}] = \text{Concentration of Protein} \times \Delta F/\Delta F_{\text{m}} \tag{3.8}$$

The concentration of free lipid is calculated as:

$$[\text{Free lipid}] = [\text{Total lipid}] - [\text{Bound lipid}] \tag{3.9}$$

K_d values can be determined by nonlinear least-squares (NLLSQ) fit of the plot of bound lipid versus free lipid.

Spectral data of fluorescence titration by the ethanol injection method were analyzed according to:

$$\varepsilon - 1 = (\varepsilon_b - 1) - K_d(\varepsilon - 1)/mn \tag{3.10}$$

where K_d is the dissociation constant of lipid–protein complex, m is the concentration of lipid, and n is the number of binding sites of lipid. The quantity ε is the spectral parameter relative value (peak emission wavelength, λ_{max}, or intensity, I) associated with lipid binding to GLTP at a concentration of lipid "m". Thus, ε can symbolize either I/I_o or $(\lambda_{max})_o/\lambda_{max}$, where the subscript "o" refers to the values in the absence of lipid. The parameter ε_b represents the spectral properties of the protein–lipid complex. As per Eq. (3.10), $\varepsilon - 1$ versus $(\varepsilon - 1)/m$ graph gives the slope as K_d/n, the reciprocal of the protein/lipid association constant.

3.3 Fluorescence Quenching Analysis

Fluorescence quenching refers to any process that involves a variety of molecular interactions that lead to a reduction in fluorescence intensity. The interactions can be energy transfer, ground state complex formation, excited state reactions, molecular collisions, etc.

Quenching is broadly classified under two categories: static and dynamic quenching. Dynamic quenching involves a collision between two molecules with fluorophore losing energy as kinetic energy. In other words, dynamic quenching occurs when the fluorophore and another molecule diffuse in solution and collide with each other. Each collisional events lead to the transfer of excitation energy to the quencher molecule. Molecules do not form complex in this case.

Static quenching occurs when two molecules interact with each other to form a complex (Fig. 3.7). The interaction between the two molecules can be monitored by means of variation in the intrinsic fluorescence of aromatic amino acids or using extrinsic fluorophores such as ANS. Changes in fluorescence parameters can be used to follow the interaction [48–50]. This technique has been employed for various applications including pathogen sensing, chemosensing, heavy metal detection, imaging, etc. The extent of quenching can be quantitatively estimated using the Stern–Volmer equation.

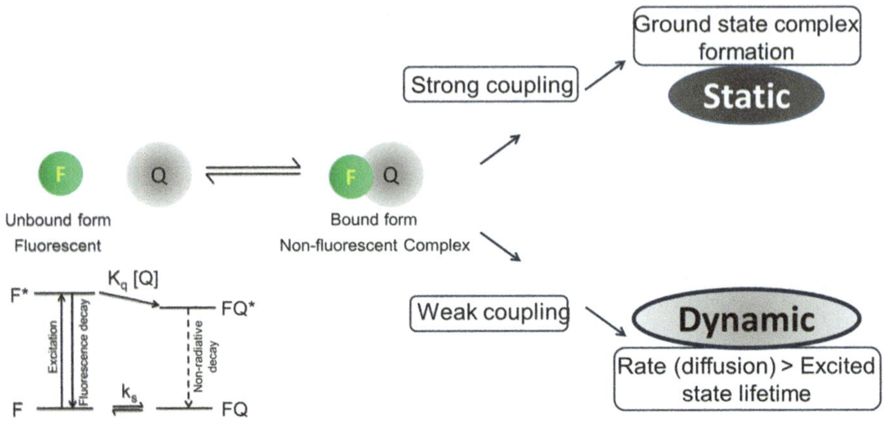

Fig. 3.7 Schematic illustration of fluorescence quenching: static and dynamic

3.3.1 Collisional Quenching: The Stern–Volmer Plot

The relationship between the ratio of fluorescence intensity in the absence and presence of quencher and the concentration of quencher is known as the Stern–Volmer equation.

$$F_o/F_c = 1 + K_{SV}[Q] \tag{3.11}$$

where F_o and F_c are the respective fluorescence intensities in the absence and presence of the quencher, K_{SV} is the Stern–Volmer (SV) quenching constant, and $[Q]$ is the concentration of the quencher.

A linear Stern–Volmer plot suggests a single class of fluorophores available to the quencher equally. SV plots often show deviation from linearity because of a variety of factors such as heterogeneity in fluorophore excited state, the mixture of quenching mechanisms, and deviation from ideal behavior.

Static quenching will dominate at higher concentrations and diffusional quenching at lower. In the case of static quenching, the slope of the graph decreases, while for dynamic quenching, it rapidly increases with temperature. In the former case, the complex formation is inhibited with an increases in temperature, while in the latter one, the number of collisions increases with temperature. The shape and position of the fluorescence spectrum change with protein concentration for static quenching, while during dynamic quenching, fluorescence intensity decreases without change in shape and position of the spectrum. Linear SV plots are obtained for dynamic quenching. An upward curvature is explained in terms of both static and dynamic quenching. A downward curvature may be indicative of static quenching and may be attributed to fractional accessibility of fluorophore to the quencher. In such situations, modified Stern–Volmer plots are used for analyzing the process.

Modified Stern–Volmer equation is expressed as:

$$F_o/\Delta F = f_{\mathrm{a}}^{-1} + (K_{\mathrm{a}}f_{\mathrm{a}})^{-1}[Q]^{-1} \tag{3.12}$$

where f_a is the portion of fluorophore sites available to the quencher, and K_a is the Stern–Volmer quenching constant. When there is an explicit interaction between protein and the quencher during complex formation, the modified Stern–Volmer plot displays linearity.

The position of the fluorophore within the protein determines its accessibility to the solvent and subsequently to the quencher. For instance, Trp present at the surface residues will have better accessibility to the solvent as compared to the buried ones. Hydrophobic regions of proteins are not freely accessible even to water-soluble quenchers, like iodide and acrylamide. The quenching constant is intensely influenced by the emission maximum. Studies reveal that a red-shifted is observed when the tryptophan residues of proteins are almost completely accessible to acrylamide quenching and blue-shifted when the tryptophan residues remain largely inaccessible to quenching by acrylamide. Being polar molecules, acrylamide and iodide cannot readily enter into the nonpolar regions of proteins. Small nonpolar oxygen molecules on the other hand can freely enter.

In the recent past, emitting semiconductor quantum dots along with quenchers were employed for detecting biologically significant molecules [51, 52]. But their applications were limited by their poor biodegradability and potential toxicity. Recently, fluorescence emission from graphene quantum dots was quenched by novel MnO_2 nanoflowers and has been demonstrated for the detection of intracellular glutathione [53]. This method follows a "fluorescence turn-on" mechanism for its detection (Fig. 3.8). The interaction between the fluorophore and the quencher was electrostatic. In the presence of glutathione, the nanoprobe MnO_2-GQDs formed by the adsorption of MnO_2 degrade to release Mn^{2+} and highly fluorescent free GQDs. This method was reported to be highly specific and sensitive for glutathione with a detection limit of 2.8 μM. This method can be extended for cellular imaging to detect GSH.

For the past few decades, ethidium bromide has been used as a luminescent probe to investigate the binding of small molecules or drugs to nucleic acids [54]. It is known that neither ethidium bromide nor the nucleic acids are fluorescent as such, but when they form a complex with each other, they become fluorescent. But ethidium bromide being a carcinogen, its use is limited as a probe. Recently, nanoceria particles were reported to form bionanoconjugates with nucleic acid

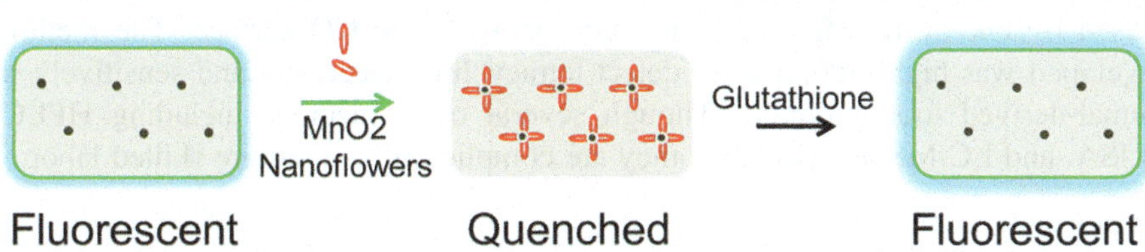

Fig. 3.8 Quenching of fluorescence from graphene quantum dots by manganese dioxide nanoflowers follows a fluorescence turn-on mechanism in the detection of intracellular glutathione

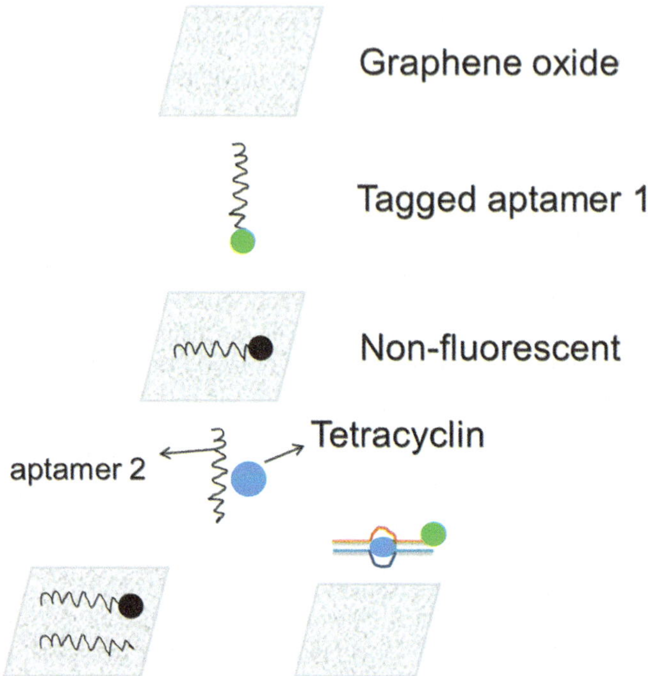

Fig. 3.9 Schematic represents the principle involved in the detection of tetracycline by fluorescence turn-on aptasensing method with graphene oxide and fluorophore-tagged aptamer

bases and were utilized as molecular nanoprobes, for the single-step investigation of oligonucleotides [55]. This method has a detection limit of 0.12 nM. This method can find a wide range of applications [56] in diagnostics, biosensors and chemosensors, gene expression analysis, cellular imaging, and forensic analysis.

The fluorescence turn-on aptasensing method has been proposed for the detection of tetracycline in animal-derived food (Fig. 3.9). Tetracycline being reported as an allergen to humans, its use in excess leaves out its residues in animal-derived food products [57]. In this method, a tetracycline binding RNA aptamer that is linked to a fluorophore is employed for detection. The tagged aptamer becomes fluorescent only upon detecting tetracycline, as this disrupts the π–π stacking between the graphene oxide (GO) surface and the tagged aptamer. This leads to the inhibition of quenching caused by energy transfer from the fluorophore to the GO surface. The method developed was highly efficient to detect tetracycline selectively and sensitively in animal-derived food products. Though several other methods including HPLC, ELISA, and LC-MS are available, they are complicated and require skilled labor.

3.4 Red-Edge Excitation Shift (REES) Analysis

It is known that the emission spectra are always independent of the excitation wavelength when the solvent relaxation times are much faster than the life time of the excited electronic state. But this is not true when the spectra are recorded in a polar environment with considerably high viscosity. In an equilibrium condition, a fraction of the ground state molecules are likely to have the same solvation sphere as the relaxed excited state. Excitation of such low-energy molecules leads to an emission spectrum that is more red-shifted. Such phenomenon is known as red-edge excitation spectrum or red-edge excitation shift (REES). REES is an optical phenomenon that can be utilized to obtain unique information on molecular conformations and the associated equilibrium. In this technique, the more stable fluorophores are excited resulting in a red shift in the maximum of the emission intensity. This phenomenon helps to identify the fluorescent conformers with varying solvation spheres and biomolecular binding interactions in the ground and excited states. This effect is due to a change in dipole moment which further induces a change in the extent of solvation at different electronic levels. Higher the solvation, longer will be the life time in that particular state than that of the environmental relaxation which results in an emission maximum independent of the excitation wavelength. In this method by following photoselection, it is possible to identify fluorophores in different solvation environments especially at the red edge of the excitation spectra that are in equilibrium. Experimentally, one observes a red shift in the emission maximum with respect to the excitation wavelength. In the case of proteins, the intrinsic fluorescence from the tryptophan and/or tyrosine varies depending on the proportions of the folded, molten globule, or unfolded states [58, 59]. Using this REES method, a novel and structurally unique lipid-binding motif has been identified and investigated in human glycolipid transfer protein [45].

There are several reports wherein the selective excitation of species in a complex environment to instigate its chemical reaction in a controlled manner was studied extensively. Similar studies were carried out for benzophenone and its coordination with phenol via hydrogen bonding in a polar aprotic solvent [60]. Antimicrobial peptides have high selectivity toward bacterial targets in various host cells and have restricted bacterial resistance. The role of pH on the activity and biophysical properties of one such peptide C18G was investigated by red-edge excitation spectroscopy (REES). It was shown that clear pH dependence toward aggregation was observed [61].

In another study, the binding interactions of malathion with human serum albumin in the presence of silver nanoparticles were investigated using various optical techniques including fluorescence quenching, fluorescence resonance energy transfer (FRET), circular dichroism, red-edge excitation shift (REES), synchronous fluorescence, and three-dimensional fluorescence spectroscopy. The results of synchronous fluorescence spectroscopy and REES indicated the changes in the microenvironment of the fluorophore [62].

3.5 Synchronous Spectroscopy

Synchronous spectroscopy is a technique that has been used for the investigation of multicomponent systems. This method is considered as a simple optical method with minimum light scattering compared to conventional steady-state fluorescence emission methods, where difficulty with the analyses of impure or mixture of samples exists. Synchronous fluorescence spectroscopy was first developed by Lloyd. Here, both the emission and excitation monochromators are scanned simultaneously during measurement to get the synchronous spectrum. The advantages include spectral simplification, bandwidth narrowing, minimized scattering, and enhanced resolution. It also provides high selectivity and sensitivity in analytical measurements of complex samples [63].

There are two types of synchronous fluorescence spectra: constant wavelength or constant energy and variable angle (Fig. 3.10). The choice of type selection depends on the flexibility and selectivity of the measurements to be made. From the above spectrum, it is possible to generate an isopotential trajectory by connecting the points of equal intensity.

Synchronous spectroscopy has been in the application of determining polycyclic aromatic hydrocarbons, benzopyrenes, benzofluoranthene, anthracene, etc. in

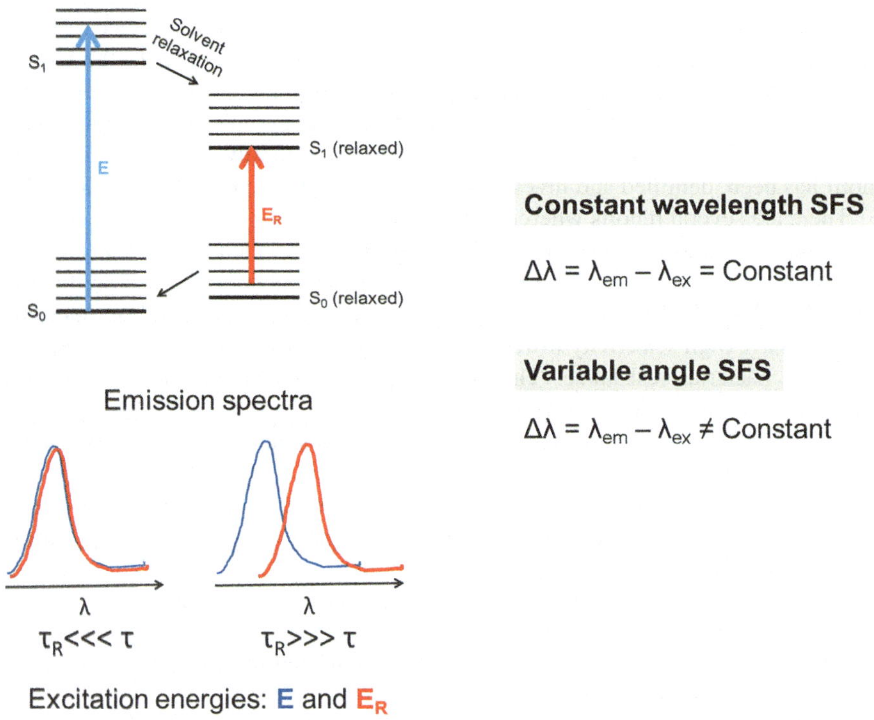

Fig. 3.10 Schematic representing the principle of synchronous fluorescence spectroscopy

foodstuffs. Moreover, these approaches are rapid, cost-effective, and efficient as an analytical tool for rapid screening of metabolized products and by-products in a large number of human samples [64] like blood, feces, and saliva. Overall, the combination of these approaches along with other techniques, such as derivative and/or low-temperature techniques and chemometrics, has shown potential performances in real-time applications. These techniques also serve as potential tools for clinical analysis and food safety evaluation [65–67].

The potential of fluorescence spectroscopy is not limited to the above discussions. With the emergence of new diseases and more virulent pathogens, investigations on novel therapeutics and diagnostic methods need to be intensified. Thermally activated delayed fluorescence (being employed for imaging applications) and circularly polarized emission (being employed as a probing tool to monitor luminescent chiral molecular systems and in particular the biological systems) are some of the advanced optical methods optical techniques that are still in their preliminary stage of development. There is still a lot more to be explored for implications of fluorescence-assisted techniques for the advent of many probing devices.

References

1. Lakowicz JR (ed) (2013) Principles of fluorescence spectroscopy. Springer Science and Business media, Berlin
2. Demchenko AP, Lakowicz JR (eds) (1991) Topics in fluorescence spectroscopy: biochemical applications. Plenum, New York, pp 65–112
3. Lakowicz JR (2002) Principles of fluorescence spectroscopy, 2nd edn. Kluwer Academics/ Plenum, NewYork, pp 445–486
4. Saxena A, Udgaonkar JB, Krishnamoorthy G (2005) Protein dynamics and protein folding dynamics revealed by time-resolved fluorescence. In: Hof M, Hutterer R, Fidler V (eds) Fluorescence spectroscopy in biology, vol 3. Springer, Berlin, Heidelberg, pp 163–179
5. Eftink MR (1994) The use of fluorescence methods to monitor unfolding transitions in proteins. Biophys J 66:482–501
6. Chen Y, Barkley MD (1998) Toward understanding tryptophan fluorescence in proteins. Biochemistry 37(28):9976–9982
7. Demchenko AP, Mély Y, Duportail G, Klymchenko AS (2009) Monitoring biophysical properties of lipid membranes by environment-sensitive fluorescent probes. Biophys J 96(9): 3461–3470
8. Martin MM, Lindqvist L (1975) The pH dependence of fluorescein fluorescence. J Lumin 10: 381–390
9. Ma LY, Wang HY, Xie H, Xu LX (2004) A long lifetime chemical sensor: study on fluorescence property of fluorescein isothiocyanate and preparation of pH chemical sensor. Spectrochim Acta A Mol Biomol Spectrosc 60(89):1865–1872
10. Varlan A, Hillebrand M (2010) Study on the interaction of 2-carboxyphenoxathiine with bovine serum albumin and human serum albumin by fluorescence spectroscopy and circular dichroism. Rev Roum Chim 55:69–77
11. Khan SN, Islam B, Khan AU (2007) Probing midazolam interaction with human serum albumin and its effect on structural state of proteins. Int J Integr Biol 1(2):102–112
12. Alarcón E, Edwards AM, Aspée A, Borsarelli CD, Lissi EA (2009) Photophysics and photochemistry of Rose Bengal bound to human serum albumin. Photochem Photobiol Sci 8:933–943

13. Weljie AM, Vogel HJ (2002) Steady-state fluorescence spectroscopy. Methods Mol Biol 173: 75–87
14. Ward LD (1985) Measurement of ligand binding to proteins by fluorescence spectroscopy. Methods Enzymol 117:400–414
15. Mocz G, Ross JA (2013) Fluorescence techniques in analysis of protein-ligand interactions. Methods Mol Biol 1008:169–210
16. Yammine A, Gao J, Kwan AH (2019) Tryptophan fluorescence quenching assays for measuring protein-ligand binding affinities: principles and a practical guide. BioProtocol 9(11):e3253
17. Williams MA, Daviter T (eds) (2013) Protein-ligand interactions: methods and applications, methods in molecular biology. Springer Science Business Media, New York, p 1008
18. Chipman DM, Grisaro V, Sharon N (1967) The binding of oligosaccharides containing N-acetylglucosamine and N-acetylmuramic acid to lysozyme. J Biol Chem 242:4388–4394
19. Roberts DD, Goldstein IJ (1982) Lectin from Lima beans (Phaseolus lunatus). J Biol Chem 257: 11274–11277
20. Komath SS, Kenoth R, Swamy MJ (2001) Thermodynamic analysis of saccharide binding to snake gourd (Trichosanthesanguina) seed lectin fluorescence and absorption spectroscopic studies. Eur J Biochem 268:111–119
21. Khan MI, Mazumder T, Pain D, Gaur N, Surolia A (1981) Binding of 4-methylumbelliferyl b-d-galactopyranoside to Momordica charantia lectin. Eur J Biochem 113:471–476
22. Bessler W, Shafer J, Goldstein IJ (1974) Spectroscopic study of the carbohydrate binding site of con a. J Biol Chem 249:2819–2822
23. Kubista M, Sjöback R, Ericsson S, Albinsson B (1994) Experimental correction for the inner-filter effect in fluorescence spectra. Analyst 119:417–419
24. Royer CA (2006) Probing protein folding and conformational transitions with fluorescence. Chem Rev 106(5):1769–1784
25. Cohen BE, McAnaney TB, Park ES, Jan YN, Boxer SG, Jan LY (2002) Probing protein electrostatics with a synthetic fluorescent amino acid. Science 296:1700–1703
26. Abou-Zied OK, Al-Shini OIK (2008) Characterization of subdomain IIA binding site of human serum albumin in its native, unfolded, and refolded states using small molecular probes. J Am Chem Soc 130(32):10793–10801
27. Er JC, Vendrell M, Tang MK, Zhai D, Chang YT (2003) Fluorescent dye cocktail for multiplex drug-site mapping on human serum albumin. ACS Comb Sci 15(9):452–457
28. Yamasaki K, Chuang VTG, Maruyama T, Otagiri M (2013) Albumin–drug interaction and its clinical implication. Biochim Biophys Acta 1830:5435–5443
29. Bagshaw C (2001) ATP analogues at a glance. J Cell Sci 114:459–460
30. Jameson DM, Eccleston JF (1997) Fluorescent nucleotide analogs: synthesis and applications. Methods Enzymol 278:363–390
31. Weisbrod SH, Marx A (2008) Novel strategies for the site-specific covalent labelling of nucleic acids. Chem Commun (Camb) 44:5675–5685
32. Morris MC, Gondeau C, Tainer JA, Divita G (2002) Kinetic mechanism of activation of the Cdk2/cyclin a complex. Key role of the C-lobe of the Cdk. J Biol Chem 277(26):23847–23853
33. Heitz F, Morris MC, Fesquet D, Cavadore JC, Dorée M, Divita G (1997) Interactions of cyclins with cyclin-dependent kinases: a common interactive mechanism. Biochemistry 36(16): 4995–5003
34. Creighton TE (2010) The biophysical chemistry of nucleic acids and proteins. Helvetian Press, Eastbourne
35. Divita G, Müller BU, Immendörfer U, Gautel M, Rittinger K, Restle T, Goody RS (1993) Kinetics of interaction of HIV reverse transcriptase with primer/template. Biochemistry 32(31): 7966–7971
36. Liu S, Harada BT, Miller JT, Le Grice SFJ, Zhuang X (2010) Initiation complex dynamics direct the transitions between distinct phases of early HIV reverse transcription. Nat Struct Mol Biol 17:1453–1460

37. Abbondanzieri EA, Bokinsky G, Rausch JW, Zhang JX, Le Grice SFJ, Zhuang X (2008) Dynamic binding orientations direct activity of HIV reverse transcriptase. Nature 453(7192): 184–189
38. Agopian A, Depollier J, Lionne C, Divita G (2007) Trp24 and Phe61 are essential for accurate association of HIV-1 reverse transcriptase with primer/template. J Mol Biol 373(1):127–140
39. Kurzawa L, Morris MC (2010) Cell-cycle markers and biosensors. Chem Biochem 11(8): 1037–1047
40. Kurzawa L, Pellerano M, Coppolani JB, Morris MC (2011) Fluorescent peptide biosensor for probing the relative abundance of cyclin-dependent kinases in living cells. PLoS One 6(10): e26555
41. Gondeau C, Gerbal-Chaloin S, Bello P, Aldrian-Herrada G, Morris MC, Divita G (2005) Design of a novel class of peptide inhibitors of cyclin-dependent kinase/cyclin activation. J Biol Chem 280(14):13793–13800
42. Pommier Y, Cherfils J (2005) Interfacial inhibition of macromolecular interactions: nature's paradigm for drug discovery. Trends Pharmacol Sci 26(3):138–145
43. Bianchi G, Carafoli E, Scarpa A (1986) Membrane pathology. Ann N Y Acad Sci 488:1–153
44. Kamlekar RK, Gao Y, Kenoth R, Molotkovsky JG, Prendergast FG, Malinina L, Patel DJ, Wessels WS, Venyaminov SY, Brown RE (2010) Human GLTP: three distinct functions for the three Tryptophans in a novel peripheral amphitropic fold. Biophys J 99:2626–2635
45. Kenoth R, Zou X, Simanshu DK, Pike HM, Malinina L, Patel DJ, Brown RE, Kamlekar RK (2018) Functional evaluation of intrinsic Tryptophans in glycolipid binding and membrane interaction by HETC2, a fungal glycolipid transfer protein. Biochim Biophys Acta Biomembr 1860(5):1069–1076
46. Santra MK, Panda D (2003) Detection of an intermediate during unfolding of bacterial cell division protein FtsZ: loss of functional properties precedes the global unfolding of FtsZ. J Biol Chem 278:21336–21343
47. Srivastava R, Ratheesh A, Gude RK, Rao KVK, Panda D, Subrahmanyam G (2005) Resveratrol inhibits type II phosphatidylinositol 4-kinase: a key component in pathways of phopsphoinositide turnover. Biochem Pharmacol 70:1048–1055
48. Eftink MR, Ghiron CA (1981) Fluorescence quenching studies with proteins. Anal Biochem 114:199–227
49. Eftink MR, Ghiron CA (1976) Exposure of tryptophyl residues in proteins. Quantitative determination by fluorescence quenching studies. Biochemistry 15:672–680
50. Eftink MR, Ghiron CA (1976) Fluorescence quenching of indole and model micelle systems. J Phys Chem 80:486–493
51. Clift MJ, Stone V (2012) Quantum dots: an insight and perspective of their biological interaction and how this relates to their relevance for clinical use. Theranostics 2:668–680
52. Holzinger M, Le Goff A, Cosnier S (2014) Nanomaterials for biosensing applications: a review. Front Chem 2:63
53. Song ZL, Dai X, Li M, Teng H, Song Z, Xie D, Luo X (2018) Biodegradable nanoprobe based on MnO_2 nanoflowers and graphene quantum dots for near infrared fluorescence imaging of glutathione in living cells. Mikrochim Acta 185:485
54. Kiran T, Vasavi CS, Munusami P, Pathak M, Balamurali MM (2017) Evaluation of DNA/protein interactions and cytotoxic studies of copper (II) complexes incorporated with N, N donor ligands and terpyridine ligand. Int J Biol Macromol 95:1254–1266
55. Bulbul G, Hayat A, Mustafa F, Andreescu S (2018) DNA assay based on Nanoceria as fluorescence quenchers (NanoCeracQ DNA assay). Sci Rep 8:2426
56. Liu B, Liu J (2015) Comprehensive screen of metal oxide nanoparticles for DNA adsorption, fluorescence quenching, and anion discrimination. ACS Appl Mater Interfaces 7:24833–24838
57. Zhang L, Wang J, Deng J, Wang S (2020) A novel fluorescent "turn-on" aptasensor based on nitrogen-doped graphene quantum dots and hexagonal cobalt oxyhydroxide nanoflakes to detect tetracycline. Anal Bioanal Chem 412:1343–1351
58. Patra SK, Pal MK (1997) Red edge excitation shift emission spectroscopic investigation of serum albumins and serum albumin-bilirubin complexes. Sectrochim Acta Pt A Mol Biomol Spectrosc 53A(10):1609–1614

59. Chattopadhyay A, Haldar S (2014) Dynamic insight into protein structure utilizing red edge excitation shift. Acc Chem Res 47:12–19
60. Venkatraman RK, Orr-Ewing AJ (2019) Photochemistry of benzophenone in solution: a tale of two different solvent environments. J Am Chem Soc 141:15222–15229
61. Hitchner MA, Santiago-Ortiz LE, Necelis MR, Shirley DJ, Palmer TJ, Tarnawsky KE, Vaden TD, Caputo GA (2019) Activity and characterization of a pH-sensitive antimicrobial peptide. Biochim Biophys Acta Biomembr 1861:182984
62. Baghaee PT, Divsalar A, Jamshikhan CJ, Donya A (2019) Human serum albumin–malathion complex study in the presence of silver nanoparticles at different sizes by multi spectroscopic techniques. J Biomol Struct Dyn 37:2254–2264
63. Rubio S, Gomez-Hens A, Valcarcel M (1986) Analytical applications of synchronous fluorescence spectroscopy. Talanta 33:633–640
64. Lecrenier MC, Baeten V, Taira A, Abbas O (2018) Synchronous fluorescence spectroscopy for detecting blood meal and blood products. Talanta 189:166–173
65. Dankowska A, Malecka M, Kowalewski W (2015) Detection of plant oil addition to cheese by synchronous fluorescence spectroscopy. Dairy Sci Technol 95:413–424
66. Durakli VS, Ercioglu E, Boyaci IH (2017) Rapid discrimination between buffalo and cow milk and detection of adulteration of buffalo milk with cow milk using synchronous fluorescence spectroscopy in combination with multivariate methods. J Dairy Res 84(2):214–219
67. Temiz HT, Tamer U, Berkkan A, Boyaci IH (2017) Synchronous fluorescence spectroscopy for determination of tahini adulteration. Talanta 167:557–562

Fluorescence Anisotropy: Probing Rotational Dynamics of Biomolecules

4

Gourab Prasad Pattnaik and Hirak Chakraborty

Abstract

Molecules present in the lowest energy state (ground state) absorb light to be raised to the higher energy state (excited state). Fluorescence is an optical phenomenon through which the excess energy is being dissipated in the form of light (emission of photons). Fluorescence can be discriminated from the other emissive deactivation pathway, phosphorescence, by the timescale of the process. The fluorescence is a phenomenon that occurs within nanosecond timescale, while phosphorescence is observed at higher time window (millisecond to second). The fluorescence is highly sensitive to the molecular as well as environmental properties, which makes fluorescence spectroscopy one of the most unique and versatile techniques utilized for chemical and biological research. It is widely used to understand the molecular structure and dynamics in complex systems. The emission properties of fluorophores and its dependence on its immediate environment are being explored by the steady-state and time-resolved fluorescence spectroscopy. Time-resolved fluorescence measurements are preferred over steady-state fluorescence measurements to obtain comprehensive information on the microenvironment of the fluorophore as it captures the competing kinetic processes that are occurring in the timescale of fluorescence lifetime and influence fluorescence, such as collisional quenching, solvent relaxation, energy transfer, and rotational reorientation.

G. P. Pattnaik · H. Chakraborty (✉)
School of Chemistry, Sambalpur University, Burla, Odisha, India
e-mail: hirak@suniv.ac.in

© The Author(s), under exclusive license to Springer Nature Singapore Pte Ltd. 2022

H. Sahoo (ed.), *Optical Spectroscopic and Microscopic Techniques*,
https://doi.org/10.1007/978-981-16-4550-1_4

4.1 Fluorescence

Molecules present in the lowest energy state (ground state) absorb light to be raised to the higher energy state (excited state). Fluorescence is an optical phenomenon through which the excess energy is being dissipated in the form of light (emission of photons). Fluorescence can be discriminated from the other emissive deactivation pathway, phosphorescence, by the timescale of the process. The fluorescence is a phenomenon that occurs within nanosecond timescale, while phosphorescence is observed at higher time window (millisecond to second). The fluorescence is highly sensitive to the molecular as well as environmental properties, which makes fluorescence spectroscopy one of the most unique and versatile techniques utilized for chemical and biological research. It is widely used to understand the molecular structure and dynamics in complex systems. The emission properties of fluorophores and its dependence on its immediate environment are being explored by the steady-state and time-resolved fluorescence spectroscopy. Time-resolved fluorescence measurements are preferred over steady-state fluorescence measurements to obtain comprehensive information on the microenvironment of the fluorophore as it captures the competing kinetic processes that are occurring in the timescale of fluorescence lifetime and influence fluorescence, such as collisional quenching, solvent relaxation, energy transfer, and rotational reorientation.

The scope of application of fluorescence spectroscopy ranges from detection of solvent properties to conformational dynamics of the macromolecules. The application of fluorescence in ion and molecular sensing has gathered tremendous momentum as the fluorescence-based assays are simple but robust. The fluorescence-based methods are very useful in detecting small molecules binding to the protein and nucleic acid, protein–protein interaction, lateral movements of lipids in membranes, and many important biological processes. With the development of green fluorescence protein (GFP) and antibody-based fluorophores, fluorescence imaging has become more versatile in its in vitro and in vivo applications. The recent progress in single-molecule fluorescence measurements, such as fluorescence correlation spectroscopy, image correlation spectroscopy, and single-photon counting methodologies, offers much precise handle to monitor a specific macromolecule under physiological conditions.

Fluorescence anisotropy is a powerful tool to characterize small molecule–protein, protein–protein, and protein–DNA interactions as the measurements are rapid, quantitative, and adaptable to a broad range of experimental conditions. In addition, fluorescence anisotropy has been demonstrated as a sensitive parameter to study membrane organization and dynamics. It has achieved an edge over other fluorescence techniques as the fluorophore concentration does not have any effect on the value of fluorescence anisotropy. In this chapter, we will discuss the theory of steady-state and time-resolved fluorescence anisotropy and their applications in biological systems.

4.2 Steady-State Fluorescence Anisotropy

Light is an electromagnetic radiation with electric and magnetic vectors aligned perpendicular to each other. Light is being absorbed by a molecule, while the molecular dipole is parallel to the electric vector of the electromagnetic radiation (light). Therefore, the orientation of the electric vector of light is important for molecular absorption. In depolarized light, the electric vectors are randomly oriented in all directions perpendicular to the direction of its propagation. On the other hand, when the majority of the electric vectors are oriented in one plane, it is known as partially polarized light, and if all the electric vectors travel in any particular plane, it is called plane polarized light. Figure 4.1 depicts the schematic representations of depolarized, partially polarized, and plane polarized light.

A polarizer is an optical device that converts a depolarized or partially polarized light into a beam polarized in a particular plane. The function of a plane polarizer is shown in Fig. 4.2. The dichroic devices and double refracting calcite crystals are being used as polarizer; however, the double refracting calcite crystal polarizer is more common. The emitted light retains the polarization against the excitation with a plane polarized light if fluorophores do not rotate in the excited state. Nevertheless, the freely rotating molecules convert the plane polarized excitation into a depolarized emission due to change in its orientation during fluorescence lifetime (Fig. 4.3). The property of polarized emission against a polarized excitation provides an opportunity to monitor the probability of molecular rotation in the experimental condition. The change in orientation of the molecule, i.e., the average angular displacement of the fluorophore that occurs within the span in excited state of the fluorophore, can be measured with the help of fluorescence anisotropy. Therefore, fluorescence anisotropy is being considered as a tool to determine the change in molecular orientation (in space) regarding the time between the absorption and emission of photons. The angular displacement of the fluorophore is directly related to its rotational diffusion, and the rate of rotational diffusion is dependent on its size, shape, and viscosity of the medium. The measurement of fluorescence anisotropy for a particular fluorophore makes the anisotropy value solely dependent on the viscosity of the medium. The emission from fluorophore in rotationally restricted environment (high viscosity) is highly polarized, whereas fluorophores emit depolarized light in low viscous medium (freely rotatable), leading to nearly zero fluorescence anisotropy value (Fig. 4.3). The fluorescence anisotropy of a sample is determined by the following equation:

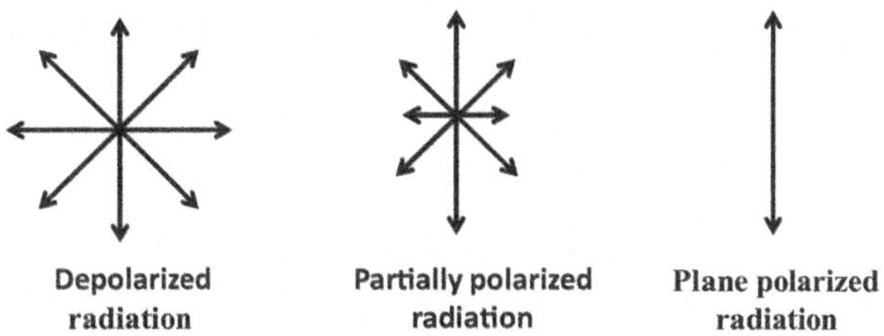

Fig. 4.1 Electric vector representations of depolarized, partially polarized, and plane polarized light

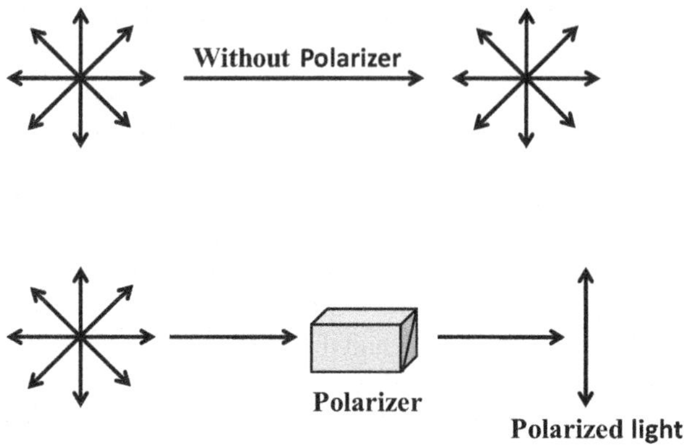

Fig. 4.2 Schematic representation of function of a plane polarizer

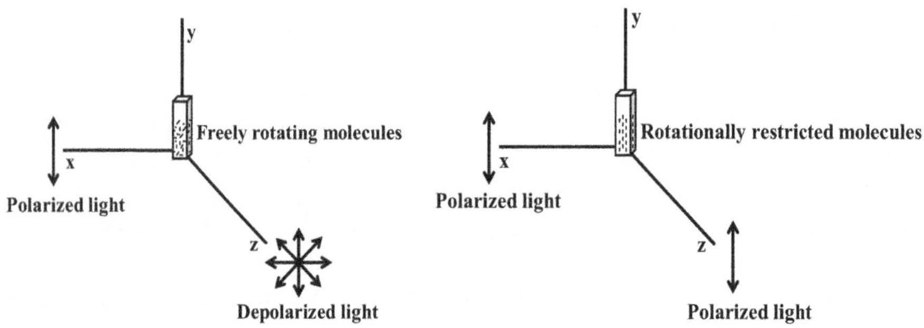

Fig. 4.3 Schematic representation of the effect of probe rotation on the polarization of emission, while the probe is being excited by plane polarized light

$$r = \frac{I_{VV} - G \times I_{VH}}{I_{VV} + 2G \times I_{VH}} \tag{4.1}$$

where G is the grating correction (G-factor) and is defined as the ratio of I_{HV} and I_{HH}. G factor corrects any dependencies upon the plane of polarization within the instrument's optical and electrical components. Here, I is the fluorescence intensity, and subscripts V and H denote the vertical and horizontal orientation of the polarizer. Further, the first and second subscripts refer to the orientation of the excitation and emission polarizer, respectively.

Fluorescence anisotropy depends upon the angle between absorption and emission transition dipoles. The fundamental anisotropy is given by the equation:

$$r_0 = \frac{2}{5}\left(\frac{3\cos^2\theta - 1}{2}\right) \tag{4.2}$$

where r_0 is the fundamental anisotropy, and θ is the angle between absorption and emission dipoles.

It has been found that the r_0 value of 1,6-diphenyl-1,3,5-hexatriene (DPH) is highest, i.e., 0.39 as the θ is minimum for DPH ($\theta \sim 7.4°$) [1]. Theoretically, maximum r_0 (0.4) can be obtained when $\theta = 0$, whereas r_0 become zero at $\theta = 54.7°$. The r_0 value becomes negative and reaches -0.20 when $\theta = 90°$. Therefore, the range of r_0 is considered to be $-0.2 \leq r_0 \leq 0.4$ for single-photon excitation of an isotropic solution. The angle θ and r_0 differ on the excitation wavelength as the orientation of the absorption dipole depends on wavelength. Since emission is mainly from the lowest singlet state, the fluorescence anisotropy value is independent of the emission wavelength. However, in solvent relaxation process within the lifetime of a fluorophore, the lowest singlet state relaxes to lower energies, and therefore, fluorescence anisotropy depends on the emission wavelength, mainly in the restricted environment. The various solvent-relaxed states are relatively more nondegenerate in motionally restricted environment.

Rotational diffusion of fluorophore within its fluorescence lifetime causes fluorescence depolarization. Therefore, the time required for the probe rotation (rotational correlation time) of the fluorophore is an important factor for fluorescence anisotropy. The rotational correlation time depends on the size of the fluorophore and the viscosity of immediate environment of the fluorophore. The relation between rotational correlation time and fluorescence lifetime is given in Perrin equation (Eq. 4.3)

$$r = \frac{r_0}{1 + \tau/\theta} \tag{4.3}$$

Here, r_0, θ, and τ are fundamental anisotropy, rotational correlation time, and fluorescence lifetime of the fluorophore, respectively. It is important to note that the fundamental anisotropy is an intrinsic property of fluorophore. When θ is greater than τ, r approaches to r_0, and when θ is less than τ, r value is close to zero. The rotational correlation time is being extensively used for binding of small molecule (e.g., drug) to their macromolecular targets (e.g., proteins). As the rotation depends on the size of the molecule, free small molecules show relatively smaller θ, and upon binding to their target macromolecules, the θ value increases. The θ value can be monitored as a function of protein concentration to obtain the binding isotherm of any small molecule (drug), and IC_{50} value can be calculated from the binding isotherm. Therefore, the timescale of θ and τ has significant impact on the fluorescence anisotropy value.

Fluorescence anisotropy can be measured either in an L- or T-format fluorescence spectrophotometer instrument. A single-emission channel is used in case of L-format instrument, whereas two emission channels, one for parallel and another for perpendicular component, are used in T-format instrument. Figure 4.4a, b represent the schematic diagram of L-and T-format fluorescence anisotropy instruments, respectively. Fluorescence anisotropy is calculated from I_{VV}, I_{VH}, I_{HH}, and I_{HV} values at a particular wavelength. However, values of I_{HV} and I_{HH} are required to measure the G-factor, which is majorly constant for a particular instrument for a certain wavelength. Therefore, I_{VV} and I_{VH} are the important determinants of fluorescence

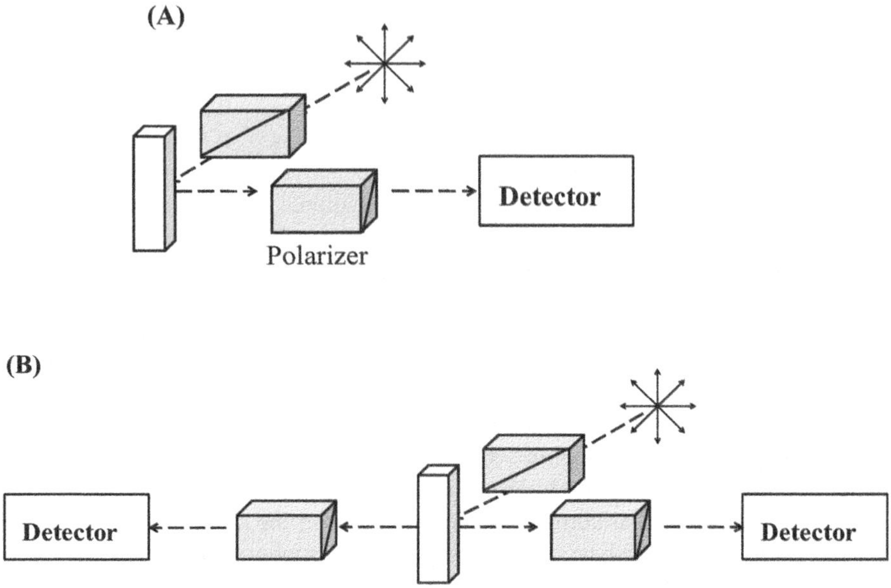

Fig. 4.4 Schematic representations of (**a**) L- and (**b**) T-format fluorescence spectrophotometers for anisotropy measurements

anisotropy value. In L-format instrument, these two components (I_{VV} and I_{VH}) are measured separately against two different excitations, whereas in T-format instrument, one can measure two emission components simultaneously against a single excitation. The simultaneous measurements provide fluctuation-free signal intensity in T-format instrument. Nevertheless, with the increment advancement in instrument, the importance of T-format instrument does not provide much edge over L-format instrument other than minimizing experimental time. In general, L-format instrument is widely used for fluorescence anisotropy measurement.

4.2.1 Factors Affecting Fluorescence Anisotropy

There are several factors that affect fluorescence anisotropy value leading to false interpretation of the results. Therefore, proper experimental set-up with an appropriate fluorescent probe is important to extrapolate anisotropy results in understanding environmental properties.

The shape (symmetry) of the molecule plays a key role in choice of fluorophore. 1,6-diphenyl-1,3,5-hexatriene (DPH, Fig. 4.5) is a widely used fluorophore for the determination of hydrophobic tail packing of the membrane because of its rod shape. The presence of C_2 axis symmetry makes the fluorescence anisotropy value independent of molecular rotation along its molecular axis, so the change in anisotropy value is mainly due to the rotation of the molecule in a different plane other than its

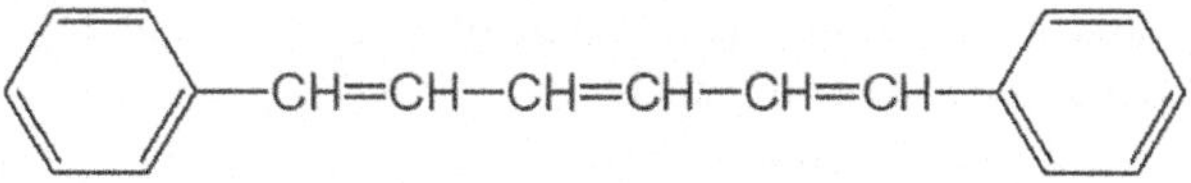

Fig. 4.5 Chemical structure of 1,6-diphenyl-1,3,5-hexatriene (DPH)

molecular axis. Therefore, change in anisotropy value of DPH in the membrane represents the change in environmental packing of the membrane as rotation in an axis other than its molecular axis requires more space. Fluorescein and its derivatives are spherical, and having minimum anisotropy (r_{min}) value of 0.021 provides a large dynamic range. In contrast, cyanine dyes do not have spherical symmetry, show higher value of its minimum anisotropy (r_{min}) of 0.17, and reduces the dynamic range of anisotropy [2, 3].

The effect of fluorescence lifetime in fluorescence anisotropy measurement is evident in Perrin equation (Eq. 4.3). The lower lifetime value in comparison to its rotational correlation time ($\tau < < \theta$) leads to completion of the emission prior to the rotation of the molecule, resulting in higher anisotropy value. On the other hand, larger lifetime value compared to its rotational correlation ($\tau > > \theta$) results in low anisotropy value. Therefore, fluorescence lifetime and rotational correlation time of the probe should be comparable for the selection of fluorescent probe for anisotropy measurement. For instance, fluorescence anisotropy of fluorescein or rhodamine, having lifetime of around 4 ns, is preferable for low molecular weight (<10 kDa) proteins having rotational correlation time less than 10 ns. Similarly, we need to select a fluorescent probe with higher lifetime for the measurement of anisotropy of proteins having higher molecular weight [1, 2].

Temperature directly affects the rotational correlation time of the fluorophore. Generally, the rotational correlation time of the probe reduces with increasing temperature resulting in lower fluorescence anisotropy value, which is obvious from the Perrin equation (Eq. 4.3). However, the reverse fluorescence anisotropy effect is found in case of coumarin or cyanine dyes as there is a substantial effect of temperature on fluorescence lifetime of these probes [4, 5]. The fluorescence lifetime of these probes decreases with increasing temperature and thereby enhances the fluorescence anisotropy value [2, 6].

Fluorescence anisotropy value is further modulated if the fluorophore displays Homo-FRET. Homo-FRET is a process through which energy is being transferred between two similar fluorophores, and generally, it is known as FRET between two matching fluorophores. Generally, fluorophore with low Stokes' shift value shows Homo-FRET [7]. The energy transfer between two dipoles in different orientation causes depolarization of the emitted light, thereby reducing the fluorescence anisotropy value [8, 9].

Scattering artifacts is detrimental in fluorescence anisotropy measurements. Biological samples such as membranes, aggregated proteins, and insoluble particles in solution cause scattering, and the scattering contribution enhances with the particle size and concentration. The fluorescence anisotropy value deviates from its original value as scattered lights are totally polarized leading to an anisotropy value of 1. Thus, any amount of scattered light results in substantial change in fluorescence anisotropy [1].

4.2.2 Importance of Apparent Rotational Correlation Time

We have discussed in earlier section that fluorescence anisotropy has a dependency on the fluorescence lifetime value. Therefore, fluorophores with lower fluorescence lifetime often demonstrate deceptive fluorescence anisotropy values. Fluorophores such as hemicyanines that demonstrate low fluorescence lifetime (≤ 1 ns) do not get enough time to rotate before the emission displays a high fluorescence anisotropy value [10]. The complexity of interpreting fluorescence anisotropy value increases in case of temperature-dependent measurement of fluorescence anisotropy. The temperature-dependent drop of fluorescence lifetime provides very short time to the fluorophore to rotate in excited state, and often this time window is smaller than its rotational correlation time, resulting in no temperature-dependent change in fluorescence anisotropy of the probes. In this situation, calculation of apparent rotational correlation time using Perrin equation provides the actual information about the rigidity of the environment as the equation considers both the fluorescence lifetime and anisotropy.

4.3 Time-Resolved Anisotropy Decays

The steady-state anisotropy measurements are being carried out in presence of continuous illumination of the source and represent the average of anisotropy decay in terms of fluorescence intensity decay. Therefore, we obtain an average information of the probe rotation from steady-state anisotropy measurements. Additional information can be obtained from time-resolved anisotropy measurements, where anisotropy is followed as function of time against pulsed excitation [1].

The trend of anisotropy decay depends on size, shape, and flexibility of the fluorophore. A spherical molecule decays with a single rotational correlation time; however, anisotropy decays are usually multiexponential for nonspherical molecules due to rotation of the fluorophore in various molecular axes. In addition to the shape, the segmental motions of the macromolecule further result in multiexponential anisotropy decay [11]. The anisotropy decay of fluorophore in membranes is extremely informative as the decay does not reach zero anisotropy value even after long time due to anisotropic nature of the environment. The residual anisotropy of the fluorophore in membrane may be extrapolated to the order parameter of the membrane.

In time-resolved anisotropy measurements, emission was collected at directions vertical or parallel ($I_{\parallel}$) and horizontal or perpendicular ($I_{\perp}$) to the polarization of the vertically polarized excitation beam. Anisotropy was calculated using the following equation [12].

$$r(t) = \frac{[I_{\text{II}}(t) - G(\lambda)I_{\perp}(t)]}{[I_{\text{II}}(t) + 2G(\lambda)I_{\perp}(t)]} \tag{4.4}$$

where $G(\lambda)$ is the geometry factor at the emission wavelength (λ). The G factor of the emission collection optics was measured from the emission collected at the parallel ($I_{\|}$) and perpendicular ($I_{\perp}$) directions against the horizontally polarized excitation beam. The time-resolved anisotropy decays are analyzed based on the following equations:

$$\begin{aligned} I_{\|}(t) &= I(t)[1 + 2r(t)]/3 \\ I_{\perp}(t) &= I(t)[1 - r(t)]/3 \end{aligned} \tag{4.5}$$

where $I_{\|}(t)$ and $I_{\perp}(t)$ are the decays of the parallel ($_{\|}$) and perpendicular ($\perp$) components of emission. The equation for time-resolved fluorescence anisotropy can be expressed as a sum to exponential decay:

$$r(t) = r_0 \sum \beta_i \exp\left(-{}^{t}/_{\phi_i}\right) \tag{4.6}$$

where ϕ_i and β_i represent the ith rotational correlation time and the corresponding pre-exponential factor in the experimental decay such that $\Sigma\beta i = 1$. r_0 represents the anisotropy at zero time (initial anisotropy). The goodness of the fit is evaluated by the reduced χ^2, which is acceptable between 1.0 and 1.15.

A characteristic time-resolved anisotropy decay of DPH in 1,2-dimyristoyl-sn-glycero-3-phosphocholine (DMPC) is shown in Fig. 4.6. DPH shows two extremely different rotational correlation time (10.26 ns and 0.68 ns) in DMPC membranes indicating differential rotation of the DPH in DMPC membranes. This information was not available in steady-state fluorescence anisotropy measurements as it provides an average information of fluorescence anisotropy.

Moreover, fitting of the time-resolved fluorescence anisotropy provides the fraction of component (pre-exponential factor) showing a particular rotational correlation time. Therefore, time-resolved anisotropy measurements offer the statistical properties of the anisotropy data, which is not possible from steady-state fluorescence measurements.

In addition, the time-resolved anisotropy data can be utilized to determine the degree of probe wobbling in any anisotropic medium such as membranes. The degree of probe wobbling is dependent on the membrane fluidity (order). Kinosita model [14, 15] considered the "probe wobbling-in-cone" and allows to calculate the semi angle (θ) of probe wobbling. The semi angle of wobbling is given by:

$$\theta = \cos^{-1}\left[1/2\left\{\left(1 + 8\left({}^{r_\infty}/_{r_0}\right)^{\frac{1}{2}}\right)^{\frac{1}{2}} - 1\right\}\right] \tag{4.7}$$

The r_0 denotes the initial anisotropy (anisotropy at zero time) in the anisotropy decay curve, and $r_\infty = r_0 \times \beta_s$, where β_s is the pre-exponential factor corresponding

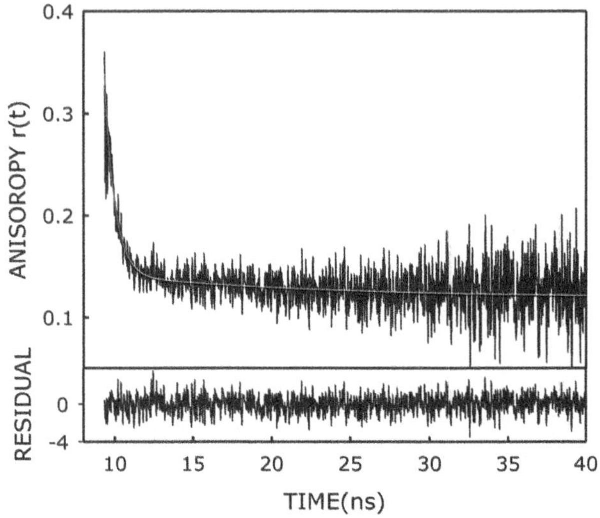

Fig. 4.6 Plot of time-resolved fluorescence anisotropy decay of DPH in the gel phase of DMPC membranes. This decay profile was best fitted in a two-exponential anisotropy decay model, and the fitted lines are shown in red. The figure has been adapted with permission from [13]

to the slower decaying component. The r_0 value for DPH was considered as 0.39 [1], while analyzing the fluorescence anisotropy decay kinetics using Eq. (4.6). The "θ" value indicates the restriction of rotation, higher θ value (~90°) represents the unhindered rotation of the probe, while lower θ value (~0°) represents the crowded or hindered rotation of the probe.

4.4 Applications of Fluorescence Anisotropy Measurements

Fluorescence anisotropy can be judiciously utilized to determine the organization and dynamics of proteins, membranes, and other biomolecules as well as their microenvironment. In the following sections, we have discussed various applications of fluorescence anisotropy (both steady-state and time-resolved) in determination of organization and dynamics of biomolecules.

4.4.1 Determination of Phase Transition Temperature of Lipid

Lipidic phase is an important aspect of biology. Lipids can exist broadly in two temperature-dependent phases, such as fluid and gel phase [16], and the study of phase transition has great importance in biology [17]. The temperature at which lipid transforms from gel to fluid phase is known as phase transition temperature [18]. Different factors such as hydrophobic chain length, number of double bonds, and its position in the hydrophobic chain affect the phase transition temperature of a lipid bilayer. Moreover, presence of external agent alters the phase transition temperature by either destabilizing or stabilizing the hydrophobic chain packing of the gel phase lipid membrane. Fluorescence probe like DPH, located with an average distance of

~7.8 Å (hydrophobic region) from the bilayer center of the lipid bilayer, is frequently used to measure the phase transition temperature of the lipid membrane [19, 20]. The rod shape geometry of DPH with C_2-axis of symmetry makes it an immensely popular fluorescent probe for anisotropy measurement as the rotation of the molecule in its molecular axis does not change the direction of the molecular dipole. Therefore, rotation of the probe other than its molecular axis is being recorded in the change in fluorescence anisotropy measurements.

The membrane with lipids in gel phase provides a motionally restricted environment, where DPH cannot rotate freely, resulting in higher fluorescence anisotropy value. On the other hand, lipids are disordered in fluid phase, where DPH rotates freely and demonstrates lower anisotropy value. Therefore, at phase transition temperature, there is a sharp fall of fluorescence anisotropy value which is easily identified through temperature-dependent fluorescence anisotropy experiment. The effect of additives on hydrophobic chain packing of lipid membrane can be easily monitored from the change in phase transition temperature. Generally, most of the organic additives destabilize hydrophobic chain packing of gel phase lipid membrane and hence reduces the phase transition temperature [21, 22]. However, there are few biomolecules like eugenol, an antimicrobial agent that enhances phase transition temperature by about 1 °C of DMPC membranes. This subtle change in phase transition temperature could also be measured by utilizing temperature-dependent steady-state fluorescence anisotropy measurement of DPH [13]. The representative plot of steady-state fluorescence anisotropy of DPH as a function of temperature in DMPC lipid membranes in the absence and presence of various concentrations of eugenol is shown in Fig. 4.7.

4.4.2 Determination of Microviscosity of the Environment

The local viscosity in biological systems provides information about crowding and packing of the local environment. Therefore, it is important to evaluate the microviscosity of proteins and membranes to understand their function from a physical perspective. Measurement of microviscosity is not possible by using methods that are used for the measurement of bulk viscosity. The molecular probes are becoming popular in measuring microviscosity of proteins and membranes as their ease of rotation depends on the microviscosity of its location.

The fluorescence anisotropy is being routinely used to monitor molecular rotation and is widely used to measure the microviscosity of the membrane [23–27] and protein. First, the anisotropy of the fluorophore is measured in different solvents of known viscosity (reference solvents) to prepare the standard calibration curve, followed by the measurement of anisotropy in the lipid membranes whose viscosity is to be known. The microviscosity of the membrane can be obtained from the standard calibration curve of fluorescence anisotropy and viscosity [1].

It has been established that the hippocampal membrane is a suitable natural source to study the interaction of neuronal receptors with membrane lipids of neuronal origin [28]. The treatment of methyl-β-cyclodextrin (MβCD) is known to

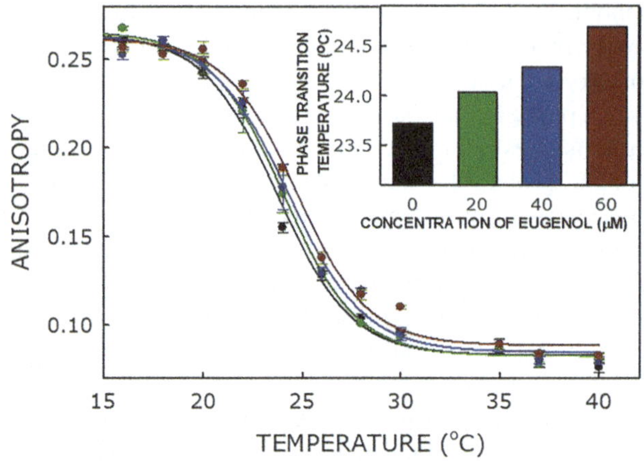

Fig. 4.7 Plot of change in fluorescence anisotropy of DPH in DMPC membranes as a function of temperature. The phase transition temperature of the lipid was determined from the inflection point of the plot. Phase transition temperature of DMPC was measured in the absence and presence of 20 μM, 40 μM, and 60 μM eugenol, and shown as black, green, blue, and red lines. The phase transition temperatures of DMPC membranes in the absence and presence of various concentrations of eugenol are shown in the inset. The figure has been adapted with permission from [13]

deplete membrane cholesterol [28], and the amount of cholesterol is being measured by standard bicinchoninic acid (BCA) assay [29]. The microviscosity of native bovine hippocampal membrane was obtained by utilizing steady-state fluorescence anisotropy of a molecular rotor, BODIPY-C_{12} [30], a meso-substituted molecular rotor based on BODIPY (4,4-difluoro-4-bora-3a,4a-diaza-s-indacene) structure [31]. This has been further shown that the microviscosity of bovine hippocampal membranes decreases with depletion of cholesterol (Fig. 4.8). The fluorescence anisotropy measurement of appropriate probe could be judiciously used to evaluate microviscosity of biological systems.

4.4.3 Study of Protein–Protein Interaction

Being the "workhorses" of the cell, proteins perform majority of the biological processes in a cell. While proteins carry out several functions independently, a considerably large number of biological activities require interaction between multiple proteins. Therefore, study of protein–protein interaction is of great importance in understanding biological processes. Several biochemical and biophysical techniques are being employed to understand the protein–protein interactions, out of which fluorescence anisotropy is an extremely convenient tool. The labeling of protein can be accomplished by the chemical addition of extrinsic fluorophores such as fluorescein, rhodamine, coumarins, BODIPY, and Alexa to the protein or biochemical tagging of the protein with genetically coded green fluorescence proteins (GFPs)

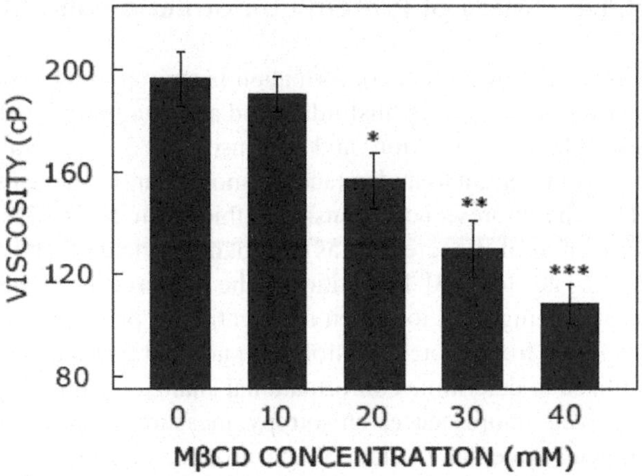

Fig. 4.8 Plot of the effect of cholesterol depletion by MβCD on the viscosity of hippocampal membranes. The figure has been adapted with permission from [31]

[32, 33]. As discussed earlier, fluorescence anisotropy depends on the rate of rotational diffusion of the fluorophore which is sensitive to the viscosity of the medium as well as the molecular size of the fluorophore. The interaction of protein of interest with other protein increases the molecular mass and hence slows down the rotational diffusion of the protein–protein complex resulting in enhancement of fluorescence anisotropy [32].

The choice of fluorescent probe is important for the evaluation of the protein–protein interaction. The fluorescence lifetime of the probe would be similar to its rotational correlation time; otherwise, protein–protein interaction would not be easily evaluated by fluorescence anisotropy measurements. For example, fluorescein or rhodamine shows fluorescence lifetime of about 4 ns, the fluorescence anisotropy value changes from 0.05 to 0.29, while the mass of the fluorescence-labeled protein (or macromolecule) changes from 100 Da to 160 kDa. Therefore, the change in molecular mass between 100 Da to 160 kDa can easily be evaluated by monitoring fluorescence anisotropy of fluorescein- or rhodamine-labeled protein. On the other hand, the same probe shows very subtle change in fluorescence anisotropy (from 0.29 to 0.30) when the molecular mass changes from 160 to 1000 kDa. Therefore, it requires fluorophore with higher fluorescence lifetime to evaluate the interaction between two large proteins [32].

Similar approach can also be applied to study protein–nucleic acid interaction. Generally, the nucleic acid is being tagged with a fluorescent probe, and the fluorescence anisotropy of the probe increases when the nucleic acid interacts with a protein. The protein–nucleic acid interaction is crucial to understand the control of genetic expression. The fluorescence anisotropy is employed to monitor equilibrium binding profiles of a bacterial repressor protein, tryptophan repressor (TR), to a fluorescently labeled oligonucleotide containing one of its target operator sequences in solution [34]. The binding constant can further be evaluated by analyzing the binding profile in relevant binding model.

4.4.4 Study of Protein Conformation and Misfolding

Evaluation of protein conformation is an important paradigm in biology, especially in the case of protein misfolding and aggregation-related neurodegenerative diseases like Alzheimer's, Prion, and Parkinson's [35–38]. Evaluation of protein conformation is important to understand the molecular mechanism of protein aggregation [39–41]. The fluorescence intensity of thioflavin T (ThT) is generally used to study the formation of β-aggregate, as the quantum yield of ThT is very high in the nonpolar aggregates [42, 43]. In addition to the fluorescence intensity, fluorescence anisotropy is also being used to screen aggregation of prion protein [39].

Apart from protein misfolding and aggregation, fluorescence anisotropy is also utilized to determine conformational changes of protein. Both steady-state and time-resolved fluorescence, anisotropy measurements are employed to determine the global/segmental motion of the protein. As fluorescence anisotropy depends on motional restriction around the fluorophore, anisotropy value is always high in globular proteins, whereas it reduces with the increase in conformational flexibility of the protein. Therefore, fluorescence anisotropy measurement can also be used to investigate the denaturation of protein as motional restriction reduces with the unfolding of the protein [44]. The conformations of κ-casein are being monitored in the presence of different denaturants, acid and base utilizing steady-state and time-resolved fluorescence anisotropy measurements [45, 46]. The steady-state fluorescence anisotropy values of tryptophan in different pHs are shown in Fig. 4.9. The fluorescence anisotropy values of κ-casein in three pHs indicate that the tryptophan environment is dynamic at acidic pH, whereas it becomes restricted in alkaline pH compared to κ-casein in neutral pH.

The time-resolved fluorescence anisotropy is used to determine the rotational correlation time of κ-casein in the presence of different concentrations of cetyltrimethyl ammonium bromide (CTAB) [46]. The results show that the κ-casein assumes a relatively compact conformation in the presence of low concentration of (0.2 mM) CTAB. The anisotropy decay profile and calculated rotational correlation time of tryptophan in κ-casein are shown in Fig. 4.10.

The tryptophan scanning mutation is further being utilized to determine to segmental motion of different region of the protein [47]. Tryptophan can be genetically inserted at various locations of the protein, and measurements of steady-state and time-resolved fluorescence anisotropy of tryptophan provide conformational information of different segments of the protein [48, 49].

4.4.5 Study of Membrane Organization

Biological membranes are complex two-dimensional, microheterogeneous, supramolecular assemblies of various lipids and proteins [50]. Therefore, understanding membrane organization and dynamics is vital to understand the function of membranes and membrane proteins. Several spectroscopic and microscopic methods have been employed to evaluate the organization and dynamics of model and cellular

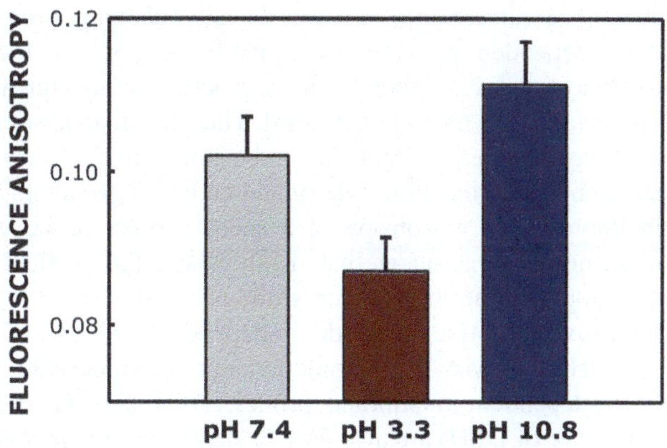

Fig. 4.9 Steady-state fluorescence anisotropy value of tryptophan in k-casein (2 μM) at three different pH of 7.4 (gray), 3.3 (red), and 10.8 (blue) at temperature 25 °C. The figure has been adapted with permission from [45]

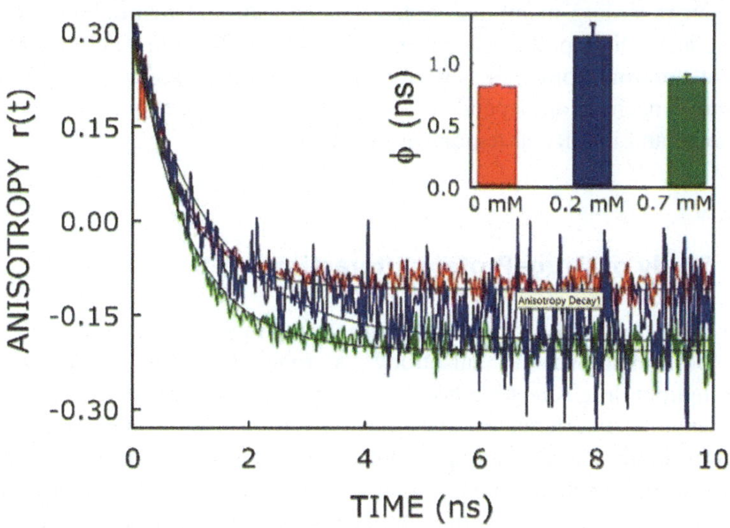

Fig. 4.10 Plot of time-resolved anisotropy decay for κ-casein in the absence (red) and presence of 0.2 mM (blue) and 0.7 mM (green) of CTAB. This decay profile was best fitted in a single exponential anisotropy decay model, and the fitted lines are shown in black through the corresponding data. Inset figure is the obtained rotational correlation times in the absence and presence of corresponding concentrations of CTAB. The figure has been adapted with permission from [46]

membranes. The organization of membranes along the bilayer normal is interesting due to the varied physical properties of the headgroup and acyl chain regions of the lipids [51–54].

Fluorescence anisotropy measurements of depth-dependent membrane probes provide insight into the molecular packing at various locations along the bilayer normal. The n-anthroyloxy stearic acids (n-AS), in which an anthroyloxy group is attached at different positions of the alkyl chain, are routinely used to determine the depth-dependent packing of the membrane [55]. The location of these probes along the bilayer normal depends on the position of attachment of the anthroyloxy group on the fatty acyl chain [14, 56–58]. Thus, the distance of the anthroyloxy group from the bilayer center is established to be almost linearly related to the number of carbon atoms between the anthroyloxy and carboxyl groups [57]. Interestingly, the change in fluorescence anisotropy of a specific n-AS probe will provide the membrane ordering information at that depth where the particular n-AS probe is situated. This approach has been successfully adopted in the study of the effect of hemagglutinin fusion peptide on the depth-dependent ordering of the membrane [59].

DPH and TMA-DPH (trimethylammonium derivative of DPH) are also used as depth-dependent membrane probes. DPH and TMA-DPH (trimethylammonium derivative of DPH) are also used as depth-dependent membrane probes [19]. Therefore, measurement of fluorescence anisotropy of TMA-DPH and DPH delivers information regarding membrane order at interfacial and hydrophobic regions of the membrane. These probes have been judiciously utilized to determine the depth-dependent ordering ability of various fusion inhibitor peptides [60]. Figure 4.11 shows the concentration-dependent effect of various fusion inhibitor peptides at hydrophobic and interfacial regions of the membrane.

4.4.6 Study of Drug–Protein Interaction

Binding efficiency and kinetics of the drug determine its efficacy as therapeutic agents [61]. Fluorescence anisotropy is routinely used as the assay for high-throughput drug screening because of its robust performance, relative insensitivity to external interferences, and sensitivity [62]. The smaller drug molecules show lower fluorescence anisotropy value compared to when it is bound to a protein or DNA (macromolecules) due to enhancement of molecular mass (Fig. 4.12). The change in fluorescence anisotropy with the addition of protein/DNA represents the binding curve for the drug molecule with its macromolecular target. The tighter binding of norharman to human serum albumin (HSA) compared to bovine serum albumin (BSA) has been shown from the anisotropy measurement of norharman in presence of BSA and HSA [25]. The major application of fluorescence anisotropy is to identify small molecular modulators of activity for a variety of enzymes. The human Myt1 kinase is a regulator of Cdk1/CycB; therefore, it is an important determinant for the G2/M transition in the cell cycle. The fluorescence anisotropy has been utilized to screen the small molecules as the inhibitor of Myt1 kinase [63]. The fluorescence anisotropy-based methodology for the high-throughput screening of small-molecule modulators of the inositol 5-phosphatases is extremely convenient [64]. Similarly, fluorescence anisotropy has been successfully utilized for identification of molecular modulators for other enzymes such as proteases [65], nucleases [66], and deacetylases [67].

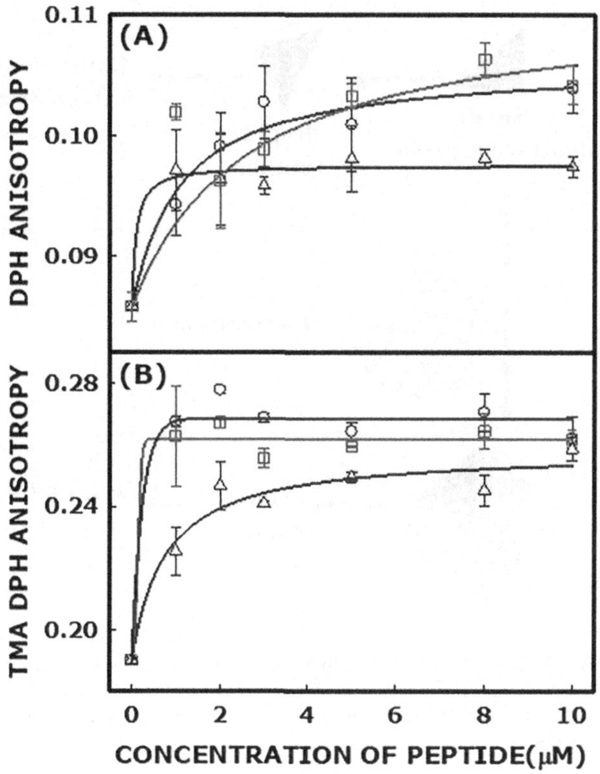

Fig. 4.11 Plot of change in fluorescence anisotropy of (**a**) DPH and (**b**) TMA-DPH in DOPC/DOPE/DOPG (60/30/10) membranes with increasing concentration of TG-23 peptide (blue, ○), GL-22 peptide (green, □), and EV-22 peptide (red, Δ). The figure has been adapted with permission from [60]

4.5 Concluding Remark and Future Perspectives

Fluorescence anisotropy remains a popular tool to study biomolecular dynamics because of its sensitivity, robustness, independence of probe concentration, and simple instrumentation. Moreover, fluorescence anisotropy can be used to study many other biomolecular processes such as drug–protein and protein–protein interaction, and protein aggregation. The advancement of instrumentation and speed of data acquisition make fluorescence anisotropy-based methodology one of the most reliable methods for high-throughput drug screening. Steady-state anisotropy measurements in tandem with time-resolved study provide a clear picture of biomolecular dynamics as the output of former measurement is an averaged information, whereas latter provides the statistical information of the system. However, further advancement of instrumentation and methodology might allow us anisotropic images of the in vivo system, similar to fluorescence lifetime images. The anisotropic images would be extremely important to understand molecular crowding and packing of the biological systems.

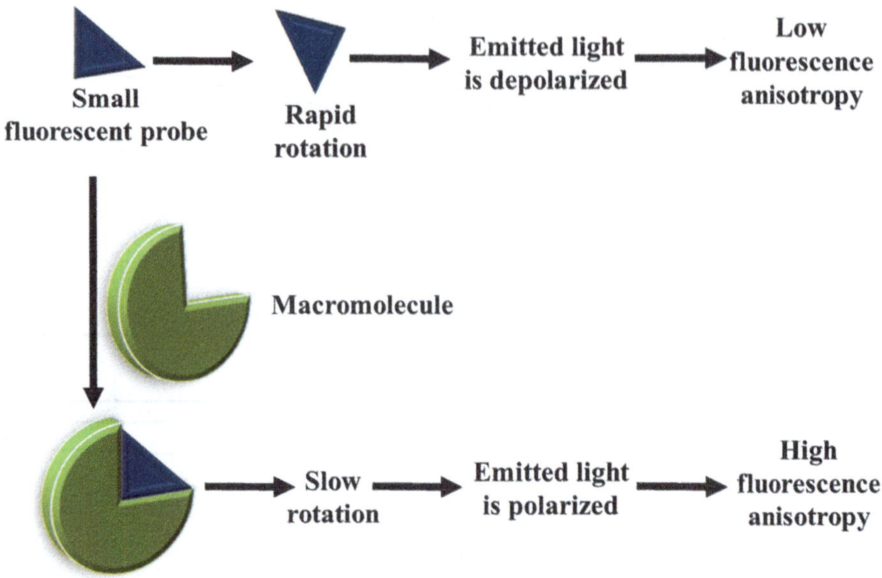

Fig. 4.12 Schematic representations of fluorescence anisotropy-based drug-protein binding assay. A rapidly rotating small fluorescent probe (drug) gives low fluorescence anisotropy. The association with a macromolecule such as a protein enhances fluorescence anisotropy due to restriction in rotation

Acknowledgments This work was supported by research grant from the Science and Technology Department, Government of Odisha. H.C. thanks the University Grants Commission for UGC-Assistant Professor position. G.P.P acknowledges Science and Technology Department, Government of Odisha, for Research Fellowship. We acknowledge Department of Science and Technology, New Delhi, and UGC for providing instrument facility to the School of Chemistry, Sambalpur University, under the FIST and DRS programs, respectively. We thank members of Chakraborty laboratory for their comments and discussions.

Note The authors declare that there is no conflict of interest.

References

1. Lakowicz JR (2006) Principles of fluorescence spectroscopy, 3rd edn. Springer
2. Zhang H, Wu Q, Berezin MY (2015) Fluorescence anisotropy (polarization): from drug screening to precision medicine. Expert Opin Drug Discov 10(11):1145–1161
3. Gustafson TP, Cao Q, Achilefu S, Berezin MY (2012) Defining a polymethine dye for fluorescence anisotropy applications in the near-infrared spectral range. ChemPhysChem 13(3):716–723
4. Zhegalova NG, Dergunov SA, Wang ST, Pinkhassik E, Berezin MY (2014) Design of fluorescent nanocapsules as ratiometric nanothermometers. Chemistry 20(33):10292–10297

5. Zhegalova NG, Aydt A, Wang ST, Berezin MY (2013) Molecular thermometers for potential applications in thermal ablation procedures, vol 8596. SPIE BiOS. SPIE
6. Kumbhakar M, Mukherjee T, Pal H (2005) Temperature effect on the fluorescence anisotropy decay dynamics of Coumarin-153 dye in Triton-X-100 and Brij-35 micellar solutions. Photochem Photobiol 81(3):588–594
7. Ganguly S, Clayton AH, Chattopadhyay A (2011) Organization of higher-order oligomers of the serotonin(1)(A) receptor explored utilizing homo-FRET in live cells. Biophys J 100(2): 361–368
8. Meher G, Bhattacharjya S, Chakraborty H (2019) Membrane cholesterol modulates oligomeric status and peptide-membrane interaction of severe acute respiratory syndrome coronavirus fusion peptide. J Phys Chem B 123(50):10654–10662
9. Chakraborty H, Chattopadhyay A (2015) Excitements and challenges in GPCR oligomerization: molecular insight from FRET. ACS Chem Neurosci 6(1):199–206
10. Mishra A, Behera GB, Krishna MMG, Periasamy N (2001) Time-resolved fluorescence studies of aminostyryl pyridinium dyes in organic solvents and surfactant solutions. J Lumin 92(3): 175–188
11. Jain N, Mukhopadhyay S (2015) Applications of fluorescence anisotropy in understanding protein conformational disorder and aggregation. In: Applied spectroscopy and the science of nanomaterials, vol 2. Springer, Singapore
12. Shrivastava S, Dutta D, Chattopadhyay A (2016) Effect of local anesthetics on the organization and dynamics in membranes of varying phase: a fluorescence approach. Chem Phys Lipids 198: 21–27
13. Meher G, Chakraborty H (2018) Influence of eugenol on the organization and dynamics of lipid membranes: a phase-dependent study. Langmuir 34(6):2344–2351
14. Chakraborty H, Haldar S, Chong PL, Kombrabail M, Krishnamoorthy G, Chattopadhyay A (2015) Depth-dependent organization and dynamics of archaeal and eukaryotic membranes: development of membrane anisotropy gradient with natural evolution. Langmuir 31(42): 11591–11597
15. Kinosita K Jr, Kawato S, Ikegami A (1977) A theory of fluorescence polarization decay in membranes. Biophys J 20(3):289–305. https://doi.org/10.1016/S0006-3495(77)85550-1
16. van Meer G, Voelker DR, Feigenson GW (2008) Membrane lipids: where they are and how they behave. Nat Rev Mol Cell Biol 9(2):112–124
17. Yeagle PL (2005) The structure of biological membranes. CRC Press
18. Chen W, Dusa F, Witos J, Ruokonen SK, Wiedmer SK (2018) Determination of the main phase transition temperature of phospholipids by nanoplasmonic sensing. Sci Rep 8(1):14815
19. Kaiser RD, London E (1998) Location of diphenylhexatriene (DPH) and its derivatives within membranes: comparison of different fluorescence quenching analyses of membrane depth. Biochemistry 37(22):8180–8190
20. Halling KK, Ramstedt B, Nystrom JH, Slotte JP, Nyholm TK (2008) Cholesterol interactions with fluid-phase phospholipids: effect on the lateral organization of the bilayer. Biophys J 95(8): 3861–3871
21. Constantinides PP, Inouchi N, Tritton TR, Sartorelli AC, Sturtevant JM (1986) A scanning calorimetric study of the interaction of anthracyclines with neutral and acidic phospholipids alone and in binary mixtures. J Biol Chem 261(22):10196–10203
22. Roy SM, Bansode AS, Sarkar M (2010) Effect of increase in orientational order of lipid chains and head group spacing on non steroidal anti-inflammatory drug induced membrane fusion. Langmuir 26(24):18967–18975
23. Shinitzky M, Barenholz Y (1978) Fluidity parameters of lipid regions determined by fluorescence polarization. Biochim Biophys Acta 515(4):367–394
24. Mallick A, Chattopadhyay N (2005) Photophysics in motionally constrained bioenvironment: interaction of norharmane with bovine serum albumin. Photochem Photobiol 81(2):419–424

25. Chakrabarty A, Mallick A, Haldar B, Das P, Chattopadhyay N (2007) Binding interaction of a biological photosensitizer with serum albumins: a biophysical study. Biomacromolecules 8(3): 920–927
26. Mallick A, Haldar B, Maiti S, Chattopadhyay N (2004) Constrained photophysics of 3-acetyl-4-oxo-6,7-dihydro-12H indolo-[2,3-a] quinolizine in micellar environments: a spectrofluorometric study. J Colloid Interface Sci 278(1):215–223
27. Mallick A, Haldar B, Maiti S, Bera SC, Chattopadhyay N (2005) Photophysical study of 3-acetyl-4-oxo-6,7-dihydro-12H-indolo[2,3-a]quinolizine in biomimetic reverse micellar nanocavities: a spectroscopic approach. J Phys Chem B 109(30):14675–14682
28. Pucadyil TJ, Chattopadhyay A (2004) Cholesterol modulates ligand binding and G-protein coupling to serotonin(1A) receptors from bovine hippocampus. Biochim Biophys Acta 1663(1–2):188–200
29. Smith PK, Krohn RI, Hermanson GT, Mallia AK, Gartner FH, Provenzano MD, Fujimoto EK, Goeke NM, Olson BJ, Klenk DC (1985) Measurement of protein using bicinchoninic acid. Anal Biochem 150(1):76–85
30. Levitt JA, Kuimova MK, Yahioglu G, Chung P-H, Suhling K, Phillips D (2009) Membrane-bound molecular rotors measure viscosity in live cells via fluorescence lifetime imaging. J Phys Chem C 113(27):11634–11642
31. Pal S, Chakraborty H, Bandari S, Yahioglu G, Suhling K, Chattopadhyay A (2016) Molecular rheology of neuronal membranes explored using a molecular rotor: implications for receptor function. Chem Phys Lipids 196:69–75
32. Gijsbers A, Nishigaki T, Sanchez-Puig N (2016) Fluorescence anisotropy as a tool to study protein-protein interactions. J Vis Exp 21(116):e54640
33. Toseland CP (2013) Fluorescent labeling and modification of proteins. J Chem Biol 6(3):85–95
34. LeTilly V, Royer CA (1993) Fluorescence anisotropy assays implicate protein-protein interactions in regulating trp repressor DNA binding. Biochemistry 32(30):7753–7758
35. De Felice FG, Vieira MN, Meirelles MN, Morozova-Roche LA, Dobson CM, Ferreira ST (2004) Formation of amyloid aggregates from human lysozyme and its disease-associated variants using hydrostatic pressure. FASEB J 18(10):1099–1101
36. Tanzi RE, Bertram L (2005) Twenty years of the Alzheimer's disease amyloid hypothesis: a genetic perspective. Cell 120(4):545–555
37. Hillger F, Nettels D, Dorsch S, Schuler B (2007) Detection and analysis of protein aggregation with confocal single molecule fluorescence spectroscopy. J Fluoresc 17(6):759–765
38. Thirunavukkuarasu S, Jares-Erijman EA, Jovin TM (2008) Multiparametric fluorescence detection of early stages in the amyloid protein aggregation of pyrene-labeled alpha-synuclein. J Mol Biol 378(5):1064–1073
39. Sabate R, Saupe SJ (2007) Thioflavin T fluorescence anisotropy: an alternative technique for the study of amyloid aggregation. Biochem Biophys Res Commun 360(1):135–138
40. Matveeva EG, Rudolph A, Moll JR, Thompson RB (2012) Structure-selective anisotropy assay for amyloid Beta oligomers. ACS Chem Neurosci 3(11):982–987
41. Zorrilla S, Rivas G, Lillo MP (2004) Fluorescence anisotropy as a probe to study tracer proteins in crowded solutions. J Mol Recognit 17(5):408–416
42. Naiki H, Higuchi K, Hosokawa M, Takeda T (1989) Fluorometric determination of amyloid fibrils in vitro using the fluorescent dye, thioflavin T1. Anal Biochem 177(2):244–249
43. Nilsson MR (2004) Techniques to study amyloid fibril formation in vitro. Methods 34(1): 151–160
44. Chakraborty A, Basak S (2008) Effect of surfactants on casein structure: a spectroscopic study. Colloids Surf B Biointerfaces 63(1):83–90
45. Mirdha L, Chakraborty H (2019) Characterization of structural conformers of kappa-casein utilizing fluorescence spectroscopy. Int J Biol Macromol 131:89–96
46. Mishra S, Meher G, Chakraborty H (2017) Conformational transition of kappa-casein in micellar environment: insight from the tryptophan fluorescence. Spectrochim Acta A Mol Biomol Spectrosc 186:99–104
47. Jain N, Bhasne K, Hemaswasthi M, Mukhopadhyay S (2013) Structural and dynamical insights into the membrane-bound α-synuclein. PLoS One 8(12):e83752

48. Kiran U, Regur P, Kreutz MR, Sharma Y, Chakraborty A (2017) Intermotif communication induces hierarchical Ca(2+) filling of caldendrin. Biochemistry 56(19):2467–2476
49. Kiran U, Kreutz MR, Sharma Y, Chakraborty A (2019) Tryptophan scanning mutagenesis of EF-hand motifs. Methods Mol Biol 1929:567–581
50. Sarkar P, Chattopadhyay A (2019) Exploring membrane organization at varying spatiotemporal resolutions utilizing fluorescence-based approaches: implications in membrane biology. Phys Chem Chem Phys 21(22):11554–11563
51. Stubbs CD, Ho C, Slater SJ (1995) Fluorescence techniques for probing water penetration into lipid bilayers. J Fluoresc 5(1):19–28
52. Chattopadhyay A (2003) Exploring membrane organization and dynamics by the wavelength-selective fluorescence approach. Chem Phys Lipids 122(1–2):3–17
53. Haldar S, Chaudhuri A, Chattopadhyay A (2011) Organization and dynamics of membrane probes and proteins utilizing the red edge excitation shift. J Phys Chem B 115(19):5693–5706
54. Seelig J (1977) Deuterium magnetic resonance: theory and application to lipid membranes. Q Rev Biophys 10(3):353–418
55. Haldar S, Kombrabail M, Krishnamoorthy G, Chattopadhyay A (2012) Depth-dependent heterogeneity in membranes by fluorescence lifetime distribution analysis. J Phys Chem Lett 3(18):2676–2681
56. Villalaín J, Prieto M (1991) Location and interaction of N-(9-anthroyloxy)-stearic acid probes incorporated in phosphatidylcholine vesicles. Chem Phys Lipids 59(1):9–16
57. Abrams FS, London E (1993) Extension of the parallax analysis of membrane penetration depth to the polar region of model membranes: use of fluorescence quenching by a spin-label attached to the phospholipid polar headgroup. Biochemistry 32(40):10826–10831
58. Abrams FS, Chattopadhyay A, London E (1992) Determination of the location of fluorescent probes attached to fatty acids using parallax analysis of fluorescence quenching: effect of carboxyl ionization state and environment on depth. Biochemistry 31(23):5322–5327
59. Chakraborty H, Lentz BR, Kombrabail M, Krishnamoorthy G, Chattopadhyay A (2017) Depth-dependent membrane ordering by hemagglutinin fusion peptide promotes fusion. J Phys Chem B 121(7):1640–1648
60. Pattnaik GP, Chakraborty H (2018) Coronin 1 derived tryptophan-aspartic acid containing peptides inhibit membrane fusion. Chem Phys Lipids 217:35–42
61. Vauquelin G (2016) Effects of target binding kinetics on in vivo drug efficacy: koff, kon and rebinding. Br J Pharmacol 173(15):2319–2334
62. Hall MD, Yasgar A, Peryea T, Braisted JC, Jadhav A, Simeonov A, Coussens NP (2016) Fluorescence polarization assays in high-throughput screening and drug discovery: a review. Methods Appl Fluoresc 4(2):022001
63. Rohe A, Henze C, Erdmann F, Sippl W, Schmidt M (2014) A fluorescence anisotropy-based Myt1 kinase binding assay. Assay Drug Dev Technol 12(2):136–144
64. Pirruccello M, Nandez R, Idevall-Hagren O, Alcazar-Roman A, Abriola L, Berwick SA, Lucast L, Morel D, De Camilli P (2014) Identification of inhibitors of inositol 5-phosphatases through multiple screening strategies. ACS Chem Biol 9(6):1359–1368
65. Vickers CJ, Gonzalez-Paez GE, Umotoy JC, Cayanan-Garrett C, Brown SJ, Wolan DW (2013) Small-molecule procaspase activators identified using fluorescence polarization. Chembiochem 14(12):1419–1422
66. Baughman BM, Jake Slavish P, DuBois RM, Boyd VA, White SW, Webb TR (2012) Identification of influenza endonuclease inhibitors using a novel fluorescence polarization assay. ACS Chem Biol 7(3):526–534
67. Haus P, Korbus M, Schroder M, Meyer-Almes FJ (2011) Identification of selective class II histone deacetylase inhibitors using a novel dual-parameter binding assay based on fluorescence anisotropy and lifetime. J Biomol Screen 16(10):1206–1216

Fluorescence Lifetime: A Multifaceted Tool for Exploring Biological Systems

5

Subhrajit Mohanty and Usharani Subuddhi

Abstract

Fluorescence is a very powerful technique widely used for the investigation of matter and living systems at a molecular and/or supramolecular level and provides important information about the structural and dynamic aspects. Steady-state and time-resolved fluorescence methods are frequently employed to analyze the emissive properties of fluorophores and their immediate environment. The advantages of using fluorescence technique, in addition to the dynamic nature of the information obtained, are its high degree of sensitivity, appropriate time-scale, noninvasive nature, and minimum perturbation. The timescale of fluorescence is best suitable to understand the processes occurring such as solvent relaxation, dynamic quenching, calculating Förster distance, and getting information regarding the dynamics of nanoenvironment surrounding excited molecule. This timescale includes the most important biologically relevant processes of fast submolecular motions at the short end of this timescale and the translational and rotational motion of macromolecules at the other extreme of this timescale. The fluorescence lifetime is one of the most important properties of a fluorophore that defines the time window of investigation of various dynamic phenomena. Fluorescence decay time is an intrinsic property of a fluorescent molecule, which is independent of the fluorophore concentration, absorption, path length, measurement technique, emission intensity, photobleaching, and/or excitation intensity. It is however influenced by various external factors, such as temperature, polarity, and the presence of fluorescence quenchers. Time-resolved fluorescence studies have got widespread applications, such as in drug discovery, medical diagnostics, and especially for exploration of biological macromolecules. Time-resolved data

S. Mohanty · U. Subuddhi (✉)
Department of Chemistry, NIT Rourkela, Rourkela, Odisha, India
e-mail: subuddhiu@nitrkl.ac.in

© The Author(s), under exclusive license to Springer Nature Singapore Pte Ltd. 2022

H. Sahoo (ed.), *Optical Spectroscopic and Microscopic Techniques*,
https://doi.org/10.1007/978-981-16-4550-1_5

provide insight into the rates and kinetics of several intramolecular and intermolecular processes that are difficult to obtain from the steady-state data. This is due to the fact that time-resolved measurements capture a range of competing or perturbing kinetic processes occurring in the timescale of the fluorophore's lifetime such as collisional quenching, solvent dynamics, energy transfer, and rotational reorientation. Therefore, understanding the dynamics of excited states of a fluorophore is highly solicited in order to decipher various photophysical, photochemical, and photobiological processes [1–3]. In this chapter, we will discuss the theory of time-resolved fluorescence and its applications in biological systems.

5.1 Introduction

Fluorescence is a very powerful technique widely used for the investigation of matter and living systems at a molecular and/or supramolecular level and provides important information about the structural and dynamic aspects. Steady-state and time-resolved fluorescence methods are frequently employed to analyze the emissive properties of fluorophores and their immediate environment. The advantages of using fluorescence technique, in addition to the dynamic nature of the information obtained, are its high degree of sensitivity, appropriate time-scale, noninvasive nature, and minimum perturbation. The timescale of fluorescence is best suitable to understand the processes occurring such as solvent relaxation, dynamic quenching, calculating Förster distance, and getting information regarding the dynamics of nanoenvironment surrounding excited molecule. This timescale includes the most important biologically relevant processes of fast submolecular motions at the short end of this timescale and the translational and rotational motion of macromolecules at the other extreme of this timescale. The fluorescence lifetime is one of the most important properties of a fluorophore that defines the time window of investigation of various dynamic phenomena. Fluorescence decay time is an intrinsic property of a fluorescent molecule, which is independent of the fluorophore concentration, absorption, path length, measurement technique, emission intensity, photobleaching, and/or excitation intensity. It is however influenced by various external factors, such as temperature, polarity, and the presence of fluorescence quenchers. Time-resolved fluorescence studies have got widespread applications, such as in drug discovery, medical diagnostics, and especially for exploration of biological macromolecules. Time-resolved data provide insight into the rates and kinetics of several intramolecular and intermolecular processes that are difficult to obtain from the steady-state data. This is due to the fact that time-resolved measurements capture a range of competing or perturbing kinetic processes occurring in the timescale of the fluorophore's lifetime such as collisional quenching, solvent dynamics, energy transfer, and rotational reorientation. Therefore, understanding the dynamics of excited states of a fluorophore is highly solicited in order to decipher various photophysical, photochemical, and photobiological processes [1–

3]. In this chapter, we will discuss the theory of time-resolved fluorescence and its applications in biological systems.

5.2 Fluorescence Lifetime: Definition and Principle

Fluorescence lifetime or the fluorescence decay time of a fluorophore is defined as the rate of depopulation of the singlet excited state (S_1) following excitation by a very short pulse of light (i.e. a δ-pulse of duration less than the reciprocal of the involved rate constants). The excited molecules return to the ground state (S_0), either by radiative or nonradiative processes, or may undergo intersystem crossing [2, 3]. The rate of change of concentration of the excited molecules, according to classical chemical kinetics, can be expressed as:

$$-\frac{\mathrm{d}[^1F^*]}{\mathrm{d}t} = (k_\mathrm{r}^\mathrm{S} + k_\mathrm{nr}^\mathrm{S})\,[^1F^*]$$

where k_r^S and k_nr^S, respectively, are the rate constants for radiative and overall nonradiative processes for $S_1 \rightarrow S_0$ transition. The above equation upon integration yields the time evolution of the concentration of the excited molecules $[^1F^*]$ of the form

$$[^1F^*] = [^1F^*]_0 \exp\left(-\frac{t}{\tau_\mathrm{S}}\right)$$

$[^1F^*]_0$ represents the excited molecule concentration at time 0 after δ-pulse excitation. In fluorescence experiments, the number of excited molecules is generally not monitored, rather the emission intensity, which is proportional to fluorophore population at a given time. Hence, the above equation is usually expressed in terms of the time-dependent intensity $I(t)$, which upon integration gives rise to the expression for the monoexponential decay:

$$I(t) = I_0 \exp\left(-\frac{t}{\tau_\mathrm{S}}\right)$$

τ_S being the fluorescence lifetime of the S_1 excited state given by: $\tau_\mathrm{S} = \frac{1}{k_\mathrm{r}^\mathrm{S} + k_\mathrm{nr}^\mathrm{S}}$, where I_0 represents the initial fluorescence intensity at $t = 0$. The fluorescence lifetime is then resolved from the slope of log $I(t)$ versus t plot, but normally it is computed by fitting the data to the presumed decay models. When there are more than one lifetime components, then the intensity is fitted to multiexponential decay functions given by:

$$I(t) = \sum_i \alpha_i \exp\left(t/\tau_i\right)$$

where α_i is the amplitude corresponding to a lifetime component τ_i, and $\sum\alpha_i$ is normalized to unity.

Both k_r^S and k_{nr}^S depend strongly on the fluorophore's environment, and factors like refractive index, polarity, viscosity, temperature, etc. have great effect on these, and as a consequence, fluorescence decay time of a fluorophore can be utilized as a probe of local environment as τ_S strongly depends on the rates of various competing decay processes. Moreover, since the fluorescence intensity is proportional to the excited state population, the fluorescence lifetime can also be defined as the time taken for the intensity to drop to $1/e$ of its initial value after the δ-pulse excitation. This is the basis of single-photon counting technique.

5.2.1 Factors Affecting Fluorescence Lifetime

When a fluorophore is in the excited state during its fluorescence lifetime, various transformations are feasible such as electron redistribution, geometrical alterations, and reorganization of the surrounding molecules. As a result of the strong influence of these factors on the nonradiative decay rates, there is an inevitable decrease in the fluorescence lifetime, and such processes are collectively known as fluorescence quenching, and these complex processes can be classified based on the way the excess energy in the excited state is dissipated. According to Levshin's classification, the excitation energy can be lost due to vibrations and rotations of the molecule thus known as energy loss *internally* or by dissipating the energy outside the molecule *externally* [2]. Processes such as molecular vibration, internal rotation, intersystem crossing, and electron and/or proton transfer in the excited state often affect the fluorescence lifetime of a fluorophore. Internal rotation of a fluorophore around some flexible bonds of the molecule leads to the nonradiative energy loss; therefore, any factor that imposes restriction of the rotation, such as change of the fluorophore microenvironment making it more compact or rigid or viscous, decreasing the temperature, or rigidification of the molecular framework, would diminish the nonradiative processes, thus increasing the fluorescence lifetime. Factors such as temperature and viscosity of the surrounding media affect the rate of rotation (k_{rot}) of a fluorophore through the rotational correlation time (θ_r) according to the Stokes–Einstein–Debye relationship [4]

$$k_{rot} = \frac{1}{\theta_r} = \frac{k_B T}{4\pi r^3 \eta}$$

where η, k_B, T, and r represent the solvent viscosity, Boltzmann constant, temperature, and radius, respectively. Since k_{rot} contributes to k_{nr}, which in turn affects the fluorescence lifetime.

Apart from this, there are certain fluorophores known as molecular rotors, whose fluorescence lifetime exhibits strong dependence on the viscosity of the medium; this dependence can be described by a molecular parameter γ that is obtained from the slope of a double logarithmic plot of fluorescence lifetime versus viscosity according to Förster–Hoffmann model, $\tau = C_{\mathrm{m}} \times \eta^{\gamma}$, where C_{m} is the concentration-temperature constants, γ is the molecular parameter, and τ represents fluorescence lifetime [2]. Fluorescence lifetime of some fluorophores also shows solvent polarity dependence. However, unlike steady-state solvatochromism [5, 6], which is a result of change in dipole moment of the excited state relative to the ground electronic states, the polarity dependence of lifetime is not properly understood [7].

Fluorescence lifetime can also be affected by modulations in the charge distribution in the excited state due to processes such as excited state electron transfer (ESET) and intersystem crossings [8–11]. In ESET, the electrons move from the donor with high electron density to the electron-deficient acceptor moiety in the excited state resulting in a nonfluorescent entity that relaxes nonradiatively to the ground state, therefore lowering the fluorescence lifetime. Similar to ESET, ESPT also is a very fast process occurring typically in a picosecond timescale and that also reduces the fluorescence lifetime. Intersystem crossing is another process that largely influences the fluorescence lifetime of a fluorophore containing Br or I like heavy atoms.

Förster Resonance Energy Transfer (FRET) is an energy transfer mechanism between two chromophores that causes decrease in the donor (D) fluorescence and can result in the gain in that of the acceptor (A) [2]. FRET efficiency can be measured in terms of the fluorescence lifetime of the donor (τ_{D}) according to the relation $E_t = 1 - \frac{\tau_{\mathrm{D}}}{\tau_{\mathrm{DA}}}$, where τ_{DA} stands for the fluorescence decay time of the donor in the presence of acceptor. τ_{D} and τ_{DA} can be related to the distance (r) between D and A as $\tau_{\mathrm{DA}} = \tau_{\mathrm{D}} \frac{r^6}{r^6 + R_0^6}$, where R_0 represents the Förster distance the fluorescence lifetime is 50% of the lifetime of the donor alone (τ_{D}) [2]. Therefore, lifetime is extremely responsive to variations in the D-A distance, which makes FRET an effective tool to explore various functional events between molecules in cells such as structural changes and interactions.

Dynamic quenching is a process, where a fluorophore loses its energy in the excited state nonradiatively due to collision with other molecules known as quenchers, and as a consequence, the lifetime of the molecule decreases with the increasing concentration of the quencher according to Stern–Volmer equation $\frac{\tau_0}{\tau} = 1 + k_q \tau_0 [Q]$, where τ_0 and τ stand for the fluorescence lifetime of the fluorophore in the absence and presence of the quencher (Q), k_{q} is the bimolecular quenching constant, and $[Q]$ represents the quencher concentration. k_{q} provides important information about the location and local environment of a fluorophore in biological macromolecules as it reflects the quenching efficiency or the accessibility of the fluorophores to the quencher. For a diffusion-controlled quenching process, k_{q} value can be up to the order of $10^{10}\ \mathrm{M}^{-1}\ \mathrm{s}^{-1}$; a smaller value of k_{q} than this indicates a low quenching efficiency due to the steric shielding of the fluorophore. And larger values of k_{q} ($>10^{10}\ \mathrm{M}^{-1}\ \mathrm{s}^{-1}$) generally point toward a probable ground state interaction between the quencher and the fluorophore [2, 3].

Fluorescence lifetime does not depend on the concentration of a fluorophore only when the molecules do not have any type of interactions either chemically or photonically, which is often the case when the concentration is kept low; at high concentrations, however, fluorescence lifetime can either increase because of photon reabsorption or decrease due to energy transfer between two close-by molecules and formation of dimers and higher aggregates that do not emit. The reabsorption process affects the fluorescence lifetime, and the measurable (apparent) fluorescence lifetime (τ_{ra}) is expressed as $\tau_{\mathrm{ra}} = \frac{\tau}{1+a\phi}$, which depends on the probability of reabsorption (a), which is a function of the extent of overlap between the emission and absorption spectra (J) and the concentration of the sample (c). τ and ϕ are the actual lifetime and the fluorescence quantum yield, respectively [12, 13]. This usually affects the fluorescence lifetime imaging especially in biological systems if by some means the local concentration of the fluorophore increases.

Another process that is of interest in the context of modulation in fluorescence lifetime is the formation of a transient excited state dimer known as excimer due to the association of an excited state fluorophore with another similar molecule in the ground state. This phenomenon is often seen in polycyclic aromatic compounds with flat π-ring structure such as pyrene. The excimer has a distinct emission spectrum and a completely different lifetime than the monomer which is extensively used in steady-state and dynamic imaging. The importance of pyrene in lifetime imaging is due to its significantly long lifetime (monomeric $\approx$680 ns and excimeric $\approx$90 ns), which is used in cell studies to eliminate interference from the background autofluorescence ($\approx$5–7 ns) [14]. Additionally, because of the long lifetime, it is extremely susceptible to quenching due to factors like presence of oxygen, free radicals, and neurotoxins, and as a result, they are widely used as probes for DNA hybridizations, mRNA recognition, and actin assembly [15]. Apart from the above, there are also several other important processes that affect excited state lifetime, e.g., the presence of paramagnetic metals, spin–orbit couplings, delayed fluorescence, etc.

5.3 Time-Resolved Fluorometry: The Technique

Pulse fluorometry and phase-modulation fluorometry are the two commonly used time-resolved techniques to measure the fluorescence lifetime; the former technique works in the time domain, whereas the later one works in the frequency domain. Theoretically, both the techniques are equivalent, but the principles are different. In the former, a short δ-pulse of exciting light is applied to the sample, which is convoluted by the instrument response. In the phase-modulation fluorometry, on the hand, the harmonic response of the sample is provided by utilizing modulated light at variable frequency, which is same as the Fourier transform of the δ-pulse response. Although both the techniques have similar capability, the time-domain methods are straightforward and hence relatively easier to comprehend. Moreover, the time-domain technology has turned out to be more preferred over the frequency domain as the instrument has become compact, versatile, less expensive, and more reliable [1–3].

5.3.1 Frequency-Domain or Phase-Modulation Method

In the frequency-domain or phase-modulation method, an intensity-modulated light is used for the excitation of the sample. This is typically carried out by a sine-wave modulation with the intensity varying at a high frequency, which is comparable to the reciprocal of the lifetime. As a result, the emission also occurs at the same modulation frequency. However, owing to the time lag between the absorption and emission, the emission shows time delays and amplitude changes relative to the excitation that are determined by the intensity decay law of the fluorophore. This delay is measured as a phase angle shift (ϕ_ω) between the excitation and emission, which can be used to calculate the decay time, and ω represents the modulation frequency in radians/s. The lifetime of the fluorophore also reduces the peak-to-peak height of the emission with respect to that of the modulated excitation and offers another independent estimation of the lifetime. ϕ_ω and the modulation (m_ω) of the emission depend on the relative values of the lifetime and the modulation frequency; these are measured over a large range of frequencies, and the frequency response is utilized to get the intensity decay of the fluorophore. The required range of modulation frequencies essential for the recovery of the intensity decay depends on the lifetime. Frequency-domain technique usually utilizes both ϕ_ω and m_ω information for the determination of lifetime. For monoexponential decay, the relations between ϕ_ω and m_ω and the decay time (τ) are given by:

$$\tan \phi_\omega = \omega\tau \quad \text{and} \quad m_\omega = \left(1 + \omega^2\tau^2\right)^{-1/2}$$

However, most of the samples exhibit multiple lifetimes, and in such cases, the lifetime values estimated from the ϕ_ω and m_ω at a specific frequency are only apparent lifetimes τ_ϕ and τ_{m}, and each signifies a different weighted average of the lifetimes displayed by the sample. Therefore, for such samples, it becomes essential to measure the phase and modulation values over the widest possible range of modulation frequencies, and the frequency response is used to resolve the intensity decay using nonlinear least-squares procedures. The calculated frequency-dependent values of the phase angle ($\phi_{c\omega}$) and the demodulation ($m_{c\omega}$) are given by:

$$\tan \phi_{c\omega} = N_\omega / D_\omega \quad \text{and} \quad m_{c\omega} = \left(N_\omega^2 + D_\omega^2\right)^{1/2}$$

where $N_\omega = \sum_i \frac{\alpha_i \omega \tau_i^2}{\left(1+\omega^2\tau_i^2\right)} / \sum_i \alpha_i \tau_i$ and $D_\omega = \sum_i \frac{\alpha_i \tau_i}{\left(1+\omega^2\tau_i^2\right)} / \sum_i \alpha_i \tau_i$.

In the frequency-domain experiment, a comparison of the sample emission and scattered light is performed similar to as that in TCSPC. The FD instruments (Fig. 5.1) generally contain two PMTs as detectors, one for the sample and other to serve as a timing reference. The sample PMT alternately receives the emission or the scattered light from the sample from the scattering solution; the sample and scattering solutions are mounted in a rotating sample holder that positions each solution exactly in the same location. Since everything else remains the same, the

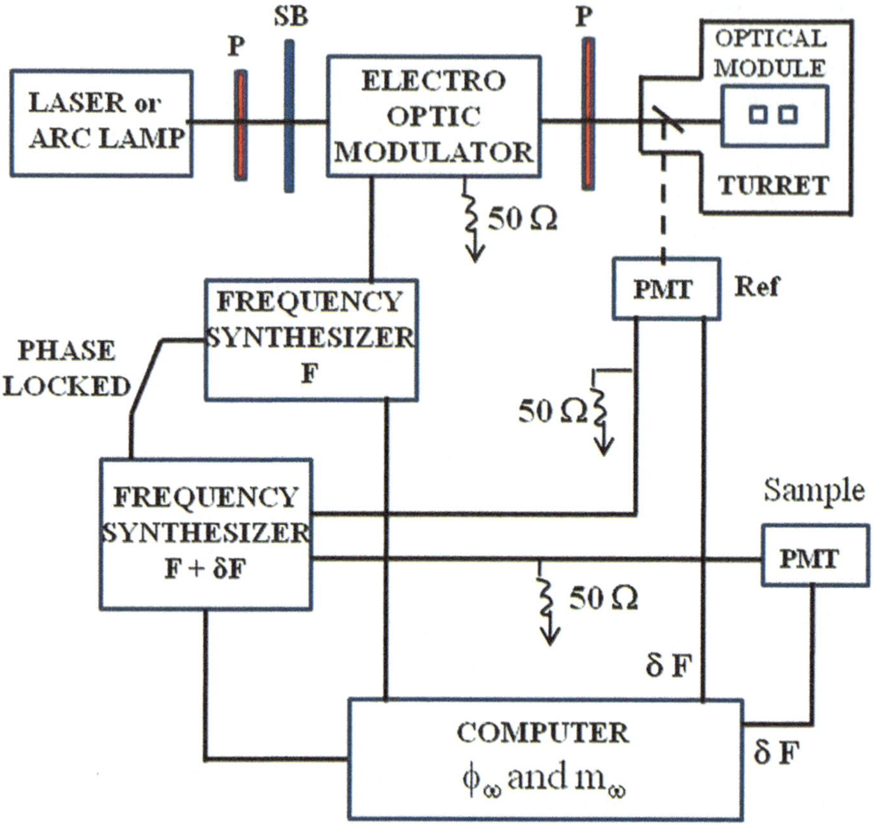

Fig. 5.1 Schematic illustration of a variable-frequency phase modulation fluorometer. *P* polarizer, *SB* Soleil–Babinet compensator, *F* frequency, *δF* cross-correlation frequency, *PMT* photomultiplier tube [2]

change in relative phase or modulation is only because of the intensity decay of the sample.

All measurements are carried with respect to the reference PMT, as the phase difference and relative modulation from the scatterer and sample cannot be measured at the same time. The phase shift between the scatterer and the reference PMT is fist determined, and the signals are shifted by an arbitrary phase angle (ϕ_1) in order to take care of the unavoidable time delays in the cables and electronic circuits. Then the phase angle of the sample relative to the reference PMT (ϕ_2) is measured, and the actual corrected phase shift (ϕ_ω) is calculated as $\phi_\omega = (\phi_2 - \phi_1)$. The modulation is determined from the AC and DC components of the sample (*B/A*) and scatterer (*b/a*) (Fig. 5.2). All the measured modulation values are corrected for modulation drifts in the instrument by dividing them with the value at the reference PMT during every measurement.

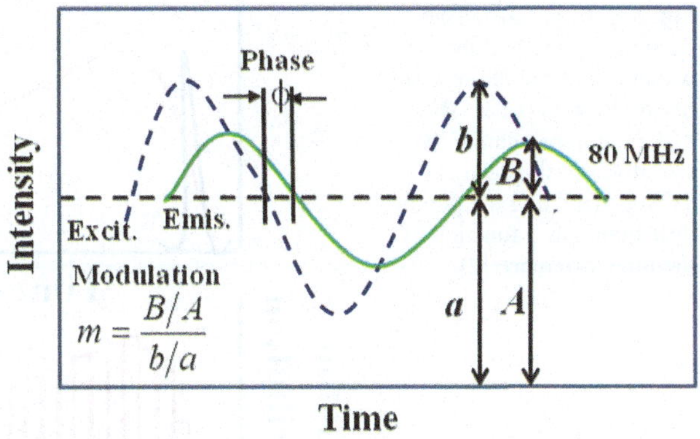

Fig. 5.2 Representation of the phase angle and modulation of fluorescence emission. The light modulation frequency is 80 MHz [2]

5.3.2 Time-Domain or Pulse Fluorometry

In time-domain or pulse fluorometry, a pulse of light from a flash lamp, pulsed laser, laser diode, or LED with sufficient delay between pulses is used for exciting the sample. The pulse width is kept much shorter than the lifetime of the sample. Then the time-dependent intensity is measured, and the fluorescence lifetime is determined from the plot of log $I(t)$ versus t, frequently by fitting the data to presumed decay models. There are various methods such as time-correlated single-photon counting (TCSPC), fluorescence upconversion, and stroboscopic method for the time-domain technique.

5.3.2.1 Time-Correlated Single-Photon Counting (TCSPC)

TCSPC is the most preferred technique for time-domain fluorescence measurements [16, 17]. In this technique, a δ-pulse of light is used to excite the sample, and the settings are adjusted in order to maintain the detection rate typically around 1 photon per 100 excitation pulses. The time lag between the excitation pulse and the arrival of the emission photon is gauged, and a histogram is constructed with the time difference in the x axis and the number of photons received during this time difference on the y axis, which denotes the waveform of the decay (Fig. 5.3).

The schematic of the specialized electronics that are generally used for the measurement of the time delay between the excitation and emission is represented in Fig. 5.4. The measurement starts with an excitation pulse exciting the sample and sending a signal that is passed through a constant function discriminator (CFD) to gauge the arrival time of the pulse. As soon as the signal passed to the time-to-amplitude converter (TAC), it starts a voltage ramp which keeps increasing till the TAC receives an additional signal from a second CFD that determines the arrival time of the emission photon. In order to minimize false reading, a fixed range of voltage is restricted for the arrival of the signal, and if the signal is not received within this range, then the event is suppressed by a window discriminator (WD). The voltage ramp thus represents the time delay between the excitation and emission signals. The voltage then is amplified by a programmable gain amplifier (PGA), and

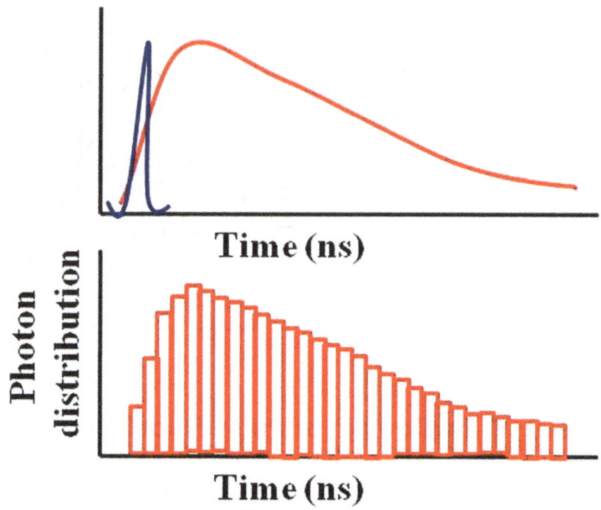

Fig. 5.3 Top panel: The excitation pulse and the waveform of the fluorescence decay. Bottom panel: The histogram representing the waveform of the decay in TCSPC depicting the photon distribution as a function of the time difference [2]

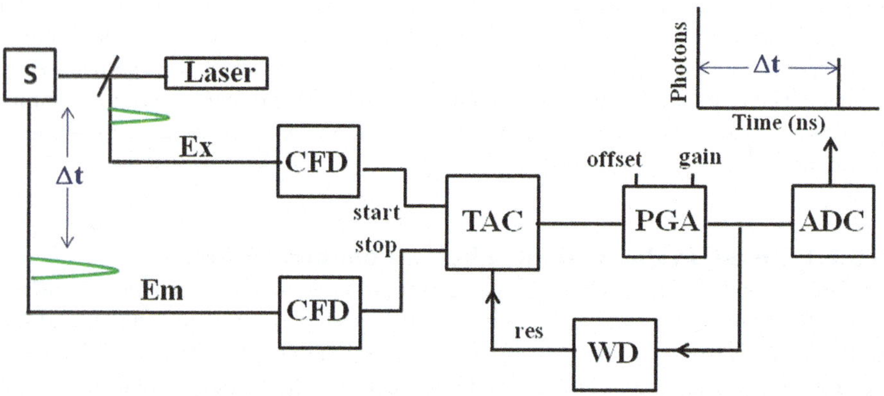

Fig. 5.4 Schematic representation of the electronics used for the measurement of the time delay between the excitation and emission in a TCSPC technique [2]

an analog-to-digital converter (ADC) is used to convert it to a numerical value, which is stored as a single event with the measured time delay. A histogram of the decay is constructed by repeating this process several times, which resembles the intensity decay (impulse response function). The height of a slab in the histogram on the *y*-axis represents the number of photons detected within a time interval of t to $t + \Delta t$, where Δt is the width of the timing channel. The time profile or the fitted function for a given intensity decay is obtained by the convolution of the instrument response function (IRF) with the impulse response function by using the nonlinear least squares analysis method, and the decay parameters (α_i and τ_i) are obtained. IRF is the response of the instrument to a zero lifetime sample and represents the shortest time profile that can be measured by the instrument. For the measurement of the IRF, generally a dilute scattering solution such as colloidal silica (Ludox) is used.

5.4 Fluorescence Lifetime-Based Applications

The applications of fluorescence lifetime measurement in biological systems can be broadly categorized as fluorescence lifetime assays, sensing, and imaging.

5.4.1 Fluorescence Lifetime Assays

Fluorescence lifetime is a robust parameter, and any change in the physicochemical environment of the fluorophore affects the lifetime; hence, it can be used in several biological assays. Fluorescence lifetime assays have been developed based on various mechanisms: simple binding assay involving change in lifetime due to the binding of two components out of which one may be intrinsically fluorescent or can be fluorescently labeled; a quench-release type assay where the fluorophore is originally in a quenched state and when the fluorescent compound gets released as a result of some enzymatic reaction or upon binding to some other analyte the lifetime of the system changes; fluorescence lifetime can also be combined with Förster resonance energy transfer (FRET) assays for the measurement of energy transfer efficiency between two fluorescent units.

5.4.1.1 Interaction of Proteins with Small and Macromolecules

Protein–Surfactant Interaction

Protein–surfactant interaction plays major roles in pharmaceutics, cosmetics, foods, etc. [18]. Fluorescence lifetime has been used extensively to distinguish and understand types of interactions between proteins and different surfactants at different concentrations. The structure and the charge of the surfactants are the two primary factors deciding the type and site of the interaction of proteins [19–22]. The synthetic surfactants are known to affect the proteins differently than the steroidal surfactants like bile salts due to their structural differences. Commonly used protein like HSA interacts with bile salts in a stepwise manner providing different stages of interaction where initially all the active sites are occupied, then cooperative binding exists and finally saturation can be observed. All these variations are well reflected by the stepwise decrease in average fluorescence lifetime as depicted in Fig. 5.5 [22]. Even bile salts with different structures interact differently with the protein, which is confirmed from the extent of change in the average fluorescence lifetime.

However, these distinct stages of interactions are absent in case of nonionic surfactants where no clear separation can be observed in the variation of fluorescence lifetime [21]. Fluorescence lifetime has also been used to understand the effect of surfactants on the thermal and chemical denaturation of bovine serum albumin

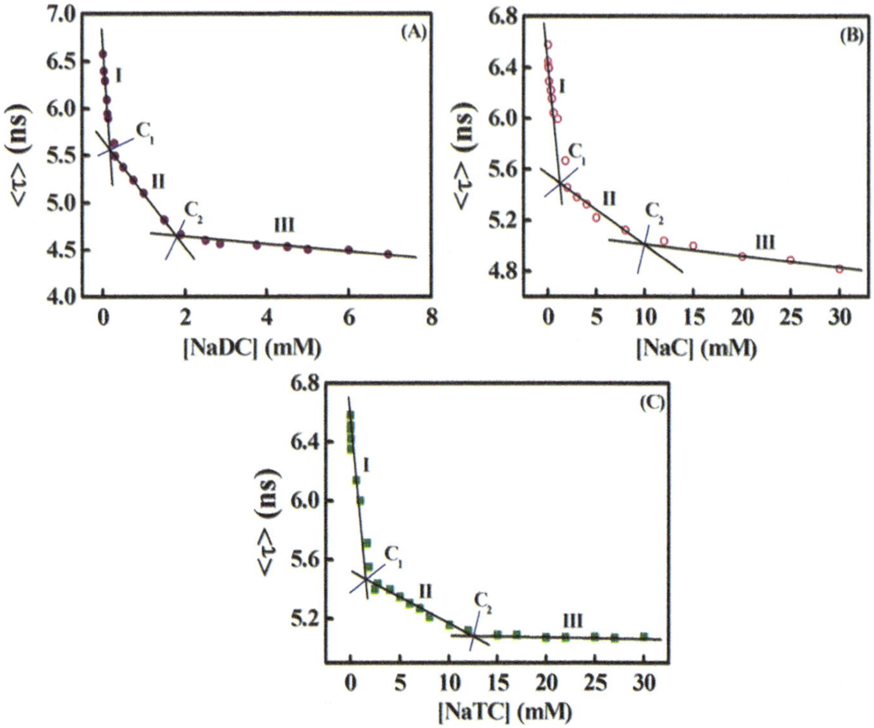

Fig. 5.5 Plot of average fluorescence lifetime of HSA against the concentrations of (**a**) NaDC, (**b**) NaC, and (**c**) NaTC. [HSA] = 10 μM. (Adapted with permission from [22]))

(BSA) during the unfolding and refolding of the proteins in the presence of different additives [23].

Protein–Drug Interaction

Proteins are highly prone to bind with variety of drugs while transporting to the targets and thus face alteration in their structure and activity which is normally observed in the quenching of the tryptophans present in the proteins. Fluorescence lifetime is a preferred technique to comprehend the type of quenching where any change in the fluorescence lifetime on addition of quencher can be monitored [24–27]. The quenching in fluorescence lifetime of proteins can also be used to calculate the FRET efficiency and the apparent distance between the donor tryptophan and the acceptor which may be any drug or external fluorophore [28, 29].

Protein–Ionic Liquid Interaction

Ionic liquids have a broad spectrum of bioapplications such as protein extraction, purification, solubilization, and stabilization, and hence, understanding their interaction with proteins becomes important [30, 31]. Fluorescence lifetime has been used popularly to identify the type of quenching of protein–ionic liquid systems such as

BSA in the presence of 1-butyl-1-methyl-2-oxopyrrolidinium bromide (BMOP), pepsin in the presence of 1-butyl-3-methylimidazolium chloride ([C4mim]Cl) and 1-octyl-3-methylimidazolium chloride ([C8mim]Cl), HSA in the presence of *N*-butyl-*N*-methyl-morpholinium bromide, and [Mor1,4] [Br] and lysozyme in the presence of 1-hexyl-3-methylimidazolium bromide ([C_6MIm][Br]) and 1-methyl-3-propylimidazolium bromide ([C_3MIm][Br]) 1,6-bis(3-methylimidazolium-1-yl) hexane bromide ([C6$(MIm)_2$][$Br]_2$) where the bimolecular quenching constant has also been calculated [32–35].

Interaction of Proteins with Macromolecules

Time-resolved fluorescence resonance energy transfer (TR-FRET) method has been implemented to monitor the interaction between two different proteins SERCA2a (sarcoplasmic reticulum Ca-ATPase) and PLB (phospholamban) where the earlier one acts as donor and the later one is the acceptor [36]. Binding of proteins with other macromolecules such as DNA and RNA plays key roles inside the cells. However, the protein-DNA/RNA stoichiometry is not fixed throughout the cell and hence the fluorescence lifetime is a useful tool for understanding the interaction process as this method is independent of the concentration of the fluorophore. The combination of both FRET and lifetime imaging technique can also be used to monitor the interaction of protein and the nucleic acid [37].

5.4.1.2 Investigation of the Conformational Changes of Proteins

The function of proteins depends on their structure, which can undergo verity of alteration depending on the surrounding media and additives. On addition of surfactant, proteins usually undergo some structural transitions depending upon the surfactant concentration which can be characterized by the average fluorescence lifetime [38]. Recently, the folding and unfolding processes of human serum albumin (HSA) and rabbit serum albumin (RSA) have been studied in the presence of different concentrations of SDS by monitoring the average fluorescence lifetime. The average fluorescence lifetime of both the proteins is found to increase slightly at the lower concentrations of SDS which then decreases up to the CMC of the surfactant indicating the unfolding, and above the CMC of SDS, the fluorescence lifetime again increases due to the refolding of the proteins [39]. Fluorescence lifetime can also be a useful technique for studying the conformational alterations in the secondary structure of proteins on their denaturation [23]. Combination of fluorescence lifetime and Three-Color Single-Molecule FRET was found to be extremely useful to understand the conformational dynamics of fast folding proteins [40]. To understand the conformational differences between Gαi1 (inhibitory G proteins) and Gαs (stimulatory G proteins) after coupling of β2AR (β2-adrenergic receptor), recently lifetime-based FRET by labeling small-sized fluorescent pair 7-HC/FlAsH ((7-hydroxycoumarin-4-yl)-ethylglycine/fluorescein arsenical hairpin) has been investigated which offers to comprehend the role of G protein in transduction [41].

5.4.1.3 Binding of DNA with Small Molecules

DNA (deoxyribonucleic acid), one of the most studied biopolymer, plays vital roles in various important biological processes such as transcription of gene, mutagenesis, gene expression, etc. and carries instructions for development and functioning of organisms [42, 43]. Binding of small molecules to DNA has been the most emerging area of the research for the formulation of new therapeutic agents and their functional efficiency in the current scenario. The binding of small molecules with DNA is trimodal in nature which includes intercalative binding, groove binding, and the electrostatic binding [44, 45]. Fluorescence lifetime provides a colossal significant method to analyze the different modes of interaction between the small molecules with DNA. The interaction of different drugs such as Ellipticine, 3-acetyl-4-oxo-6,7-dihydro-12H-indolo-[2,3-a]-quinolizine (AODIQ), and β-carboline analog with calf-thymus DNA (ctDNA) has been carried out using fluorescence lifetime to elucidate the binding sites and mechanism [46–48]. The multimodal interaction of acridine orange with DNA provides a modulation of the properties of the dye which can be monitored by fluorescence lifetime studies [49]. By analyzing the modulation in the lifetime values and their components, it is possible to figure out the different binding modes of small molecules with DNA. The average lifetime values of the drug increase on binding with DNA, which can be ascribed to the decrease in the nonradiative rate constant because of motional restriction provided by DNA to the drug; however, the change in radiative rate constant is insignificant. Ionic strength of the medium is known to modulate the binding mode of the drug and DNA. On increasing the salt concentration of NaCl, the amplitude as well as the magnitude of the highest lifetime component, which corresponds to the intercalation, decreases due to drop in the electrostatic contribution toward the binding process. Whereas, with an increase in the ionic strength, there is an enhancement of the intermediate lifetime component value which signifies the preferring of groove binding. However, the overall drug binding ability increases on increasing the ionic strength which is suggested from the drop in population of free drug [50].

The distance between the nucleic acids can be determined using FRET and fluorescence lifetime where the double-stranded DNA and RNA have been internally labeled with Alexa488 and Cy5 as donor and acceptor moiety [51]. FRET has also been used to detect the RNA degradation by labeling fluorescein to the end of RNA [52]. Using time-resolved FRET, the conformational analysis and calculation of the distance between quantum dot and DNA have been done [53]. Recently, FRET has been constructed in a double-stranded DNA using fluorescent thymidine and cytidine analogs and calculated the FRET efficiency using fluorescence lifetime. The calculated FRET efficiency follows a periodical variation suggesting the sensitivity of FRET toward the orientation of the fluorophore [54].

5.4.1.4 Elucidating the Self Assembly of Bile Salts and Their Interaction with Drugs

Fluorescence lifetime is helpful in elucidating the aggregation behavior and properties of different biosurfactants such as bile salts. As the aggregation of bile salt proceeds through different types and size of aggregates, the contribution of these

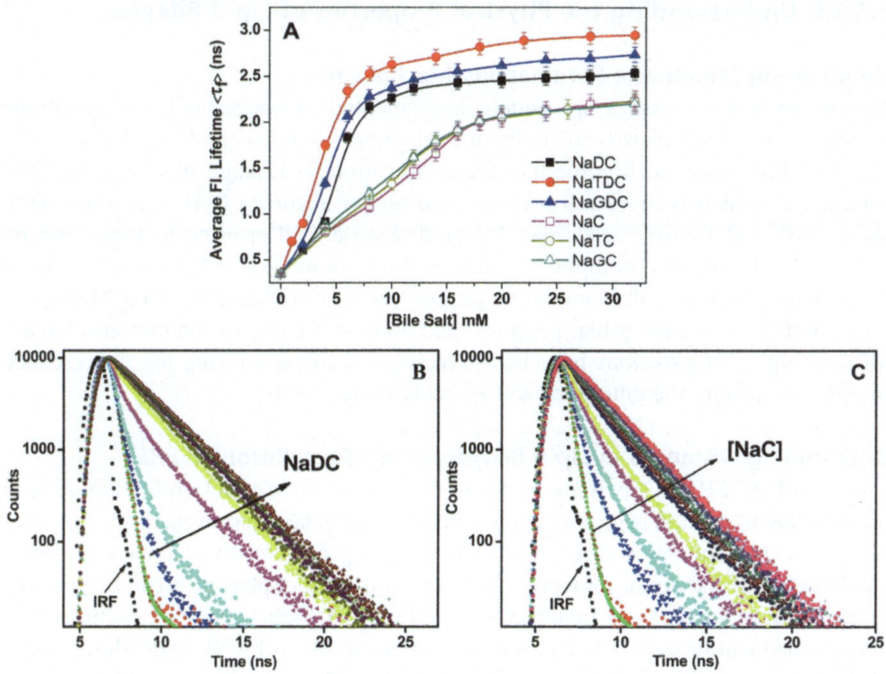

Fig. 5.6 (**a**) Plot of average fluorescence lifetime against the concentration of bile salts (standard deviation is shown as error bars). The fluorescence lifetime decay transients at different concentrations of (**b**) NaDC and (**c**) NaC. [Bile salt] = 0–32 mM, $\lambda_{ex} = 370$ nm. (Adapted with permission from [57])

aggregates in the process of micellization can be estimated using different components of lifetime decay of Dansyl-labeled cholic acid through speciation diagram [55]. The CMC of the bile salts can also be calculated by monitoring the changes in the decay pattern and average lifetime of the extrinsic fluorophores like DPH and Coumarin 1 whose lifetimes are highly sensitive toward a small change in the microenvironment surrounding them [56, 57]. Recently, Coumarin 1 has been employed to understand the aggregation behavior of six different bile salts where a significant increase in average lifetime values of Coumarin 1 was observed due to its conversion of TICT (Twisted Intramolecular Charge Transfer) state to ICT (Intramolecular Charge Transfer) state (Fig. 5.6) [57]. Owing to the biodegradable and biocompatible nature of bile salts, they are becoming most suitable candidate for drug delivery, and hence, their interaction with several fluorescent drugs such as curcumin, ellipticine, piperine, and berberine has been in the focus recently, with bile salts also being investigated using fluorescence lifetime [58–61]

5.4.1.5 Understanding the Physical Properties of Lipid Bilayers

Determining Membrane Permeability Mechanisms

Membrane leakage assays are widely implemented to understand the membrane permeability mechanisms in both model lipid systems and in living cells [62, 63]. Fluorescence lifetime-based quantification of leakage has been reported for calcein, which is more informative with higher accuracy [64]. Calcein usually shows different lifetime responses for graded versus all-or-none leakage, and by monitoring the lifetime components, it is possible to understand the mechanism of the leakage. Recently, the membrane permeabilization induced by SMAMPs (synthetic mimics of antimicrobial peptides) and their selectivity for membranes having different lipid compositions have been investigated by monitoring the fluorescence lifetime of calcein, the entrapped self-quenching dye [65].

Determining Distances in Lipid Bilayers Using Time-Resolved FRET

Time-resolved FRET is extremely useful in the domain size determination of lipid bilayers due to its high resolution [66]. This method usually monitors the affinities of donor–acceptor moieties inside the bilayer. Using time-resolved data as a linear combination of the donor molecule decays in two distinct phases, it can be possible to access the domain sizes quantitatively [67–69]. A higher precised estimation of domain size can be achieved by combining time-resolved FRET with Monte Carlo (MC) simulations [70, 71]. By analyzing the interleaflet contribution to FRET, it can be possible to determine the depth of molecules in membranes.

Interaction of Liposomes with Small and Macromolecules

Fluorescence lifetime has been used extensively to investigate the location and distribution of extrinsic probes and effects of additives such as surfactants, drugs, and polymers on lipid bilayers. By monitoring the two different emitting species, i.e., neutral and anionic of 1-naphthol through fluorescence lifetime, water permeation introduced by bile salts and wetting of the membrane at the interface and at the core of the DPPC (dipalmitoylphosphatidylcholine) bilayer in the submicellar concentration range of bile salts has been observed [72]. A similar kind of study has been performed by using ANS as a fluorescent probe, where a decrease in average lifetime was seen because of the hydration of the lipid bilayer membrane induced by bile salts (Fig. 5.7). Using the lifetime of ANS, it was found that on increasing temperature, a drop in the average lifetime can be observed due to increase in the fluidity inside the liposomes as the solid gel phase transforms into the liquid phase [73].

The fluorescence lifetime of 1-naphthol has also been studied to examine the effect of aggregation of FDA-approved oral drug molecule FTY720 on DMPC (dimyristoylphosphatidylcholine) [74]. In a recent work, a contrasting interaction of PEG on the properties of different phases of 1,2-dimyristoyl-*sn*-glycero-3-phosphocholine (DMPC) liposomes has been found out, where both the phases are affected differently by the addition of PEG which is well reflected in the unequal change in the lifetime of probe in different phases [75]. Two-dimensional fluorescence lifetime correlation spectroscopy can be used to understand and quantify the

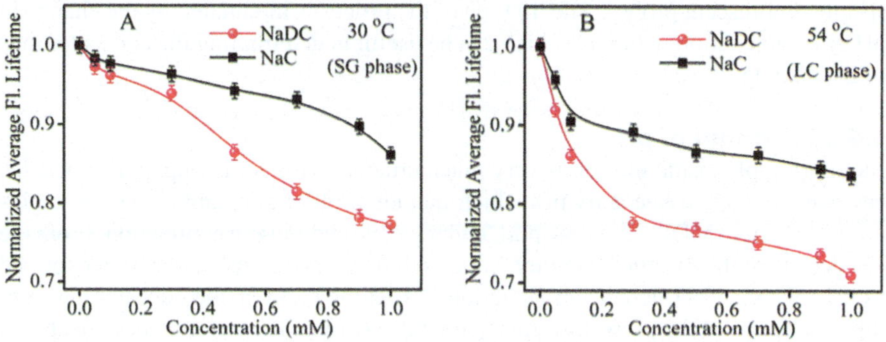

Fig. 5.7 Variation of the average fluorescence lifetime of ANS with bile salt concentration in DPPC vesicles at (**a**) 30 °C (SG phase) and (**b**) 54 °C (LC phase) ($\lambda_{ex} = 370$ nm, $\lambda_{em} = 475$ nm). [DPPC] = 0.4 mM and [ANS] = 4 μM. (Adapted with permission from [73])

diffusion of lipids in the proximal and distal leaflets of an SLB-supported lipid bilayer [76]. Recently, the photophysical properties of a cationic fluorophore, lumichrome (LC) in the presence of a small unilamellar vesicle of 1,2-dimyristoyl-*sn*-glycero-3-phosphocholine (DMPC), has been studied using fluorescence lifetime and time-resolved area-normalized emission spectra (TRANES), where, more than one prototropic forms of LC are found in the DMPC liposome [77].

Interaction of Liposomes with Proteins and DNA

Interaction of drug probed in different liposomal carriers on addition of human serum albumin (HSA) and hen egg white lysozyme has been investigated through fluorescence lifetime [78, 79]. The migration of the drug from liposomes to the albumin was found owing to the penetration of HSA into the liposomes due to hydrophobic interaction which has been confirmed by analyzing different components of the fluorescence lifetime. Effect of DNA on liposome-encapsulated ellipticine can be monitored using fluorescence lifetime studies which revealed the decrease in fluorescence lifetime components of the drug in presence of DNA due to the DNA-induced hydration and rupture of bilayers [80].

5.4.2 Fluorescence Lifetime Sensing

Fluorescence lifetime has been widely utilized for sensing of H^+, Ca^{2+}, K^+, glucose, and other metabolites in biological samples because of high sensitivity. Time resolution of the fluorescence signal enables to separate the population of fluorophores bound to nonspecific locations, therefore displaying high spatial resolution even inside the living cells. Recently, application of lifetime-based sensing in tissues and living cells has gained pace using fluorophores in the near-infrared region. Fluorescence lifetime sensing is considered to be one of the most preferred techniques for chemical sensing owing to the surge in the numbers of sensitive and specific fluorescent probes. The lifetime of the probes is modulated by the change in pH or complexation with ions which can be useful in determining the concentrations of the analyte.

5.4.2.1 Sensing of pH

Monitoring pH inside the cell is very much crucial owing to its importance in cells, enzymes, tissues, tissue activities which include proliferation and apoptosis, multi-drug resistance (MDR), ion transport, endocytosis, and muscle contraction. From the last decade of the twentieth century, various fluorescence probes have been explored as pH sensor. A pH-dependent alteration in lifetime requires the contribution from different forms of the probe toward apparent decay time. If one form does not absorb at the excitation wavelength, then the observed emission results from just one form of species and, thus, independent of pH [81]. The contribution of each species toward the emission can also be modified by selecting the excitation or emission wavelength. For physiological pH, the most common sensors are fluorescein-based indicator BCECF and fluorescent proteins. BCECF has a smaller lifetime of 2.75 ns at pH 4.5 in its protonated state, which increases to 3.9 ns at pH 8 in its deprotonated state [15]. Another organic molecule, acridine, is also known to have a remarkable change in fluorescence lifetime values (20 ns) in the pH range of 4–7 [82]. Currently, quantum dot nanosensors, different functionalized materials such as functionalized Tristyrylbenzene, and functionalized perylene bisimide (PBI) were also synthesized and used as lifetime-based pH sensor [83–85]. The fluorescence lifetime response of functionalized PBI to pH variation can be related to the photo-induced electron transfer (PET) which is attributed to the protonation–deprotonation of the amine groups present in the side chains and resulted in a ten times increase in average fluorescence lifetime on decreasing the pH from 10.2 to 5.2 [84]. Recently, Pt-Nanocluster has been developed as a fluorescence lifetime-based pH sensor to understand the physiological systems such as blood and culture media due to change in its surface states depending on the pH of the medium. Pt-Nanocluster is known to have two different lifetime components, out of which the faster component results from the metal core signifying the intrinsic quantization effect, and the slower component is attributed to the interaction of the metal core with surface ligands [86]. The slower component and the average lifetime of the nanocluster drop down on increasing pH of the medium (Fig. 5.8). The slower lifetime component and the average lifetime of the nanocluster are responsive toward the pH change, i.e., both decrease on increasing the pH. Due to deprotonation, the structure of surface ligand is modified, which might affect the metal–ligand binding, and thus results in the modulation of lifetime on changing the pH.

5.4.2.2 Sensing of Glucose

Sensing and quantification of glucose in blood and urine has been a major interest in the medicinal world due to its long-term health effects of diabetes that include blindness and neuropathies. One of the most common glucose sensors is NADH (dihydro nicotinamide adenine dinucleotide) which is the reduced from NAD (nicotinamide adenine dinucleotide). In the presence of glucose and glucose

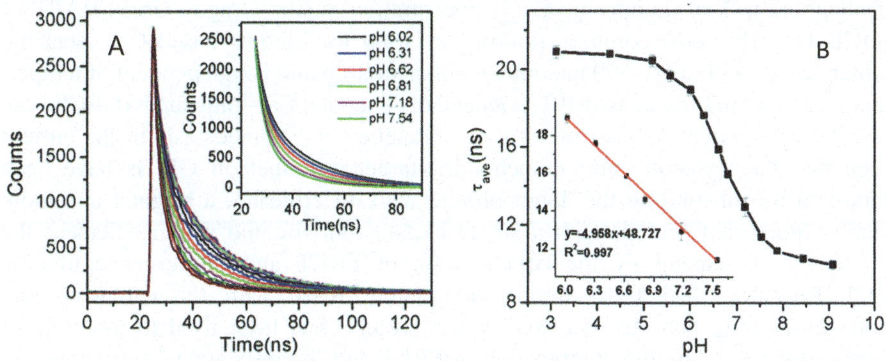

Fig. 5.8 (**a**) The variation of fluorescence decay lifetime of Pt NCs in PBS with the pH 3.12–9.09. Inset: The nonexponential fluorescence decay fitting lines from the pH 6.02 to 7.54. (**b**) The variation of average fluorescence lifetimes of Pt NCs with pH. Inset: The linear calibration plot of pH 6.02–7.54. (Adapted with permission from [86])

dehydrogenase, NAD is reduced, whereas glucose is oxidized to gluconolactone. The free state of NADH has two short lifetime components with a small average lifetime value of around 0.43 ns. However, NADH bound to glucose dehydrogenase has a significantly higher lifetime component of 2.9 ± 0.2 ns which is significantly higher than that of the free state. When the average lifetime value of NADH is around 0.43 ns, the concentration of NADH is higher than the coenzyme and hence most of the molecules of NADH are considered to be in free state, and when the average lifetime value is around 2.90 ns, the bound form is considered to be dominant. Thus, monitoring the average lifetime values of free and bound state of NADH, not only the glucose concentration in the body, can be estimated, but the effect of external factors like temperature can also be understood [87]. The derivative of NADH, i.e., Carba nicotinamide adenine dinucleotide (cNAD) which is more stable toward hydrolytic cleavage, has also been used in sensing of glucose through lifetime measurement [88]. The fluorescence lifetime of badan-labeled glucose/galactose-binding protein (GBP) has also been used for glucose sensing having the concentration in the micromolar range [89]. The development of genetically encoded fluorescent biosensor has introduced real-time visualization of dynamics of glucose not only from individual neurons but also from astrocytes. Fluorescence lifetime studies of iGlucoSnFR-TS can also be helpful in monitoring the quantity of glucose concentration up to 10 mM deep within ex vivo or in vivo tissue [90].

5.4.2.3 Sensing of Different Ions

Various ions such as Ca^{2+}, Na^{+}, and K^{+} play vital part in living cells and organisms; therefore, monitoring the ion concentrations as well as the fluctuations is very important. Modulation in intracellular Ca^{2+} concentration plays a key role in signaling cascades in brain cells, including axonal release of neurotransmitters, dendritic control of synaptic plasticity, and intracellular communication in electrically passive astroglia. Fluorescence lifetime provides an edge over other methods in the challenging task of monitoring Ca^{2+} concentration in situ. Oregon Green BAPTA1 (OGB-1) is the most common fluorophore used for lifetime-based Ca^{2+} sensing, which works within 10–500 nanomolar concentration and is independent of temperature, pH, and microviscosity [91]. Genetically encoded Ca^{2+} indicators (GECIs) can also be used extensively to investigate organelle Ca^{2+} homeostasis in the current scenario. The cytosolic and mitochondria-targeted Cameleon GECIs have been modified by substituting the donor protein with mCerulean3, a brighter and more stable fluorescent protein with a single exponential lifetime which enhances the efficiency of sensing by the collaboration of FRET and fluorescence lifetime [92]. Recently, Twitch-2B fluorescent protein fused with the cellulose- and collagen-binding peptides (ColBD-Twitch protein) has been used to monitor the extracellular Ca^{2+} by the combination of FRET and fluorescence lifetime imaging. The ColBD-Twitch protein displays well-resolved modulation in the fluorescence lifetime of the FRET donor part (blue emitting Cerulean) for changing Ca^{2+} concentrations up to 4 mM [93]. 1,2,3-Triazol fluoroionophore was used to detect Na^+ in the millimolar range where the average decay time of the fluorophore increases with a simultaneous increase in contribution of slower lifetime component and decrease in the contribution of faster component [94]. Two photon microscopy-based time-correlated fluorescence lifetime measurements of four different dyes—SBFI (sodium-binding benzofuran isophthalate), CoroNaGreen, Sodium Green, and Asante Na-TRIUM Green-2 (ANG-2)—have been carried out to quantify Na^+ dynamics [95]. In a recent work, specifically CoroNaGreen lifetime has been used to quantify intracellular Na^+ concentration [96]. The average fluorescence lifetime of CoroNaGreen was found to increase significantly from 480 to 1030 ps on increasing Na^+ concentration up to 150 mM inside HEK (human embryonic kidney, 96,121,229) cells. The alteration in the fluorescence lifetime of borondipyrromethenes (BODIPY) has been analyzed to detect K^+ ion [97]. Combination of fluorescence lifetime and FRET can also be used for sensing application inside the cells. In a recent work, ionic strength sensors mCerulean3 (a cyan fluorescent protein) and mCitrine (a yellow fluorescent protein) have been used, which act as donor–acceptor pair undergoing FRET to sense Hofmeister series of salts [98]. Nowadays, fluorescence lifetime imaging microscopy (FLIM), a more advanced technique, is being used to detect and quantify different ions present in tissues and living cells.

5.4.3 Fluorescence Lifetime Imaging

The concept of lifetime imaging is a very recent technique developed only in the end of the twentieth century [99]. This technique evaluates the spatial distribution of fluorescence lifetime at every pixel of an image. As time-resolved measurements can be realized with different techniques, e.g., single-and two-photon microscopy, endoscopy, and tomography, the number of applications has increased in diverse areas of biological imaging. Since the lifetime of a fluorophore is independent of the concentration and depends only on its molecular environment, it has been applied successfully in fluorescence microscopy, where it is difficult to have a control on the local concentration of the fluorophore. Fluorescence lifetime imaging microscopic (FLIM) technique has been used for monitoring dynamic cellular processes in real time by employing an array of fluorescence lifetime-sensitive fluorophores [100]. It has been extensively applied for the measurement of molecular environment parameters, biomolecular interactions in cell biology, the metabolic condition of cells and tissues, in vivo drug delivery and release, tissue imaging, etc. [15, 101]

5.4.3.1 FLIM for Mapping Viscosity

Fluorophores with high γ-value (viscosity sensitive) known as molecular rotors (Scheme 5.1) have been widely used in lifetime imaging for the intracellular viscosity mapping and to recognize different organelles in the cell due to heterogeneity in the viscosities of their local environments by means of multiexponential fluorescence decays [102–108].

Various *meso*-substituted BODIPY-based fluorophores have been developed for this purpose since the fluorescence lifetime of these molecules shows remarkable sensitivity to the viscosity of their microenvironment. FLIM images for live SK-OV-3 human ovarian carcinoma cells incubated with BODIPY-phenyl-C12 are presented in Fig. 5.9; two high viscosity domains characterized by apparent viscosities of 160 ± 20 and 260 ± 40 cP were detected in terms of short and long lifetime components representing low viscous endosomes (or lipid droplets) and high viscous other hydrophobic domains, respectively [106].

Recently, a distorted conformation-induced viscosity-sensitive fluorescent molecular rotor BODIPY–P has been developed by Cui et al. for quantitative monitoring of lysosomal viscosity in autophagy and apoptosis [109]. Fluorescence lifetime image of MCN, a new two-photon fluorescent probe, stained living HeLa cells suggested the intracellular average viscosity value of 73.45 ± 21.55 cP inside the cytosol. MCN was found to work even in the deep tissue (~130 μM) under two-photon excitation and showed the ability for in vivo imaging of viscosity in zebrafish. The intracellular viscosity measurements have also been accomplished by ratiometric measurements by using a rigid reference whose lifetime is independent of the viscosity while that of the molecular rotor is sensitive to the viscosity variation [103, 107, 108, 110]. The ratiometric fluorescence detection from a mixture of two cyanine dyes cy5 and cy3 has been utilized for establishing the viscosity in the aqueous phase of the cellular cytoplasm; cy3 showed strong dependence of fluorescence lifetime on viscosity, while cy5 was insensitive to viscosity [111]. A conjugated butadiyne-linked porphyrin dimer has been applied simultaneously as a molecular rotor to measure intracellular viscosity to image viscosity in cells during photo-induced cell death and as a PDT photosensitizer. This dimer acts as a molecular rotor owing to the existence of planar and twisted conformations which are interconvertible due to rotation around the butadiyne linker which results in the viscosity-dependent excited state properties of the porphyrin dimer, and ratio of emission intensity of the twisted and the planar conformations could be calibrated in a large viscosity range between 0.6 and 1000 cP.

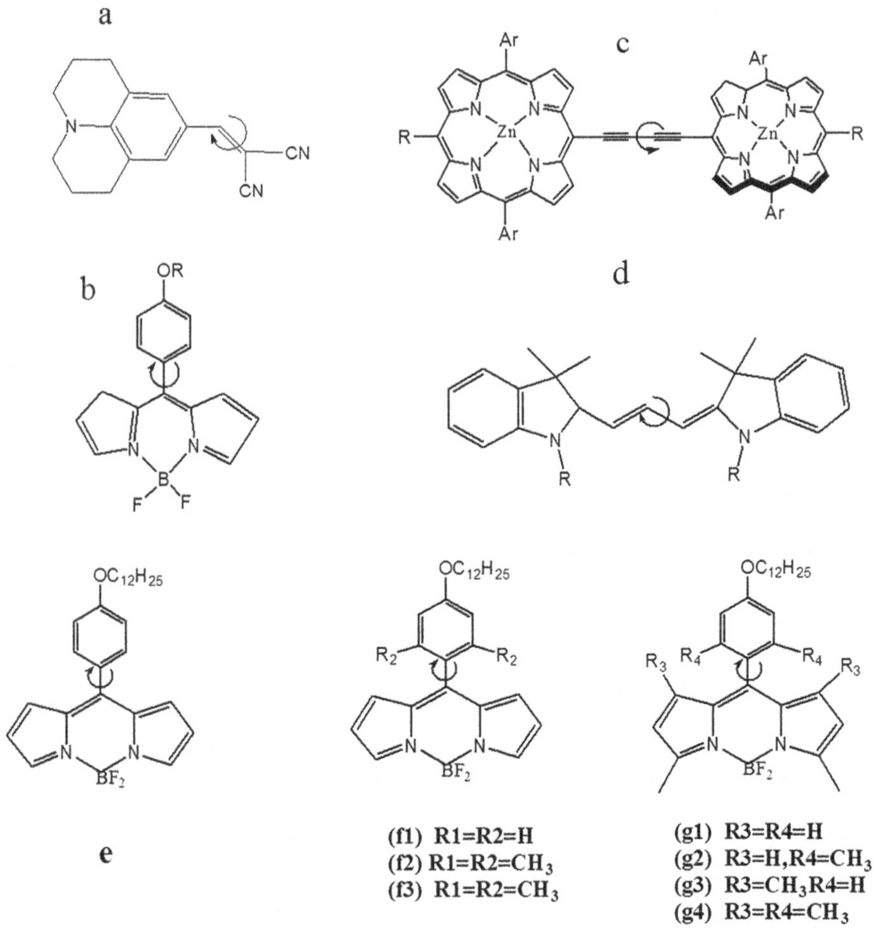

Scheme 5.1 Structure of different molecular rotors (**a**) 9-(dicyanovinyl)julolidine (DCVJ); (**b**) *p*-phenyl-substituted 4,4′-difluoro-4-bora-3a,4adiaza-s-indacene, BODIPY-phenyl-C12; (**c**) conjugated porphyrin dimer; (**d**) substituted indocyanine dye (cy3); and (**e–g**) structures of different parasubstituted BODIPY dyes having various degrees of constraint on rotation of a meso-phenyl group. (Adapted from [105])

5.4.3.2 FLIM for Mapping Intracellular Temperature

Recently, FLIM has been used to map temperature in the living cells with the help of fluorescent molecular thermometers, wherein temperature-sensitive polymers have been conjugated with certain fluorophores which display a large fluorescence lifetime variation near the physiological temperature of 37 °C [112–116]. The thermoresponsive polymers undergo some structural or configuration phase change with temperature, which in turn induces an alteration in the local environment of the fluorophore as a result of its fluorescence quantum yield and lifetime changes. Using these fluorescent polymeric thermometers (FPT), the intracellular temperature

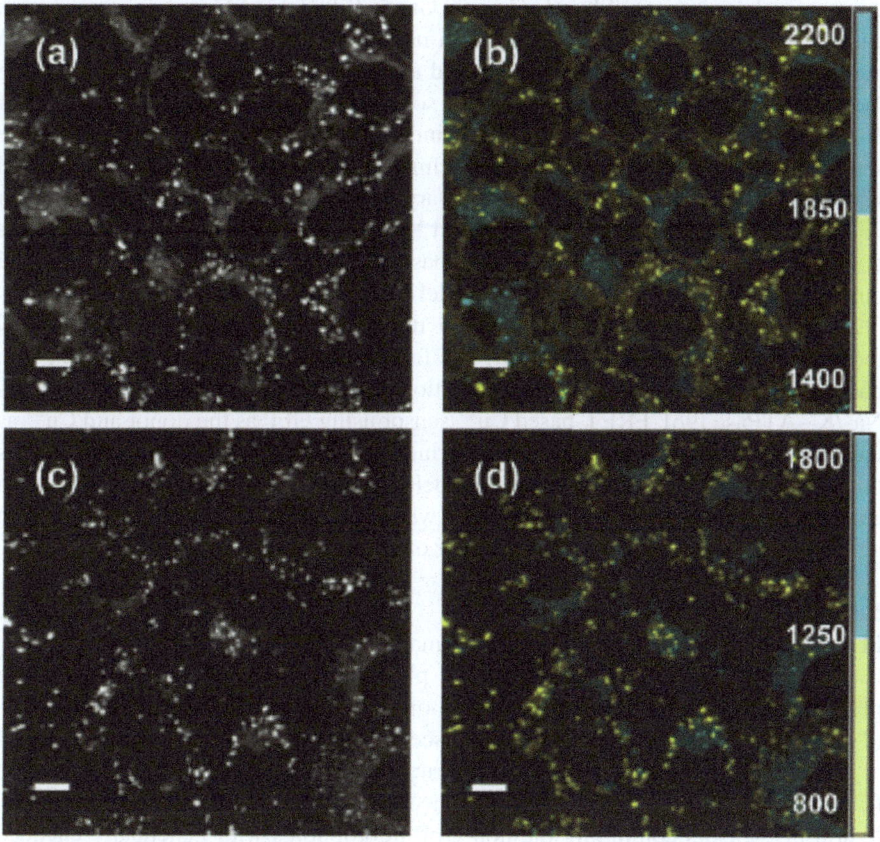

Fig. 5.9 Fluorescence intensity (**a**, **c**) and FLIM images (**b**, **d**) of live SK-OV-3 human ovarian carcinoma cells incubated with BODIPY-phenyl-C12 (**a**, **b**) and BODIPY phenyl-farnesyl (**c**, **d**). The color scale signifies shorter lifetimes and longer lifetimes in yellow and blue, respectively. (**b**) for BODIPY phenyl-C12: 1350–1850 ps in yellow and 1850–2200 ps in blue color (**d**) for BODIPY phenyl-farnesyl: 800–1250 ps in yellow and 1250–1800 ps in blue color. (Adapted with permission from [106])

mapping in various mammalian cell lines such as HeLa, COS7, and NIH/3T3 cells has been carried out, and the FLIM illustrated a temperature difference of around 1 °C between the nucleus and the cytoplasm and also the thermogenesis in the mitochondria [114, 116].

5.4.3.3 FLIM to Map Ion Concentrations

Fluorescent dyes, which change their fluorescence lifetime upon binding with these ions, are chosen for FLIM to map intracellular ion concentrations. The advantage of mapping ion concentrations with FLIM is because of the fact that this technique is independent of the fluorophore concentration and also it is unaffected by variation in incident intensity and/or photobleaching. Various dyes such as Quin-2, calcium crimson, calcium green, Fluo-3, Oregon Green BAPTA1-AM, magnesium green, and Oregon Green BAPTA5N have been used for mapping Ca^{2+} concentration in biological samples such as brain slices and astrocytic calcium homeostasis in living mouse models of Alzheimer's disease and so on [117–121]. Fluorescence lifetime of *N*-(ethoxycarbonylmethyl)-6-methoxy-quinolinium bromide (MQAE) is found to be sensitive to Cl^- due to collisional quenching and has been used for FLIM-based mapping of Cl^- concentrations in salivary acinar cells of cockroach [122], in mammalian olfactory sensory neurons [123], and in nociceptors neurons in the dorsal root ganglion [124, 125]. FLIM-based analysis of [Na^+] has been carried out by using benzofuran isophthalate in HeLa cells [126] and ANG-2 in living cells [127]. Recently, CoroNaGreen has been utilized for detection of somatic [Na^+] increase in mouse brain tissues [128] and for monitoring alterations in the Na^+ ion concentration induced by metabolic inhibition or inhibition of the cellular ubiquitous Na^+/K^+-ATPase [96]. FRET-based Cu^{2+} sensor using GFP as the donor and Cu^{2+} as the acceptor has been utilized for monitoring Cu^{2+} uptake and release in plant cells through FLIM [129]. Oregon Green-labeled carbonic anhydrase has also been employed for Cu^{2+} sensing in live cells, where the fluorescence of Oregon Green gets quenched due to FRET upon binding of Cu^{2+} to the carbonic anhydrase [130].

5.4.3.4 FLIM for Mapping pH

FLIM technique has been widely used for mapping the pH or H^+ ion concentration in live cells and monitoring the effect of pH on various cellular processes. The pH-dependent fluorescence lifetime ratio of the two prototropic forms of 2,7-bis-(2-carboxyethyl)-5-(and-6) carboxyfluorescein (BCECF) has been used to image pH at various depths of the skin stratum corneum [131]. Various fluorescent protein (FP)-based pH indicators have been designed for intracellular pH sensing; moreover, hybrid pH sensors combining the non-GFP fluorophores with genetically encoded tags have been proven to be effective noninvasive tools for selective targeting to different cellular locations [132–134]. Diazaoxatriangulenium (DAOTA) dye-based pH indicators that feature pH-dependent long fluorescence lifetimes (17–20 ns) are found suitable for studying 2D pH distributions with the help of FLIM [135]. Mercaptopropionic acid-capped quantum dots have been developed for pH mapping in cells; the advantage of these quantum dots is their long fluorescence lifetimes, which facilitates to discriminate their signal from cellular autofluorescence [83]. Recently, a DNA-based FLIM reporter containing a Ca^{2+} specific probe assembled onto a DNA nanostructure along with pH-responsive and lysosome-targeting molecules has been developed for real-time imaging and simultaneous quantification of pH and Ca^{2+} in lysosomes, and it was found that the pH and Ca^{2+} concentration are interrelated, and together, they regulate the autophagy and also the transformation of NSCs, resulting in Rapamycin-induced antiaging [136].

5.4.3.5 FLIM for Mapping Glucose

Most of the glucose sensors are based on glucose/galactose-binding protein (GBP), which alters its conformation on binding to glucose, and can be detected by FRET and also by labeling the glucose binding site with environmental sensitive probes

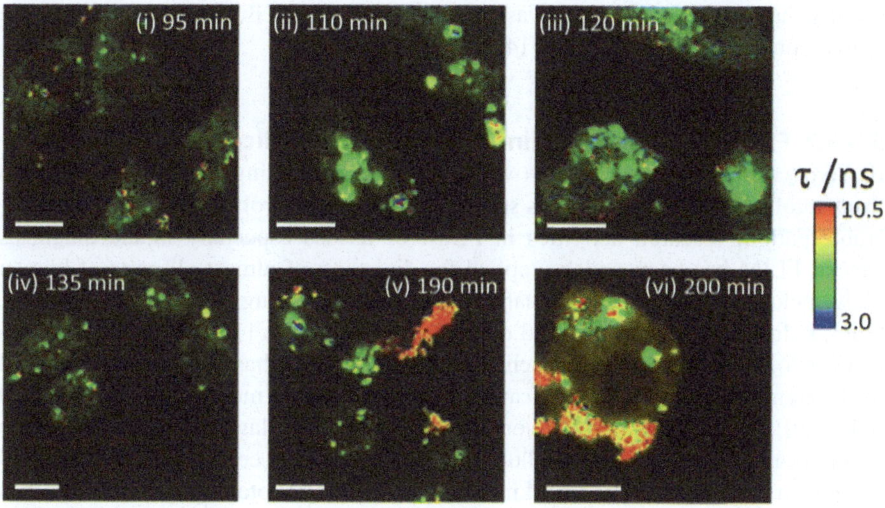

Fig. 5.10 FLIM images of QD-APBA conjugates into MDA-MB-231 cells in PBS pH 8.0 before and after addition of a solution of 50 mM glucose to the extracellular medium: (i) 95 min, (ii) 110 min, (iii) 120 min, (iv) 135 min, (v) 190 min, and (vi) 200 min. (Adapted from open access article [139])

like badan [89, 137]. The N- and C-terminals of a mutant GBP have respectively been tagged with a FRET pair AcGFP1 as donor and a mCherry as acceptor to develop AcGFP1-GBP-mCherry biosensor protein for continuous monitoring [138]. Recently, CdSe/ZnS quantum dots modified with aminophenylboronic acid (APBA) showing long fluorescence decay times have been developed, which act as nanosensors for detection of glucose in MDA-MB-231 cells via FLIM (Fig. 5.10). The long lifetime of these nanosensors made it easy for demarcating the cell autofluorescence, thereby improving the sensitivity and selectivity [139].

5.4.3.6 FLIM for Mapping Oxygen

Complexes of transition metals such as ruthenium, iridium, osmium, or rhenium are generally used as optical oxygen sensors as their fluorescence quantum yield and lifetime is very sensitive to collisional quenching by oxygen [140]. TCSPC-based FLIM study has revealed spatial pericellular oxygen gradients around isolated viable chondrocytes [141]. A frequency-domain FLIM set-up has been reported for oxygen sensing using ruthenium lifetimes in cells [142]. Using one-photon phosphorescence lifetime imaging microscopy (PLIM), a low-molecular-weight phosphorescent platinum(II) complex has been used for real-time oxygen mapping [143]. By employing a two-channel FLIM/PLIM system, the intrinsic oxygen partial pressure has been monitored that provided complex metabolic and redox imaging of living cells [144]. Recently, a FRET-based oxygen sensor, Myoglobin-mCherry, suitable for FLIM has been developed, which presents important information about metabolic activity through real-time, noninvasive, cell-by-cell intracellular pO_2 and coenzyme status monitoring in living cells [145].

5.4.3.7 FLIM for Tissue Imaging and Medical Applications

Ex vivo (cells, tissue) and in vivo (whole body) imaging using fluorescence lifetimes from autofluorescence and tissues stained with extrinsic probes comprises a considerable part of research activities in FLIM of tissues especially in the diagnosis cancer. FLIM has been widely applied for diagnosis of almost all parts of human body including the gastrointestinal (GI) tract, bronchi/lung, skin, head and neck, brain, kidney, cardiovascular, and ophthalmic pathologies [146]. Autofluorescence from endogenous fluorophores such as NAD(P)H or FAD has been extensively used to differentiate cancer from noncancerous tissue and to monitor cell metabolism [147–150]. The decrease in the fluorescence lifetime of collagen due to its disruption by tumor-associated matrix metalloproteinases from basal cell carcinomas provides a platform for early detection of skin cancer [151, 152]. Protoporphyrin IX, which is a metabolic end product of aminolevulinic acid (5-ALA), a naturally occurring intermediate of hemebiosynthesis, has been used to differentiate malignant from healthy tissue in brain and oral mucosa due to its relatively long fluorescence lifetime and preferential accumulation in tumor cells compared to normal tissue [153, 154]. By using the autofluorescence of cardiac myocytes, it has been shown that it is possible to distinguish necrosis from apoptosis [155]. The difference in the scatter plots of short and long fluorescence lifetime components between a healthy retina and that of retinas at the onset of age-related macular degeneration (AMD) is used for the diagnosis of ophthalmic pathologies [156, 157]. FLIM studies using autofluorescence of teeth can be utilized for early detection of dental caries by monitoring the difference between the lifetime decays of the carious regions and the healthy, hard dental tissues owing to the presence of metal-free porphyrin monomers with long fluorescence lifetime [158].

5.4.3.8 FLIM for Tracking Drug Delivery and Release

Fluorescence lifetime measurement and imaging techniques have been widely used for monitoring the drug delivery kinetics and the degradation of drug carriers. The increase in fluorescence lifetime from 1.1 ns of free doxorubicin to 2.4 ns after binding to DNA has been utilized to track release of the drug from a polymeric capsule [159]. The enhancement in fluorescence lifetime of an encapsulated NIR dye has also been employed to examine the biodegradability of the polymeric nanoparticle carrier in vivo [160]. FLIM of NIR-emissive polymersomes labeled dendritic cells has been utilized for in vivo tracking of cells in live animals [161]. FLIM technique has also been applied for monitoring drug trafficking and release human hair and skin from the effect of drugs on their autofluorescence lifetime [162–164].

5.5 Concluding Remarks and Future Perspectives

Fluorescence spectroscopic techniques are expanding rapidly, and their applications in diverse areas of physical, chemical, and life sciences are growing exponentially. The ever-growing interest in time-resolved fluorescence spectroscopy is due to advances in the time resolution, improved data analysis methods, and advanced instrumentation. With these technological advancements, time-resolved fluorescence measurement has become a robust and powerful multifaceted tool with excellent resolution that has enabled to evaluate the structural and dynamic aspects of various biological macromolecules such as proteins, membranes, and nucleic acids. The enhanced data collection rates and improved spatial resolution in fluorescence lifetime measurement have been expanded for cellular imaging and single-molecule detection in real time. Advances in lasers and detector technology are decreasing the cost and complexity of earlier complex instruments and hence have paved path for application of these techniques for clinical and analytical chemistry. Development of optical fiber sensors for time-resolved fluorescence spectral analysis could further expand the horizon of this technique for potential applications in chromatography to identify elements of similar retention time. Many new innovative ideas and methodologies are envisaged to be developed by inputs from diverse areas such as organic chemistry, biophysical chemistry, semiconductor physics, nonlinear optics, molecular biology, biotechnology, and nanotechnology so that time-resolved fluorescence spectroscopy will continue to contribute rapid advancements in decoding complex physiological and biochemical processes in biology.

Acknowledgments SM thanks National Institute of Technology, Rourkela, for institute fellowship. The authors also acknowledge Binita Tirkey and Kuldeep for their help in the preparation of the book chapter.

References

1. Birch DJ, Imhof RE (2002) Time-domain fluorescence spectroscopy using time-correlated single-photon counting. In: Lakowicz JR (ed) Topics in fluorescence spectroscopy. Volume 1 techniques. Kluwer Academic Publishers, Amsterdam
2. Lakowicz JR (2006) Principles of fluorescence spectroscopy, 3rd edn. Springer, New York
3. Valeur B, Berberan-Santos M (2002) Molecular fluorescence: principles and applications. Wiley-VCH, Weinheim
4. Jue T (2009) Fundamental concepts in biophysics. Springer, Humana
5. Bertolino CA, Ferrari AM, Barolo C, Viscardi G, Caputo G, Coluccia S (2006) Solvent effect on indocyanine dyes: a computational approach. Chem Phys 330:52–59
6. Cramer CJ, Famini GR, Lowrey AH (1993) Use of calculated quantum chemical properties as surrogates for solvatochromic parameters in structure-activity relationships. Acc Chem Res 26:599–605
7. Pavlovich VS (2006) Solvent polarity effect on excited-state lifetime of carotenoids and some dyes. Biopolymers 82:435–441
8. Hsieh CC, Cheng YM, Hsu CJ, Chen KY, Chou PT (2008) Spectroscopy and femtosecond excited state proton transfer induced charge transfer reaction. J Phys Chem A 112:8323

9. Syage JA, Steadman J (1991) Picosecond measurements of phenol excited-state proton transfer in clusters. I. Solvent basicity and cluster size effects. J Chem Phys 95:2497
10. Chattoraj M, King BA, Bublitz GU, Boxer SG (1996) Ultra-fast excited state dynamics in green fluorescent protein: multiple states and proton transfer. Proc Natl Acad Sci U S A 93: 8362
11. Dance ZEX, Mickley SM, Wilson TM, Ricks AB, Scott AM, Ratner MA, Wasielewski MR (2008) Intersystem crossing mediated by photoinduced intramolecular charge transfer: Julolidine-anthracene molecules with perpendicular π systems. J Phys Chem A 112:4194
12. Birks JB (1970) Photophysics of aromatic molecules. Wiley-Interscience, London
13. Rinaldo C, Gianfranco C, Antonio P, Paola T, Gianluca V (1997) Fluorescence lifetime imaging of experimental tumors in hematoporphyrin derivative-sensitised mice. Photochem Photobiol 66:229
14. Birks JB, Dyson DJ, Munro IH (1963) Proc R Soc London Ser A 275:575
15. Berezin MY, Achilefu S (2010) Fluorescence lifetime measurements and biological imaging. Chem Rev 110:2641–2684
16. Becker W (2005) Advanced time-correlated single photon counting techniques. Springer, Berlin
17. Becker W (2008) The TCSPC handbook, 3rd edn. Becker & Hickl Gmbh, Berlin
18. Otzen D (2011) Protein-surfactant interactions: a tale of many states. BBA 1814:562–591
19. Gelamo EL, Silva CHTP, Imasato H, Tabak M (2002) Interaction of bovine (BSA) and human (HSA) serum albumins with ionic surfactants: spectroscopy and modelling. BBA 1594:84–99
20. Maurya JK, Mir MUH, Maurya N, Ali A, Patel R (2016) A spectroscopic and molecular dynamic approach on the interaction between ionic liquid type gemini surfactant and human serum albumin. J Biomol Struct Dyn 34:2130–2145
21. Ghosh N, Mondal R, Deshmukh A, Dutta S, Mukherjee S (2015) Weak interactive forces govern the interaction between a non-ionicsurfactant with human serum albumin. Chem Phys Lett 634:77–82
22. Ghosh N, Mondal R, Mukherjee S (2015) Hydrophobicity is the governing factor in the interaction of human serum albumin with bile salts. Langmuir 31:1095–1104
23. Malarkania K, Sarkar I, Selvam S (2018) Denaturation studies on bovine serum albumin–bile salt system: bile salt stabilizes bovine serum albumin through hydrophobicity. J Pharm Anal 8: 27–36
24. Liu J, He Y, Liu D, He Y, Tang Z, Lou H, Huo Y, Cao X (2018) Characterizing the binding interaction of astilbin with bovine serum albumin: a spectroscopic study in combination with molecular docking technology. RSC Adv 8:7280
25. Singh I, Luxami V, Paul K (2020) Spectroscopy and molecular docking approach for investigation on the binding of nocodazole to human serum albumin. Spectrochim Acta A Mol Biomol Spectrosc 235:118289
26. Singh N, Kumar N, Rathee G, Sood D, Singh A, Tomar V, Dass SK, Chandra R (2020) Privileged scaffold chalcone: synthesis, characterization and its mechanistic interaction studies with BSA employing spectroscopic and chemoinformatics approaches. ACS Omega 5:2267–2279
27. Zhang J, Gao X, Huang J, Wang H (2020) Probing the interaction between human serum albumin and 9-hydroxyphenanthrene: a spectroscopic and molecular docking study. ACS Omega 5:16833–16840
28. Ling I, Taha M, Al-Sharji NA, Abou-Zied OK (2018) Selective binding of pyrene in subdomain IB of human serum albumin: combining energy transfer spectroscopy and molecular modelling to understand protein binding flexibility. Spectrochim Acta A Mol Biomol Spectrosc 194:36–44
29. Wang Q, Xiao Y, Huang Y, Li H (2016) An important prerequisite for efficient Forster resonance energy transfer (FRET) from human serum albumin to alkyl gallate. RSC Adv 6: 36146

30. Fujita K, MacFarlane DR, Forsyth M (2005) Protein solubilising and stabilising ionic. Liquids Chem Commun 38:4804–4806
31. Egorova KS, Gordeev EG, Ananikov VP (2017) Biological activity of ionic liquids and their application in pharmaceutics and medicine. Chem Rev 117:7132–7189
32. Fan Y, Zhang S, Wang LJ, Fan H, Shan D (2013) Investigation of the interaction of pepsin with ionic liquids by using fluorescence spectroscopy. J Appl Spectrosc 67:648–655
33. Kumari M, Maurya JK, Singh UK, Khan AB, Singh P, Patel R (2014) Spectroscopic and docking studies on the interaction between pyrrolidinium based ionic liquid and bovine serum albumin. Spectrochimica Acta A Mol Biomol Spectrosc 124:349–356
34. Kumari M, Singh UK, Singh P, Patel R (2017) Effect of N-butyl-N-methyl-morpholinium bromide ionic liquid on the conformation stability of human serum albumin. ChemistrySelect 2:1241–1249
35. Islam MM, Barik S, Preeyanka N, Sarkar M (2020) Interaction of lysozyme with monocationic and dicationic ionic liquids: toward finding a suitable medium for biomacromolecules. J Phys Chem B 124:961–973
36. Stroik DR, Yuen SL, Janicek KA, Schaaf TM, Ji L, Ceholski DK, Hajjar RJ, Cornea RL, Thomas DD (2018) Targeting protein-protein interactions for therapeutic discovery via FRET-based highthroughput screening in living cells. Sci Rep 8:12560
37. Camborde L, Brière C, Deslandes L, Jauneau A, Dumas B, Gaulin E (2017) Detection of nucleic acid–protein interactions in plant leaves using fluorescence lifetime imaging microscopy. Nat Protoc 12:193350
38. Mandal B, Ghosh S, Moulik SP (2016) Detailed characterization of lysozyme (Lyz)-surfactant (SDDS) interaction and the structural transitions. New J Chem 40:4617–4624
39. Srivastava R, Alam MS (2020) The multi-spectroscopic approach on the interaction between rabbit serum albumin and cationic surfactant: investigation on the formation and solubilization of the protein aggregation. Spectrochimica Acta A Mol Biomol Spectrosc 239:118542
40. Yoo J, Louis JM, Gopich IV, Chung HS (2018) Three-color single-molecule FRET and fluorescence lifetime analysis of fast protein folding. J Phys Chem B 122:11702–11720
41. Shi P, Zhang Y, Lv P, Wang W, Ling S, Guo X, Li D, Liu S, Sun D, Zhang L, Liu D, Zheng JS, Tian C (2020) A genetically encoded small-size fluorescence pair reveals allosteric conformational changes of G proteins upon its interaction with GPCR by fluorescence lifetime based FRET. Chem Commun 56:6941–6944
42. Fei Y, Lu G, Fan G (2009) Spectroscopic studies on the binding of a new quinolone antibacterial agent: sinafloxacin to DNA. Anal Sci 25:1333–1338
43. Ma Y, Zhang G, Pan J (2012) Spectroscopic studies on the interaction of sodium benzoate, a food preservative, with calf thymus DNA. J Agric Food Chem 60:10867–10875
44. Saenger W (1983) Principles of nucleic acid structure. Springer Verlag, New York
45. Afzal M, Ghosh S, Das S, Chattopadhyay N (2016) Endogenous activation-induced delivery of a bioactive photosensitizer from a micellar carrier to natural DNA. J Phys Chem B 120: 11492–11501
46. Sengupta A, Koninti R, Gavvala K, Nirmalya B, Partha H (2013) An anticancer drug to probe non-specific protein-DNA interactions. Phys Chem Phys Chem:1–28
47. Kundu P, Chattopadhyay N (2017) Interaction of a bioactive pyrazole derivative with calf thymus DNA: deciphering the mode of binding by multi-spectroscopic and molecular docking investigations. J Photochem Photobiol B Biol 173:485–449
48. Kundu P, Ghosh S, Chattopadhyay N (2015) Exploration of the binding interaction of a potential nervous system stimulant with calf-thymus DNA and dissociation of the drug–DNA complex by detergent sequestration. Phys Chem Chem Phys 17:17699–17709
49. Sayed M, Krishnamurthy B, Pal H (2016) Unraveling multiple binding modes of acridine orange to DNA using a multispectroscopic approach. Phys Chem Chem Phys 18:24642–24653

50. Ganguly A, Ghosh S, Guchhait N (2015) Spectroscopic and viscometric elucidation of the interaction between a potential chloride channel blocker and calf-thymus DNA: the effect of medium ionic strength on the binding mode. Phys Chem Chem Phys 17:483–492
51. Sindbert S, Kalinin S, Nguyen H, Kienzler A, Clima L, Bannwarth W, Appe B, Muller S, Seidel CAM (2011) Accurate distance determination of nucleic acids via Förster resonance energy transfer: implications of dye linker length and rigidity. J Am Chem Soc 133:2463–2480
52. Hirata R, Hirakawa K, Shimada N, Watanabe K, Ohtsuki T (2021) Fluorescence lifetime probes for detection of RNA degradation. Analyst 146:277
53. Guo J, Qiu X, Mingoes C, Deschamps JR, Susumu K, Medintz IL, Hildebrandt N (2019) Conformational details of quantum dot-DNA resolved by Förster resonance energy transfer lifetime Nanoruler. ACS Nano 13:505–514
54. Hirashima S, Sugiyama H, Park S (2020) Construction of a FRET system in a double-stranded DNA using fluorescent thymidine and cytidine analogs. J Phys Chem B 124:8794–8800
55. Mendoza MG, Marin ML, Miranda MA (2012) Dansyl-labeled cholic acid as a tool to build speciation diagrams for the aggregation of bile acids. J Phys Chem B 116:14776–14780
56. Subuddhi U, Mishra AK (2007) Micellization of bile salts in aqueous medium: a fluorescence study. Colloids Surf B Biointerfaces 57:102–107
57. Mishra SS, Mohanty S, Mishra J, Subuddhi U (2019) Photophysical properties of coumarin 1 in bile salt aggregates: an insight into the role of bile salt structure on the aggregation behavior. Langmuir 35:16555–16567
58. Mandal S, Ghosh S, Banik D, Banerjee C, Kuchlyan J, Sarkar N (2013) An investigation into the effect of the structure of bile salt aggregates on the binding interactions and ESIHT dynamics of curcumin: a photophysical approach to probe bile salt aggregates as a potential drug carrier. J Phys Chem B 117:13795–13807
59. Thakur R, Das A, Adhikari C, Chakraborty A (2014) Anjan Chakraborty, partitioning of prototropic species of an anticancer drug ellipticine in bile salt aggregates of different head groups and hydrophobic skeletons: a photophysical study to probe bile salts as multisite drug carriers. Phys Chem Chem Phys 16:15681
60. Debnath S, Mishra J (2020) Understanding the intrinsic fluorescence of piperine in microheterogeneous media: partitioning and loading studies. New J Chem 44:831
61. Sen S, Paul BK, Guchhait N (2017) Differential interaction behaviors of an alkaloid drug berberine with various bile salts. J Colloid Interface Sci 505:266–277
62. Imura Y, Choda N, Matsuzaki K (2008) Magainin 2 in action: distinct modes of membrane permeabilization in living bacterial and mammalian cells. Biophys J 95:5757–5765
63. Patel H, Tscheka C, Edwards K, Karlsson G, Heerklotz H (2011) All-or-none membrane permeabilization by fengycin-type lipopeptides from Bacillus subtilis QST713. Biochim Biophys Acta Biomembr 1808:2000–2008
64. Patel H, Heerklotz H (2009) Characterizing vesicle leakage by fluorescence lifetime measurements. Soft Matter 5:2849–2851
65. Stulz A, Vogt A, Saar JS, Akil L, Lienkamp K, Hoernka M (2019) Quantified membrane permeabilization indicates the lipid selectivity of membrane-active antimicrobials. Langmuir 35:16366–16376
66. Amaro M, Sachl R, Jurkiewicz P, Coutinho A, Prieto M, Hof M (2014) Time-resolved fluorescence in lipid bilayers: selected applications and advantages over steady state. Biophys J 107:2751–2760
67. De ARFM, Loura MS, Fedorov A, Prieto M (2005) Lipid rafts have different sizes depending on membrane composition : a time-resolved fluorescence resonance energy transfer study. J Mol Biol 346:1109–1120
68. Loura LMS, Prieto M (2000) Resonance energy transfer in heterogeneous planar and bilayer systems: theory and simulation. J Phys Chem B 104:6911–6919
69. Loura LMS, Fedorov A, Prieto M (2001) Fluid–fluid membrane microheterogeneity: a fluorescence resonance. Biophys J 80:776–788

70. Engström S, Lindberg M, Johansson LBÅ (1988) Monte Carlo simulations of electronic energy transfer in three-dimensional systems: a comparison with analytical theories. J Chem Phys 89:204
71. Štefl M, Cebecauer M, Machan R, Sachl R, Humpolickova J, Kolarova M, Johansson LBA, Hof M (2012) Dynamics and size of cross-linking-induced lipid nanodomains in model membranes. Biophys J 102:2104–2113
72. Mohapatra M, Mishra AK (2010) 1-Naphthol as a sensitive fluorescent molecular probe for monitoring the interaction of submicellar concentration of bile salt with a bilayer membrane of DPPC, a lung surfactant. J Phys Chem B 114:14934–14940
73. Mohapatra M, Mishra AK (2013) Photophysical behavior of 8-anilino-1-naphthalenesulfonate in vesicles of pulmonary surfactant dipalmitoylphosphatidylcholine (DPPC) and its sensitivity toward the bile salt–vesicle interaction. Langmuir 29:11396–11404
74. Swain J, Mohapatra M, Santosh R, Aidhen IS, Mishra AK (2013) Study of aqueous phase aggregation of FTY720 (fingolimod hydrochloride) and its effect on DMPC liposomes using fluorescent molecular probes. Phys Chem Chem Phys 15:17962
75. Sahu S, Talele P, Patra B, Verma RS, Mishra AK (2020) A multiparametric fluorescence probe to understand the physicochemical properties of small unilamellar lipid vesicles in poly (ethylene glycol)-water medium. Langmuir 36:4842–4852
76. Otosu T, Yamaguchi S (2018) Quantifying the diffusion of lipids in the proximal/distal leaflets of a supported lipid bilayer by two-dimensional fluorescence lifetime correlation spectroscopy. J Phys Chem B 122:10315–10319
77. Bapli A, Gautam RK, Jana R, Seth D (2019) Investigation of different prototropic forms of biologically active flavin lumichrome in the presence of liposome. Photochem Photobiol 95: 1151–1159
78. Thakur R, Das A, Chakraborty A (2014) The fate of anticancer drug, ellipticine in DPPC and DMPC liposomes upon interaction with HSA: a photophysical approach. J Photochem Photobiol B 130:122–131
79. Das A, Thakur R, Dagar A, Chakraborty A (2014) A spectroscopic investigation and molecular docking study on the interaction of hen egg white lysozyme with liposomes of saturated and unsaturated phosphocholines probed by an anticancer drug ellipticine. Phys Chem Chem Phys 16:5368
80. Das A, Thakur R, Chakraborty A (2013) A steady-state and time-resolved fluorescence study on liposome-calf thymus DNA interaction: probed by an anticancer drug ellipticine. RSC Adv 3:19572
81. Szmacinski H, Lakowicz JR (1993) Optical measurements of pH using fluorescence lifetimes and phase-modulation fluorometry. Anal Chem 65:1668–1674
82. Totland C, Thomas PJ, Holst B, Akhtar N, Hovdenes J, Skodvin T (2019) A broad-range fluorescence lifetime pH sensing material based on a single organic fluorophore. J Fluoresc 29: 1125–1131
83. Orte A, Alvarez-Pez JM, Ruedas-Rama MJ (2013) Fluorescence lifetime imaging microscopy for the detection of intracellular pH with quantum dot nanosensors. ACS Nano 7:6387–6395
84. Pacheco-Liñán PJ, Garzon A, Tolosa J, Bravo J, Canales-Vazquez J, Rodriguez-Lopez J, Albaladejo J, Garcia-Martinez JC (2016) pH-sensitive fluorescence lifetime molecular probes based on functionalized tristyrylbenzene. J Phys Chem C120:18771–18779
85. Huang M, Liang X, Zhang Z, Wang J, Fei Y, Ma J, Qu S, Mi L (2020) Carbon dots for intracellular pH sensing with fluorescence lifetime imaging microscopy. J Nanomater 10:604
86. Jin L, Shi L, Shi W et al (2019) Fluorescence lifetime-based pH sensing by platinum nanoclusters. Analyst 144:3533–3538
87. Von KA, Herten D, Petrich W (2012) Fluorescence lifetime-based glucose sensing using NADH. Proc SPIE 8229:82290
88. Von KA, Herten D, Petrich W (2012) Fluorescence properties of carba nicotinamide adenine dinucleotide for glucose sensing. Chem Phys Chem 13:1302–1306

89. Pickup JC, Khan F, Zhi ZL, Coulter J, Birch DJS (2013) Fluorescence intensity- and lifetime-based glucose sensing using glucose/galactose-binding protein. J Diabetes Sci Technol 7:62–71
90. Díaz-García CM, Lahmann C, Martínez-François JR, Li B, Koneal D, Nathwani N, Rahman M, Keller JP, Marvin JS, Looger LL, Yellen G (2019) Quantitative in vivo imaging of neuronal glucose concentrations with a genetically encoded fluorescence lifetime sensor. J Neurosci Res 97:946–960
91. Zheng K, Jensen TP, Rusakov DA (2018) Monitoring intracellular nanomolar calcium using fluorescence lifetime imaging. Nat Protoc 13:581–597
92. Greotti E, Bozio R, Greotti E, Ferrante C, Galla L, Zentilin L, Giacca M, Kaludercic N, Di Sante M, Mariotti L, Lia A, Gómez-Gonzalo M, Sessolo M, Di Lisa F, Carmignoto G, Bozio R, Pozzan T (2019) Dynamics in vivo mCerulean3-based Cameleon sensor to explore mitochondrial Ca^{2+}. iScience 19:340–355
93. Okkelman IA, Mcgarrigle R, Carroll SO, Berrio DC, Schenke-Layland K, Hynes J, Dmitriev RI (2020) Extracellular Ca^{2+}-sensing fluorescent protein biosensor based on a collagen-binding domain. ACS Appl Bio Mater 3:5310–5321
94. Schwarze T, Müller H, Steinbrück SAD, Eidner S, Geißler F, Kumke MU, Holdt JH (2014) Fluorescence lifetime-based sensing of sodium by an optode. Chem Commun 50:14167
95. Naumann G, Lippmann K, Eilers J (2018) Photophysical properties of Na^+ indicator dyes suitable for quantitative two-photon fluorescence-lifetime measurements. J Microsc 272(2): 136–144
96. Meyer J, Untiet V, Fahlke C, Gensch T, Rose CR (2019) Quantitative determination of cellular [Na^+] by fluorescence lifetime imaging with CoroNaGreen. J Gen Physiol 151:1319–1331
97. Müller BJ, Borisov SM, Klimant I (2016) Red- to NIR-emitting, BODIPY-based K^+-selective fluoroionophores and sensing materials. Adv Funct Mater 26:7697–7707
98. Miller RC, Aplin CP, Kay TM, Leighton R, Libal C, Simonet R, Cembran A, Heikal AA, Boersma AJ, Sheets ED (2020) FRET analysis of ionic strength sensors in the Hofmeister series of salt solutions using fluorescence lifetime measurements. J Phys Chem B 124:3447–3458
99. König K, Uchugonova A (2009) In: Periasamy A, Clegg R (eds) FLIM microscopy in biology and medicine. Chapman & Hall/CRC, Boca-Raton
100. Sarder P, Maji D, Achilefu S (2015) Molecular probes for fluorescence lifetime imaging. Bioconjug Chem 26:963–974
101. Becker W (2012) Fluorescence lifetime imaging techniques and applications. J Microsc 247: 119–136
102. Battisti A, Panettieri S, Abbandonato G, Jacchetti E, Cardarelli F, Signore G, Beltram F, Bizzarri R (2013) Imaging intracellular viscosity by a new molecular rotor suitable for phasor analysis of fluorescence lifetime. Anal Bioanal Chem 405:6223–6233
103. Haidekker MA, Nipper M, Mustafic A (2010) Dyes with segmental mobility: molecular rotors. In: Demchenko AP (ed) Advanced fluorescence reporters in chemistry and biology. I. Fundamentals and molecular design. Springer, Berlin, pp 267–308
104. Jüngst C, Klein M, Zumbusch A (2013) Long-term live cell microscopy studies of lipid droplet fusion dynamics in a dipocytes. J Lipid Res 54:3419–3429
105. Kuimova MK (2012) Mapping viscosity in cells using molecular rotors. Phys Chem Chem Phys 14:12671–12686
106. Levitt JA, Kuimova MK, Yahioglu G, Chung PH, Suhling K, Phillips D (2009) Membrane-bound molecular rotors measure viscosity in live cells via fluorescence lifetime imaging. J Phys Chem C 113:11634–11642
107. Peng X, Yang Z, Wang J, Fan J, He Y, Song F, Wang B, Sun S, Qu J, Qi J, Yan M (2011) Fluorescence ratiometry and fluorescence lifetime imaging: using a single molecular sensor for dual mode imaging of cellular viscosity. J Am Chem Soc 133:6626–6635

108. Wandelt B, Cywinski P, Darling GD, Stranix BR (2005) Single cell measurement of micro-viscosity by ratio imaging of fluorescence of styrylpyridinium probe. Biosens Bioelectron 20: 1728–1736
109. Cui J, Nie H, Zang S, Su S, Gao M, Jing J, Zhang X (2021) A conformational transition based fluorescent probe for mapping lysosomal viscosity fluctuations by fluorescence lifetime imaging. Sens Actuators B Chem 331:129432
110. Nipper ME, Dakanali M, Theodorakis EA, Haidekker MA (2010) Detection of liposome membrane viscosity perturbations with ratiometric molecular rotors. Biochimie 93:988–994
111. Luby-Phelps K, Mujumdar S, Mujumdar RB, Ernst LA, Galbraith W, Waggoner AS (1993) A novel fluorescence ratiometric method confirms the low solvent viscosity of the cytoplasm. Biophys J 65:236–242
112. Uchiyama S, Gota C, Tsuji T, Inada N (2017) Intracellular temperature measurements with fluorescent polymeric thermometers. Chem Commun 53:10976–10992
113. Gota C, Okabe K, Funatsu T, Harada Y, Uchiyama S (2009) Hydrophilic fluorescent nanogel thermometer for intracellular thermometry. J Am Chem Soc 131:2766–2767
114. Okabe K, Inada N, Gota C, Harada Y, Funatsu T, Uchiyama S (2012) Intracellular temperature mapping with a fluorescent polymeric thermometer and fluorescence lifetime imaging microscopy. Nat Commun 3:705
115. Tsuji T, Yoshida S, Yoshida A, Uchiyama S (2013) Cationic fluorescent polymeric thermometers with the ability to enter yeast and mammalian cells for practical intracellular temperature measurements. Anal Chem 85:9815–9823
116. Hayashi T, Fukuda N, Uchiyama S, Inada N (2015) A cell-permeable fluorescent polymeric thermometer for intracellular temperature mapping in mammalian cell lines. PLoS One 10:1–18
117. Lakowicz JR, Szmacinski H, Nowaczyk K, Lederer WJ (1994) Fluorescence lifetime imaging of intracellular calcium in COS cells using Quin-2. Cell Calcium 15:7–27
118. Herman B, Wodnicki P, Kwon S, Periasamy A, Gordon GW, Mahajan N, Wang XF (1997) Recent developments in monitoring calcium and protein interactions in cells using fluorescence lifetime microscopy. J Fluoresc 7:85–92
119. Celli A, Sanchez S, Behne M, Hazlett T, Gratton E, Mauro T (2010) The epidermal Ca^{2+-} gradient: measurement using the phasor representation of fluorescent lifetime imaging. Biophys J 98(5):911–921
120. Wilms CD, Schmidt H, Eilers J (2006) Quantitative two-photon Ca^{2+} imaging via fluorescence lifetime analysis. Cell Calcium 40:73–79
121. Kuchibhotla KV, Lattarulo CR, Hyman BT, Bacskai BJ (2009) Synchronous hyperactivity and intercellular calcium waves inastrocytes in Alzheimer mice. Science 323:1211–1215
122. Lahn M, Dosche C, Hille C (2011) Two-photon microscopy and fluorescence lifetime imaging reveal stimulus-induced intracellular Na^+ and Cl^- changes in cockroach salivary acinar cells. Am J Physiol Cell Physiol 300:C1323–C1336
123. Kaneko H, Putzier I, Frings S, Kaupp UB, Gensch T (2004) Chloride accumulation in mammalian olfactory sensory neurons. J Neurosci 24(7931):7938
124. Funk K, Woitecki A, Wurtz FC, Gensch T, Möhrlen F, Frings S (2008) Modulation of chloride homeostasis by inflammatory mediators in dorsal root ganglion neurons. Mol Pain 4:32
125. Gilbert D, Wurtz FC, Funk K, Gensch T, Frings S, Möhrlen F (2007) Differential maturation of chloride homeostasis in primary afferent neurons of the somatosensory system. Int J Dev Neurosci 25(7):479–489
126. Despa S, Steels P, Ameloot M (2000) Fluorescence lifetime microscopy of the sodium indicator sodium-binding benzofuran isophthalate in HeLa cells. Anal Biochem 280:227–241
127. Roder P, Hille C (2014) ANG-2 for quantitative Na^+ determination in living cells by time-resolved fluorescence microscopy. Photochem Photobiol Sci 13:1699–1710
128. Rungta RL, Choi HB, Tyson JR, Malik A, Lasse Dissing-Olesen L, Lin PJC, Cain SM, Cullis PR, Snutch TP, MacVicar BA (2015) The cellular mechanisms of neuronal swelling underlying cytotoxic edema. Cell 161:610–621

129. Hötzer B, Ivanov R, Brumbarova T, Bauer P, Jung G (2012) Visualization of Cu^{2+} uptake and release in plant cells by fluorescence lifetime imaging microscopy. FEBS J 279:410–419
130. McCranor BJ, Szmacinski H, Zeng HH, Stoddard AK, Hurst T, Fierke CA, Lakowicz JR, Thompson RB (2014) Fluorescence lifetime imaging of physiological free Cu(II) levels in live cells with a Cu(II)-selective carbonic anhydrase-based biosensor. Metallomics 6:1034–1042
131. Hanson KM, Behne MJ, Barry NP, Mauro TM, Gratton E, Clegg RM (2002) Two-photon fluorescence lifetime imaging of the skin stratum corneum pH gradient. Biophys J 83:1682–1690
132. Poëa-Guyon S, Pasquier H, Mérola F, Morel N, Erard M (2013) The enhanced cyan fluorescent protein: a sensitive pH sensor for fluorescence lifetime imaging. Anal Bioanal Chem 405: 3983–3987
133. Schmitt FJ, Thaa B, Junghans C, Vitali M, Veit M, Friedrich T (2014) eGFP-pHsens as a highly sensitivefluorophore for cellular pH determination by fluorescence lifetime imaging microscopy (FLIM). Biochim Biophys Acta 1837:1581–1593
134. Martynov VI, Pakhomov AA, Deyev IE (1862) Petrenko AG (2018) genetically encoded fluorescent indicators for live cell pH imaging. Biochim Biophys Acta Gen Subj 12:2924–2939
135. Dalfen I, Dmitriev RI, Holst G, Klimant I, Borisov SM (2019) Background-free fluorescence-decay-time sensing and imaging of pH with highly photostable diazaoxotriangulenium dyes. Anal Chem 91:808–816
136. Zhang Z, Liu Z, Tian Y (2020) A DNA-based FLIM reporter for simultaneous quantification of lysosomal pH and ca^{2+} during autophagy regulation. iScience 23(9):101344
137. Saxl T, Khan F, Ferla M, Birch D, Pickup J (2011) A fluorescence lifetime-based fibre-optic glucose sensor using glucose galactose-binding protein. Analyst 136:968–972
138. Veetil JV, Jin S, Ye K (2012) Fluorescence lifetime imaging microscopy of intracellular glucose dynamics. J Diabetes Sci Technol 6:1276–1285
139. Ripoll C, Orte A, Paniza L, Rueda-Rama MJ (2019) A quantum dot-based FLIM glucose nanosensor. Sensors 19:4992
140. Wang XD, Wolfbeis OS (2014) Optical methods for sensing and imaging oxygen: materials, spectroscopies and applications. Chem Soc Rev 43:3666–3761
141. Hosny NA, Lee DA, Knight MM (2012) Single photon counting fluorescence lifetime detection of pericellular oxygen concentrations. J Biomed Opt 17(1):016007
142. Choi H, Tzeranis DS, Cha JW, Clémenceau P, de Jong SJG, Geest LKV, Moon JH, Yannas IV, So PTC (2012) 3D-resolved fluorescence and phosphorescence lifetime imaging using temporal focusing wide-field two-photon excitation. Opt Express 20:26219–26235
143. Raza A, Colley HE, Baggaley E, Sazanovich IV, Green NH, Weinstein JA, Botchway SW, MacNeil S, Haycock JW (2017) Oxygen mapping of melanoma spheroids using small molecule platinum probe and phosphorescence lifetime imaging microscopy. Sci Rep 7:1–9
144. Kalinina S, Schaefer P, Breymayer J, Bisinger D, Chakrabortty S, Rueck A (2018) Oxygen sensing PLIM together with FLIM of intrinsic cellular fluorophores for metabolic mapping. In: Imaging, manipulation, and analysis of biomolecules, cells, and tissues. SPIE BiOS 10497
145. Penjweini R, Roarke B, Alspaugh G, Gevorgyan A, Andreoni A, Pasut A, Sackett DL, Knutson JR (2020) Single cell-based fluorescence lifetime imaging of intracellular oxygenation and metabolism. Redox Biol 34:101549
146. Marcu L (2012) Fluorescence lifetime techniques in medical applications. Ann Biomed Eng 40:304–331
147. Ghukasyan VV, Kao FJ (2009) Monitoring cellular metabolism with fluorescence lifetime of reduced nicotinamide adeninedinucleotide. J Phys Chem C 113:11532–11540
148. Horilova J, Mateasika A, Revilla-i-Domingoc R, Raiblec F, Chorvat D Jr, Marcek Chorvatova A (2015) Finger printing of metabolic states by NAD(P)H fluorescence lifetime spectroscopy in living cells: a review. Med Photon 27:62–69
149. Krause GH, Weis E (1991) Chlorophyll fluorescent and photosynthesis: the basics. Annu Rev Plant Physiol Plant Mol Biol 42:313

150. Skala MC, Riching KM, Gendron-Fitzpatrick A, Eickhoff J, Eliceiri KW, White JG, Ramanujam N (2007) In vivo multiphoton microscopy of NADH and FAD redox states, fluorescence lifetimes, and cellular morphology in precancerous epithelia. Proc Natl Acad Sci U S A 104:19494–19499
151. Galletly NP, McGinty J, Dunsby C, Teixeira F, Requejo-Isidro J, Munro I, Elson DS, Neil MAA, Chu AC, French PMW, Stamp GW (2008) Fluorescence lifetime imaging distinguishes basal cell carcinoma from surrounding uninvolved skin. J Dermatol 159:152
152. Laiho LH, Pelet S, Hancewicz TM, Kaplan PD, So PTC (2005) Two-photon 3-D mapping of ex vivo human skin endogenous fluorescence species based on fluorescence emission spectra. Biomed Opt 10:024016
153. Kantelhardt SR, Diddens H, Leppert J, Rohde V, Hüttmann G, Giese A (2008) Multiphoton excitation fluorescence microscopy of 5-aminolevulinic acid induced fluorescence in experimental gliomas. Lasers Surg Med 40:273–281
154. Kantelhardt SR, Leppert J, Krajewski J, Petkus N (2007) Imaging of brain and brain tumor specimens by time-resolved multiphoton excitation microscopy ex vivo. Neuro-Oncology 9: 103–112
155. Chorvat D, Chorvatova A (2006) Spectrally resolved time-correlated single photon counting: a novel approach for characterization of endogenous fluorescence in isolated cardiac myocytes. Eur Biophys J Biophys Lett 36:73–83
156. Schweitzer D, Hammer M, Schweitzer F, Anders R, Doebbecke T, Schenke S, Gaillard ER, Gaillard ER (2004) In vivo measurement of time-resolved autofluorescence at the human fundus. J Biomed Opt 9:1214–1222
157. Schweitzer D, Schenke S, Hammer M, Schweitzer F, Schweitzer F, Jentsch S, Birckner E, Becker W, Bergmann A (2007) Towards metabolic mapping of the human retina. Microsc Res Tech 70:410–419
158. McConnell G, Girkin JM, Ameer-Beg SM, Barber PR, Vojnovic B, Ng T, Banerjee A, Watson TF, Cook RJ (2007) Time-correlated single-photon counting fluorescence life-time confocal imaging of decayed and sound dental structures with a white-light supercontinuum source. J Microsc Oxf 225:126–136
159. Dai X, Yue Z, Eccleston ME, Swartling J, Slater NKH, Kaminski CF (2008) Fluorescence and lifetime imaging of free and micellar-encapsulated doxorubicin in living cells. Nanomedicine 4:49
160. Almutairi A, Akers WJ, Berezin MY, Achilefu S, Fréchet JMJ (2008) Monitoring the biodegradation of dendritic near-infrared Nanoprobes by in vivo fluorescence imaging. Mol Pharm 5:1103
161. Christian NA, Benencia F, Milone MC, Li G, Frail PR, Therien MJ, Coukos G, Hammer DA (2009) In vivo dendritic cell tracking using fluorescence lifetime imaging and near-infrared-emissive polymersomes. Mol Imaging Biol 11:167
162. Ehlers A, Riemann I, Stark M, König K (2007) Multiphoton fluorescence lifetime imaging of human hair. Microsc Res Tech 70:154
163. Bird DK, Schneider AL, Watkinson AC, Finnin B, Smith TA (2008) Navigating transdermal diffusion with multiphoton fluorescence lifetime imaging. J Microsc 230:61
164. Roberts MS, Roberts MJ, Robertson TA, Sanchez W, Thörling C, Zou Y, Zhao X, Becker W, Zvyagin AV (2008) In vitro and in vivo imaging of xenobiotic transport in human skin and in the rat liver. J Biophotonics 1:478

From Ensemble FRET to Single-Molecule Imaging: Monitoring Individual Cellular Machinery in Action

6

Farhana Islam, Manali Basu, and Padmaja Prasad Mishra

Abstract

Fluorescence resonance energy transfer is a suitable approach for observing conformational changes in biomolecules, as it is a "spectroscopic molecular ruler" sensitive to nanometer-scale distances. The experimental outcomes of FRET measurements in ensemble studies provide averaged overpopulation that could hide the details of dynamics for individual molecules. Though ensemble measurements have much application and have their advantages, including easy optical setup, easy sample preparation, etc., single-molecule methods have appeared as popular and powerful tools for the understanding of complex biophysical processes. This is because of its unique abilities to probe the molecular structure, function, and dynamics from information obtained from every individual molecule. At present, several laboratories around the world employ this technique for studying various biological systems such as cells, nucleosomes as well as important biomolecular interactions between DNA, RNA, and proteins. Single-molecule FRET assays are also widely used for studying DNA hybridization, interactions between DNA–protein, protein–protein, and protein folding, etc. The dynamic changes induced on DNA/protein substrate can easily be monitored in real-time observations.

Farhana Islam and Manali Basu contributed equally with all other contributors.

F. Islam · M. Basu · P. P. Mishra (✉)
Chemical Sciences Division, Saha Institute of Nuclear Physics, Kolkata, West Bengal, India

HBNI, Mumbai, India
e-mail: padmaja.mishra@saha.ac.in

© The Author(s), under exclusive license to Springer Nature Singapore Pte Ltd. 2022
H. Sahoo (ed.), *Optical Spectroscopic and Microscopic Techniques*,
https://doi.org/10.1007/978-981-16-4550-1_6

6.1 Introduction

Fluorescence resonance energy transfer is a suitable approach for observing conformational changes in biomolecules, as it is a "spectroscopic molecular ruler", sensitive to nanometer-scale distances. The experimental outcomes of FRET measurements in ensemble studies provide averaged overpopulation that sometimes hide the details of dynamics for individual molecules. Though ensemble measurements have much application and have their advantages, including easy optical setup, easy sample preparation, etc., single-molecule methods have appeared as popular and powerful tools for the understanding of complex biophysical processes. This is because of its unique abilities to probe the molecular structure, function, and dynamics from information obtained from every individual molecule. At present, several laboratories around the world employ this technique for studying various biological systems such as cells, nucleosomes as well as important biomolecular interactions between DNA, RNA, and proteins. Single-molecule FRET assays are also widely used for studying DNA hybridization, interactions between DNA–protein, protein–protein, and protein folding, etc. The dynamic changes induced on DNA/protein substrate can easily be monitored in real-time observations.

6.2 Fluorescence

Do we ever wonder why Cyalume sticks glow after they are bent, or how do fireflies glow in the dark? Well, all these processes are collectively termed luminescence. But is the glowing of a Cyalume stick the same as the glowing of fireflies? Definitely, no! Then, how are they different? To answer these questions, let us first understand luminescence, termed by Eilhardt Weidemann in 1888. Luminescence is a collective term that is used to describe the emission of light by a substance, an atom, or any molecule that is not a result of high temperature. In luminescence when incident radiation is absorbed by some material, the wavelength of light that is re-emitted is of longer wavelength as compared to the wavelength of incident light (Stokes law) [1, 2].

Luminescence can be classified into two types based on the time takes for emission, after absorption of incident radiation (τc):

- Fluorescence (temperature-independent): where τ_c is less than 10^{-8} s [2].
- Phosphorescence (temperature-dependent): where τ_c is greater than 10^{-8} s.

Luminescence can further be classified according to the source of energy used for excitation into the following:

- Chemiluminescence: Release of light energy from a chemical reaction, for example, luminol.
- Bioluminescence: a type of chemiluminescent reaction in cells that leads to emission of light from bodies or special glands of living organisms. For example,

in fireflies (*Photinus sp.*), the reaction of luciferin with ATP in presence of luciferase enzyme, jellyfish, squid, glow worms, sea salps, and Angler fish.

- Cathodoluminescence: Light is emitted as a result of excitation with electrons as in Scanning electron microscopes (SEM) and Cathode ray tubes.
- Photoluminescence: The source of excitation is electromagnetic radiation such as gamma rays, x-ray, ultraviolet rays, and visible rays. It encompasses both phosphorescence and fluorescence. It is commonly used in X-ray diagnostics, watches, and clocks.
- Electroluminescence: Benjamin Franklin in 1752 first identified lightning as a type of luminescence caused due to electric discharge. Examples are modern fluorescent lamps, zinc sulfide doped with manganese, silver, or copper.
- Sonoluminescence: It is produced by excitation with ultrasonic sound waves and first discovered in 1934 while working on sonar.
- Radioluminescence: A radioactive substance is used to produce luminescence as in self-luminous paints. A striking example of radioluminescence is the aurora borealis.

The electronic dynamics behind the phenomena of luminescence and fluorescence or phosphorescence can be best understood with the help of an energy diagram known as the Jablonski diagram, formulated by Polish physicist Aleksander Jablonski [3]. A Jablonski diagram represents the various molecular electronic states upon light absorption and also the transition between two particular states (Fig. 6.1).

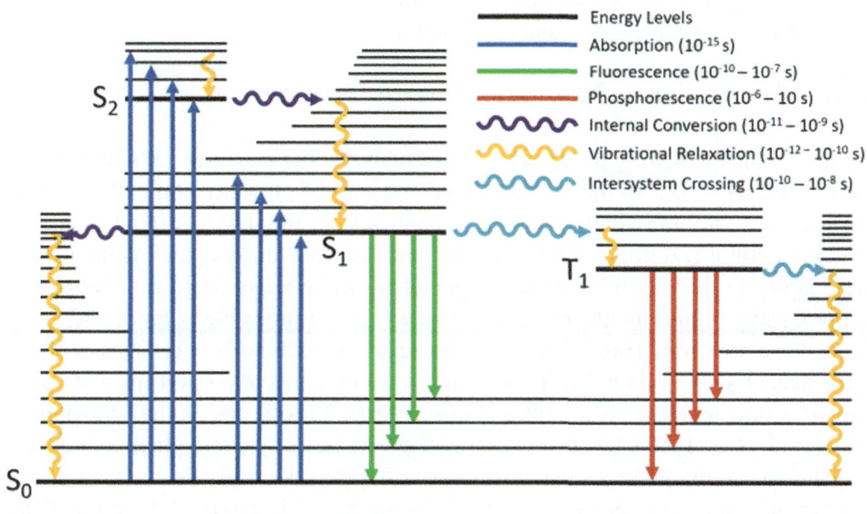

Fig. 6.1 A typical Jablonski diagram representing possible electronic transitions in a molecule when excited with a light of a specific wavelength

It is a representation of how an electron absorbs energy from a photon of a specific wavelength and then dissipates it in one of several ways.

The ground state, first, second singlet electronic state, and the triplet electronic state are represented by S0, S1, S2, and T1. It is important to mention here that upon absorbance of energy (photon), the nuclei of the molecule do not experience any significant change. This is because an atomic nucleus is heavier as compared to an electron, and hence, its transition cannot be afforded in instantaneous time scales at which absorbance and electronic transitions happen (Franck–Condon principle).

6.2.1 Fluorescence Quenching

Many processes can decrease or quench fluorescence intensity. Though it acts as a major predicament of fluorescence application as it interferes and diminishes fluorescence intensity, yet it is something that can be resolved by using appropriate quenchers. Fluorescence quenching can occur through several mechanisms [4]. Following are the three mechanisms of quenching:

- Collisional Quenching by heavy atoms such as halogens promotes intersystem crossing
- Dexter interactions between a donor and an acceptor similar to that of Resonance energy transfer (RET)
- Photoinduced electron transfer (PET) in which a complex is being formed between an electron donor and electron pair

6.2.2 Fluorescence Resonance Energy Transfer

Among all the quenching processes discussed above, fluorescence resonance energy transfer (FRET) is one of the most explored and studied phenomenon in fluorescence that involves the nonradiative energy transfer due to the dipole–dipole interaction between a donor (*D*) fluorophore in an excited state and an acceptor (*A*) fluorophore while being in the ground state (or a quencher) when appropriate spectral overlap and proximity requirements are satisfied between two molecules (donor and acceptor) in close proximity (Fig. 6.2a) [5, 6]. As a part of the donor's energy is transferred to the acceptor, there is a fall in the donor's fluorescence emission intensity and a corresponding increase in acceptor fluorescence intensity. From the viewpoint of classical physics, FRET has been described as an electrodynamic phenomenon. The mechanism usually involves two molecules, namely, a donor in the excited state that transfers its energy to another molecule, known as an acceptor in the ground state that absorbs this energy to get excited. However, the acceptor need not be fluorescent. If fluorescent, then an enhanced acceptor emission occurs due to light energy transfer from donor to acceptor. This transfer of energy is due to a long-range interaction like dipole–dipole coupling, and as it usually occurs without the emission of photons, so instead of FRET, this process is also known as RET.

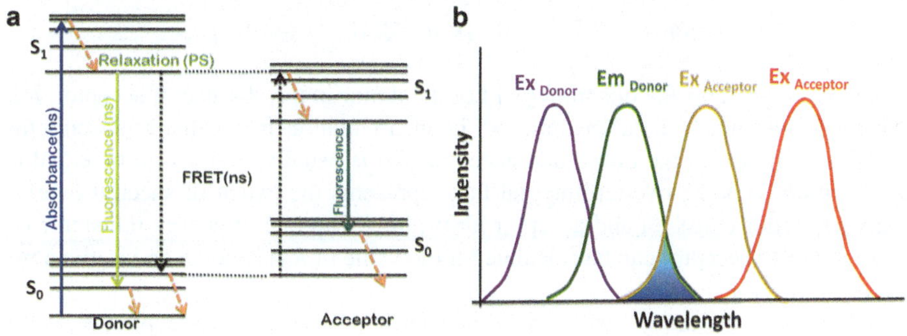

Fig. 6.2 (**a**) A schematic representation of the molecular mechanism leading to FRET, and (**b**) the spectral overlap between donor and acceptor molecule an important prerequisite of FRET

$$D + h\vartheta \rightarrow D*$$

$$D * + A \rightarrow D + A * \, [D \rightarrow \text{Donor}, A \rightarrow \text{Acceptor}]$$

$$A* \rightarrow A + h\vartheta'$$

The FRET efficiency depends upon several factors, the most integrals of which are described below:

Spectral overlap integral, $J(\lambda)$: There must be a significant overlap between the donor emission spectrum of the donor and the absorption spectrum of the acceptor. Thus, more the overlap of spectra, the better is the energy transfer from donor to the acceptor (Fig. 6.2b). This phenomenon of energy complementarity is called resonance [7]. The overlap integral, $J(\lambda)$, is given by the following equation: $J(\lambda) = \int_0 F_D(\lambda)\varepsilon_A(\lambda)\lambda^4 d\lambda$

where F_D = normalized emission spectra of the donor, ε_A = molar absorption of the acceptor, and λ = wavelength.

Distance between *D* and *A*: For FRET, the two fluorophores (donor and acceptor) must be in close vicinity of each other having a distance range of 10 to 80 Å [8, 9]. The distance between donor and acceptor corresponding to 50% transfer of energy is known as the Forster radius (R_0) which has a characteristic value for a specific donor–acceptor pair and is typically in the range of 20 to 60 angstrom. The value of R_0 is calculated from the spectral properties of the donors and acceptors, by using equations given as: $R_0 = 9.78 \times 10^3 \left(\kappa^2 \eta^{-4} Q_D J(\lambda)\right)^{1/6}$ (in Å)

$$(R_0)^6 = 8.79 \times 10^{-25} \left(\kappa^2 \eta^{-4} \text{QDJ}(\lambda)\right) \left(\text{in cm}^6\right)$$

$$R_0 = 8.79 \times 10^{23} \left(\kappa^2 \eta^{-4} \text{QDJ}(\lambda)\right) 1/6 \, (\text{in Å})$$

$$(R_0)^6 = 8.79 \times 10^{23} \left(\kappa^2 \eta^{-4} \mathrm{QDJ}(\lambda)\right) \left(\text{in } \mathring{\mathrm{A}}^6\right)$$

where Q_D represents the quantum yield of the donor in the absence of acceptor, *n* is the refractive index of the medium, κ^2 is an orientation factor that represents the transition dipoles of the donor and acceptor and is usually assumed to be equal to 2/3, and $(J(\lambda))$ is the overlap integral that represents the extent of spectral overlap between donor emission and acceptor absorption. The rate of energy transfer $k_T(r)$ from donor to acceptor can be calculated if the value of R_0 is known and is given by: $k_T = 1/\tau_D(R_0/r)^6$

where τ_D is the decay time of the donor in the absence of an acceptor, R_0 is the Förster distance, and *r* is the distance between donor and acceptor fluorophore. The value of R_0 directly depends on the overlapping of the emission spectrum of the donor with the absorption spectrum of the acceptor.

6.2.3 FRET Efficiency

The ratio between the photons absorbed by the donor to that transferred to the acceptor is known as the energy transfer efficiency and is represented as:

$$E_{\text{fret}} = \frac{R_0^6}{R_0^6 + r^6}$$

where *r* is the distance between the *D* and *A*, and R_0, known as Förster radius, is the distance between them at which the *D* transfers half of its excitation energy to the acceptor chromophore, i.e., efficiency of energy transfer is 50% [10]. So this is inversely proportional to the sixth power of the distance separating the donor and acceptor molecules, and it strongly becomes distance-dependent (Fig. 6.3a, b). Thus, FRET is an exclusively distance-dependent phenomenon and occurs efficiently when the distance between the molecules is in the range between 1 and 10 nm, highlighting its most common application being the measurement of the distances between two sites on a macromolecule justifying itself as an effective proximity indicator phenomenon. Hence, it is often termed as a "spectroscopic" or "molecular ruler" as it can measure distances either between two bio-molecules or between two sites on the same biomolecule labeled appropriately with donor and acceptor fluorophores, thus monitoring the conformational changes through the efficiency of energy transfer between the donor and acceptor fluorophores [12].

However, the transfer efficiency becomes zero if the value of *r* is greater than R_0. Definitive values of distance (*r*) are obtained when *r* is less than or equal to a factor of 2 of the R_0 value. By calculating, it can be deduced that when *r* is double the Förster distance ($r = 2R_0$), then the transfer efficiency is 1.54%, while if $r = 0.5R_0$, then the efficiency becomes 98.5%. Thus, it is practical to use RET to study the

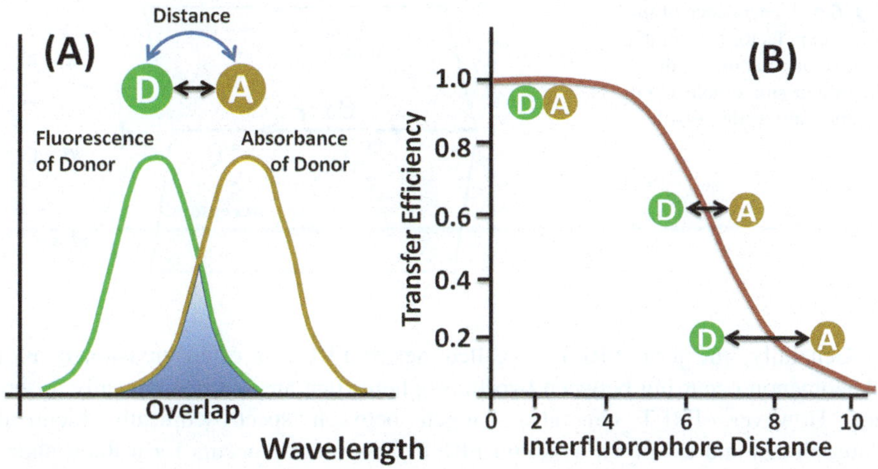

Fig. 6.3 (**a**) Fluorescence and absorption and spectra of an ideal *D–A* pair. The shaded region is the spectral overlap between the fluorescence spectra of the donor and the absorption spectra of the acceptor. (**b**) FRET efficiency is a function of interdye distance *r* [11]

conformational dynamicity of biomolecules only when the distance between donor and acceptor moieties are in the range of $r = 0.5R_0$ to $r = 2R_0$.

The Förster radius (R_0) depends on the fluorescence quantum yield of the donor in the absence of acceptor, the refractive index of the medium (*n*), the spatial relative orientation between the transition dipoles of the donor and acceptor (K^2), and the spectral overlap integral of the donor-–acceptor pair (*J*) and is given by

$$R_0 = 0.21\left(\text{Jqdn-}4\kappa^2\right)1/6$$

FRET efficiency also depends on the dipole orientation of the donor and acceptor. The value of orientation factor K^2 varies from 0 (when two dipoles are perpendicular) to 4 (when two diploes are collinear) and = 1 when they are parallel (Fig. 6.4). However, as the donor and acceptor are randomly orientated and for loosely bound fluorophores, the free rotational motion around single bonds should enable the use of an average orientation value [13]. κ^2 is generally assumed to have a value of 2/3 if the fluorescence anisotropy (*r*) of both fluorophores is less than 0.2 [14].

FRET has been widely explored as a novel technique to study a wide range of biological phenomena, such as for studying molecular interactions inside living cells with improved spatial (angstrom) and temporal (nanosecond) resolution, sensitivity, and distance range. Being a ratiometric method, this technique is advantageous as it allows measuring of the internal distance in the molecular frame than in the laboratory frame, thus making it largely immune to instrumental noise and drifts. Also, the measurement of FRET in freely diffusing single molecules is rather simple to implement and is powerful in enlightening population distributions of interdye distances.

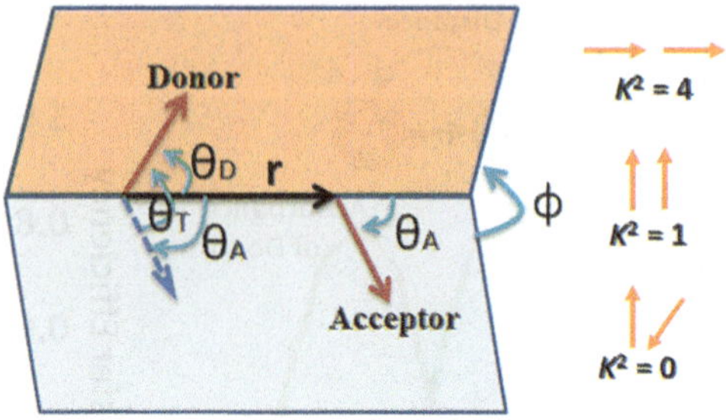

Fig. 6.4 Dependence of the orientation factor (κ^2) on the relative orientations of the donor emission dipole and the acceptor absorption dipole

Generally, the term FRET or called hetero FRET is often envisioned as a phenomenon occurring between two fluorophores that are spectroscopically different. However, FRET can also happen between spectroscopically identical fluorophores and is known as homo FRET. This usually occurs for a fluorophore with a small Stokes shift. It has been observed that when the two molecules in close proximity are fluorescent species, any changes in their spectroscopic properties that would occur due to FRET are reflected in terms of their fluorescence intensity, quantum efficiency, fluorescence lifetime, and anisotropy.

6.3 Single-Molecule FRET

Though FRET has already been a well-established method for studying conformations, as well as dynamics of biological systems at both ensemble and single-molecular levels, ensemble measurements provide only bulk averaging properties that could hide the basic molecular structure and conformation of biomolecules. By watching the motion of single molecules, all details can be seen that would have been averaged out in the ensemble measurement [15, 16]. In the conventional ensemble FRET measurements, synchronization appears as a major issue in monitoring the conformational changes or detecting the short-lived conformers of biomolecules, due to simultaneous involvement and interrogation of a large number of molecules. Single-molecule FRET (smFRET) opens up new opportunities with extremely clear and often astounding views of these molecules by real-time probing of the structural changes of individual biological molecules. A few advantages of single-molecule techniques are:

- The energy transfer of single donors and acceptors pair is monitored.
- Individual hidden kinetic pathways, the population distribution of each state, and short-lived rare intermediate states can be recognized.
- Opens the possibility of studying heterogeneity dynamics between molecules.

- Works at very low concentration (i.e. ~pM range) and hence, less amount of sample is needed. This is very useful for some biological samples; those are prone to aggregate at higher concentrations.
- Fluorophore molecules are excited using an evanescent wave generated due to total internal reflection (TIR) for better imaging that reduces instant photobleaching and background noise.
- The need for synchronization during a measurement eliminates due to synchronization before data analysis.

Therefore, in single-molecule FRET experiments, a pair of donor and acceptor FRET are excited and detected simultaneously, which sheds light on their in vivo and in vitro dynamics which in turn provides kinetics and conformational changes associated with these pair of individual molecules. However, due to the presence of Avagadro's number of molecules, conventional ensemble-averaged spectroscopic imaging methods provide results that are inhomogeneous and enormously complicated. This difficulty could be eliminated by single-molecule techniques. The apparent FRET efficiency in the case of a single-molecule study is calculated as $E_{app} = I_A/(I_A + I_D)$, where I_A and I_D are the fluorescence intensities of acceptor and donor, respectively. The E_{app} only provides an estimated sign of the interdye distance presuming the uncertainty in the orientation factor κ^2 between the two fluorophores and the required instrumental corrections. The efficiency of energy transfer that is the projection of the relative intensities of an individual donor–acceptor pair provides a way to predict the conformational alternations and dynamics of the target molecules. Correlation function analysis also can reveal the information related to dynamic equilibrium and is usually deliberated based on the sizes and structural features of individual molecules. Hitherto, the smFRET method has been extensively used to study protein–nucleic acid interactions, folding–unfolding kinetics/dynamics and conformational alternations of proteins, receptor–ligand interactions, ion channel dynamics, nucleic acid structure [9] and conformation, vesicle fusion, etc.

6.3.1 Essentials for Experimental Designing

To execute the smFRET experiment, preferably the donor-labeled or both the donor–acceptor-labeled molecule need to be surface-immobilized to restrict their free diffusion in solution [17]. The donor molecule is then excited using an evanescent wave generated due to total internal reflection (TIR) from preferably a LASER source. A high numerical aperture (NA) objective is then used to collect the resultant emission signal from the donor and/or acceptor, which is eventually projected to be recorded by the detector. This is conventionally done after splitting the donor and acceptor signal using a dichroic mirror, that helps in monitoring the individual intensities. The stored data are further analyzed by using either Matlab, IDL, or C ++ based code to extract the legitimated information. The factors that need to be maintained while designing any smFRET experiments are described below [17, 18]. Three major criteria for designing smFRET experiments are:

1. The molecules under study should be site-specifically labeled with *D* and *A* fluorophores and are located within characteristic Förster distance proximity.
2. A low concentration of labeled molecules is considered so that only one molecule is present in the observation volume at a time. The observation volume is determined and controlled by the focus laser beam.
3. Keeping a small observation volume is an important criterion for smFRET experiments. Ideally, the observed volume surrounding a single fluorophore is mentioned about 1 femtoliter (10^{-15} L), in order to avoid the intrinsic Raman scattering as well as to increase the signal-to-noise ratio (S/N). Depending on the experimental purposes, molecules of interest are either immobilized on a surface or dissolved in a solution.

6.3.2 Selecting the Right Fluorophore Pair

The photophysical properties of fluorescent dyes and the choice of appropriate FRET pair are important parameters/limitations for the smFRET experiment. As described in the earlier part, the molecules under investigation are labeled with specific *D* and *A* fluorophores in order to enumerate their conformational alternations or molecular interactions using smFRET [19]. Ideal dyes especially used to probe nonfluorescent molecules preferably need to possess the following characteristics listed below to be for a single-molecule fluorescence study. The organic dyes should be [12, 20–22]:

1. small in size (~1 nm) so that they introduce minimum perturbation with the molecules they are labeled with
2. highly water-soluble
3. high extinction coefficients 20,000 M^{-1} cm^{-1} and high quantum yields
4. commercially available in a form that can be either chemically modified or conjugated to molecules of interest (either NHS ester modified or maleimide functionalized)
5. able to excite and emit in the visible wavelengths
6. stable in terms of intensity fluctuation, at least, within the time scale of the events under investigation
7. short-lived in the excited state
8. photostable, thus capable of emitting a substantial number of photons before they photobleach and
9. large Stoke's shift to reduce scattered stray light

In addition to the nature of the fluorophore, an ideal FRET pair for single-molecule experiments should possess (1) substantial overlapping between the emission spectrum of donor with the absorption spectrum of acceptor as shown in Fig. 6.3 for an ideal commercial *D*–*A* pair, (2) significant spectral separation between the emission spectra of donor and acceptor, as it minimizes the donor emission leakage into the spectral range of acceptor emission (Fig. 6.3), (3) well-distinguished and

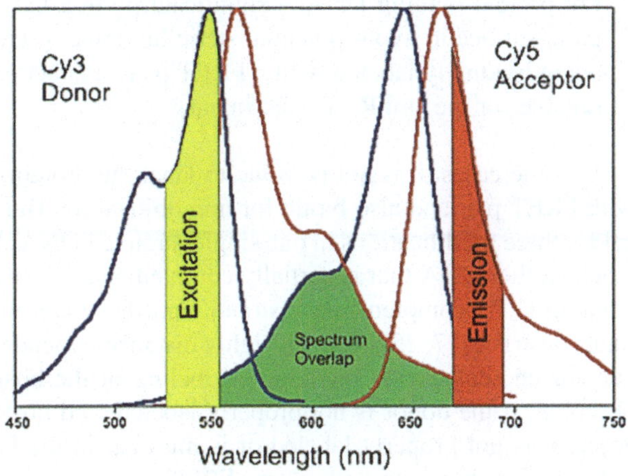

Fig. 6.5 Excitation and Emission Spectra of Cy3 and Cy5

longer wavelength excitation spectrum of the acceptor than the excitation source to avoid direct excitation, and (iv) a comparable fluorescence quantum yield of both *D* and *A*. Cyanine dyes, mostly Cy3, Cy5, and Cy7 (Fig. 6.5), the widely used donors and acceptors, though Alexa Fluor and ATTO dye series, are gaining increased interest in this kind of experiments due to their individual unique characteristic properties [23].

6.3.3 Sample Design

Labeling or alteration with *D–A* pairs in a suitable location of any complex system is decisive for successfully carrying out smFRET experiments. FRET, being an essentially distance-dependent phenomenon, the sensitivity is even higher when the donor and acceptor are placed about R_0 apart from each other. Hence, on the molecule(s) of interest, they are usually placed in such a way that the interdye distance alternates around R_0 during any interaction or conformational change. Labeling on DNA using FRET pair preserves the base–pairing interaction, and the dsDNA backbone is not disrupted by these modifications that are needed to be cross-checked with biochemical ensemble data.

Some general rules have to be followed for selecting the labeling site of biomolecules like DNA, RNA, or proteins to carry out a smFRET experiment. These are as follows:

- Altering of the position of the labeling site should be considered if the fluorescence of either the donor or acceptor is quenched due to interaction with its protein/DNA/molecule of interest that it is labeled at.
- For end or internal labeling of DNA, it is recommended to keep the backbone continuity of DNA, and for some special cases, nucleotide labeling has been considered [24]

- For protein-binding DNA, a significant portion has to be placed away from the glass surface to avoid potential steric hindrances. However, if protein and DNA are individually labeled with a FRET pair, it is better for understanding and also reliable for the smFRET experiment.

In some cases, it is not possible to label the protein, then DNA must be labeled with FRET pair and also biotin for immobilization. But it is economically impractical for three modifications in one single-stranded DNA. In this situation, it's better to label another DNA that is partially complementary to each other or to distribute the labeling to its complementary strand, and those are annealed later to generate the final construct [17, 25]. Being a highly distance-dependent phenomenon, smFRET is very much sensitive to incomplete labeling of the host molecule. Nothing can be observed, if the donor is not properly labeled, and no FRET will be detected if the acceptor is not properly labeled or is missing. In the latter case, an intense donor-only species shows up with a zero-FRET event.

6.3.4 Slide Preparation and Surface Immobilization

One of the popular methods for monitoring conformational changes of individual molecules using smFRET for an extended period is to localize in space. To do so, the immobilization of the sample to the slide or the coverslip by a biotin–streptavidin linkage has been extensively used in most single-molecule studies. Moreover, biotinylated bovine serum albumin (BSA) is also a good choice for studies involving only nucleic acids, as biotinylated BSA can easily be adsorbed to glass or quartz surfaces and subsequently binds to the biotinylated biomolecules to achieve immobilization at the essential single-molecule density through the multivalent streptavidin protein, neutravidin, or streptavidin. However, for studies involving protein, an additional passivating agent, usually polyethylene glycol (PEG), is applied between biotin and the surface-activated aminosilanized slide to suppress the nonspecific binding of protein to the slide (Fig. 6.6) [26].

Now, to get high-quality data in single-molecule experiments, one of the vital requirements is to prevent the fluorescent dyes from reacting in the excited state to form permanently nonfluorescent compounds or to enter to a non-fluorescent states. The two major determinants are photobleaching (permanent loss fluorescent property due to photon-induced chemical damage) and photoblinking (no fluorescent signal due to the dye being trapped in a state that does not allow fluorescence, usually triplet state). The most common agent resulting in both photobleaching and blinking is excited state oxygen. When the oxygen species collide with the excited state fluorophore, they can permanently react with it or temporarily send it to a triplet state, but the removal of oxygen molecules perpetuates the dark triplet state lifetime causing millisecond or longer fluorescence intermittency or early onset of the signal. Therefore, a triplet state quencher is necessary for the absence of oxygen, the most commonly used being vitamin E analog Trolox that serves as an excellent triplet-state quencher (another being β-mercaptoethanol (βME) as the triplet state

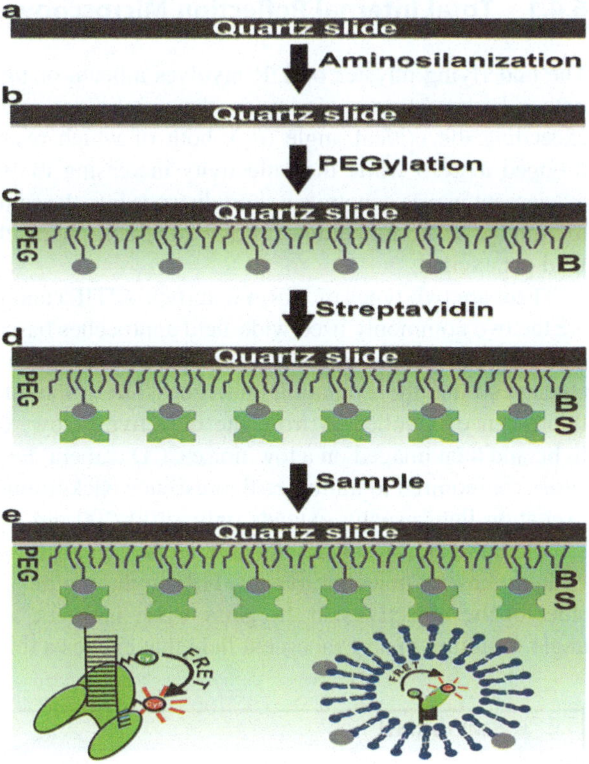

Fig. 6.6 Systematic representation of the protocol for surface passivation and sample immobilization. (**a**) The quartz slide is cleaned (protocol described later). (**b**) The slide is then aminosilanized to make it ready for pegylation. (**c**) a coat of PEG and biotin-PEG (in 100:1 ratio) to the aminosilanized surface. (**d**) Streptavidin is added to the immobilized biotin-PEG. (**e**) The sample can either have biotin to be tethered directly with attached biotin or can be encapsulated in a biotinylated lipid vesicle that can be immobilized. The slide becomes ready for use after subsequent washing to remove any unbound sample present [26]

quencher), suppressing blinking and thereby stimulating long-lasting emission of the most of the cyanine dyes in combination with an enzyme-based oxygen-scavenging system and reducing agents described as follows: 0.8% w/v D-glucose, 1 mg/ml (165 U/ml) glucose oxidase, and 0.04 mg/ml (2170 U/ml) catalase.

The resultant reactions occurring are:

$$\mathbf{Glucose + O_2 + H_2O \underset{glucose\ oxidase}{\rightarrow} gluconic\ acid + H_2O_2}$$

$$\mathbf{H_2O_2 \overset{catalase}{\rightarrow} {}^1\!/_2O_2 + H_2O}$$

6.4 Single-Molecule Detection

For monitoring of biological systems at single molecular resolution by using smFRET, two popular types of experimental setups are usually followed. One is the widely used total-internal reflection (TIRF) microscopy, and another setup uses a continuous wave (CW) or a pulsed laser as the light source for sample excitation in a confocal microscopy system.

6.4.1 Total Internal Reflection Microscopy

The underlying physics of TIR involves a beam of light encountering an interface with a lower refractive index on the distal side, and the angle of incidence (θ_I) exceeding the critical angle (θ_C), both of which result in the transmission being dropped to zero while the reflectivity increasing to 100%. When TIR occurs, the evanescent waves penetrate a short distance into the sample, thereby providing a way to limit the observed volume, as only the fluorophores that lie within this evanescent field get excited.

There are two types of TIR, prism-type (PTIR) and objective-type (OTIR); these are the two commonly used wide-field approaches based on an inverted microscope. In PTIR, the sample is excited through a fused silica prism on top of the sample channel so that the excitation undergoes TIR at the quartz–water interface, and the fluorescence is collected from the objective below, filtered to remove unwanted light, and then imaged on a low-noise CCD camera. Expensive yet recyclable quartz slides are required to minimize fluorescence background, and the evanescent field of excitation light usually extends only ~100–200 nm from the surface to which the sample is bound, which also greatly reduces background fluorescence (Fig. 6.7).

In the case of objective-type (OTIR) configuration, light is illuminated around the sides of the objective, on the glass–water interface at an angle above the critical angle, resulting in an evanescent field that excites a thin plane of the sample next to

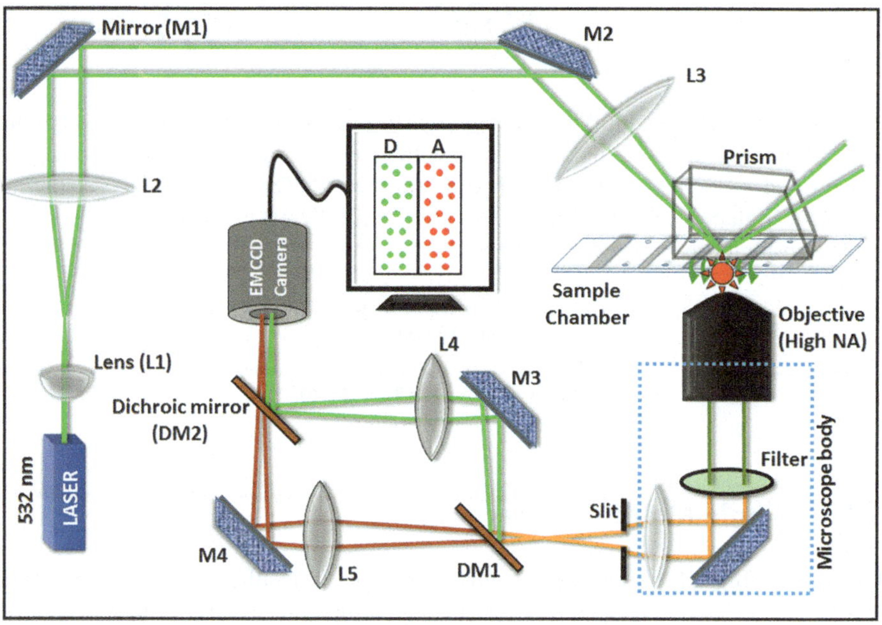

Fig. 6.7 Schematic representation of PTIR types of total internal reflection spectroscopy and the ray diagram of a typical smFRET setup

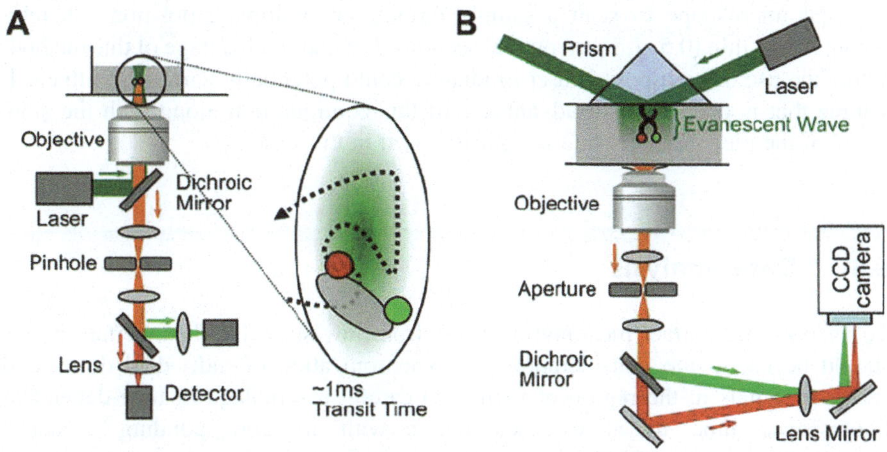

Fig. 6.8 Schematic representation of confocal microscopy (**a**) and total internal reflection microscopy (**b**) setup for FRET detection of single molecules diffusing in solution. Image courtesy: https://doi.org/10.1007/978-0-387-76497-9_5

the interface. Since the excitation and emission pass through the same objective, in this case, the excitation light is always aligned with the collection optics. In addition to higher photon-collection efficiency, it also provides additional sample control by freeing up space above the sample and has wider commercial options.

6.4.2 Confocal Microscopy

Unlike TIRF microscopy setup, which fails to image beyond 100 nm from the interface between the sample and coverslip, confocal microscopy uses a continuous wave (CW) or a pulsed laser light source excitation of target molecules that helps to acquire image beyond 100 nm. In confocal microscopy, the incident light excites all the fluorophores within the illuminated cone that is collected back through the objective, and this phenomenon is referred to as epifluorescence configuration. Usually, a dichroic filter is used in this case that separates the emission from the incident wavelength. A dichroic filter is an optical device that transmits some wavelengths while reflecting others; therefore, in this configuration, wavelengths that are longer than the incident light are reflected to a detector, that is, transmitting excitation wavelength thereby reflecting the emission wavelengths (Fig. 6.7). However, the undesirable signals from the solvent can be eliminated to some extent since in this confocal optics, a small pinhole aperture is placed at a focal point in the light path. Therefore, the light above or below the focal plane is not focused on the pinhole and is eliminated from the detector which increases the S/N ratio. Also, here the effective depth of the sample could be reduced to about twice the incident wavelength, and the resolution of the *z*-axis is usually several-fold less than the resolution in the *x*–*y* focal plane (Fig. 6.8). However, the ability of laser scanning confocal microscope to scan a sample throughout multiple z-positions, thereby giving rise to thin (0.5–1.5 μm) optical sections, is a major advantage of this method. While the use of high-power laser irradiation could potentially damage the live cell sample that is one of the disadvantages of this configuration along with the slow speed of the piezo scanner that is typically used in this case.

6.5 Data Analysis

To analyze and extract meaningful results from raw smFRET data, certain points should be taken into consideration: (a) synchronization of individual donor and acceptor signals in the region of focus; (b) careful assemblage of time-dependent fluorescence time traces of each donor with its corresponding acceptor; (c) generation of the FRET efficiency histogram from the time trace for the target molecule by elimination of photobleaching state; and (d) investigating conformational dynamics of biomolecules from the autocorrelation and cross-correlation analysis. Before data analysis, mapping is done using a high-order polynomial transformation for the correct assignment of donor–acceptor pairs. Since the donor and acceptor channels of the EMCCD image usually do not have a linear correspondence, so fluorescent beads, having a wide range of emission spectrum, are used as a reference and aids in the calibration of both the channels. Now, having known the value of Förster radius, E_{FRET} can be used to calculate absolute distances using the eq. $E = 1/[1 + (R_0/R)^6]$, assuming anisotropy of both donor and acceptor fluorophores is zero. But the apparent FRET efficiency in this is calculated as $E_{app} = I_A/ (I_A + I_D)$, where I_A and I_D represent acceptor and donor intensities, respectively, and also to determine actual FRET efficiency, one has to determine the correction factor, γ, which accounts for the differences in quantum yield and detection efficiency between the donor and the acceptor.

Using the time traces for hundreds of molecules under the same condition, the FRET efficiency is calculated, and then it is fitted into a histogram using a single or multiple peak Gaussian function which provides an estimation about the number of states that are possibly involved in these conformational fluctuations. The signals collected from the experiments are described as a series of discrete states governed by an underlying physical property of the molecule(s) being interrogated. These discrete states are often obscured by background noises that are inherent to every experimental technique, making impossible the identification and characterization of these states. To facilitate the assignment of FRET states, Hidden Markov Modeling (HMM), a probabilistic model, has been widely applied for data analysis. This model has been well suited for single-molecule analysis because of its ability to find usually stable discrete states and biologically relevant conformations, within noisy time series data, to reliably obtain the most probable path through these states, and providing information regarding the equilibrium properties along with the conformation states of the molecule could be deciphered.

6.6 Visualization of Dynamical Conformations of Biomolecules Using Single-Molecule Fluorescence Resonance Energy Transfer

The power of single-molecule fluorescence resonance energy transfer (smFRET) in biophysical applications has been proven as an excellent tool to study conformational dynamics to decipher coordinated large-scale motions within biomolecules. smFRET methods, in recent years, have geared toward an understanding of the time scales and amplitudes of function-related motions, and analyzing the smFRET temporal trajectories provides direct dynamic information on transitions between conformational states of proteins and nucleic acids. Here, we provide a few prominent examples where the single-molecule FRET method has been successfully and dominantly employed to investigate the interactions as well as the dynamical motions of biomolecules.

6.6.1 Initiation and Reinitiation of DNA Unwinding by the *Escherichia coli* Rep Helicase

In 2002, Ha et al. [27] using single-molecule techniques have visualized the motion of a highly processive helicase, *Escherichia coli* Rep helicase, a motor protein that uses ATP hydrolysis to move along DNA track using ATP as fuel, where it proceeds toward the junction between single-stranded and double-stranded DNA, resulting in various conformational states due to partial dissociation of the functional complex during unwinding results installing that would lead either to duplex rewinding upon complete dissociation of the complex or reinitiation of unwinding upon reformation of the functional helicase. A single-molecule FRET-based helicase assay has been developed to monitor the DNA unwinding by a helicase of less than 10 bp, in addition to other underlying mechanisms like unwinding stalling, rewinding, and unwinding restarting. In this study, the fluorescent probes are attached to the DNA from where the unwinding activity begins. Therefore, the energy transfer would be almost 100% efficient before unwinding and gradually decreases as unwinding progresses, thereby providing a direct measure of the extent of unwinding. They reported stalling of the unwinding by Rep helicase for a few seconds, either leading to DNA rewinding that could be easily inferred from FRET values returning to 100% or unwinding restart and completion, although the monomeric form undergoes rapid ATP-dependent conformational fluctuations upon encountering the single-stranded/double-stranded DNA junction, but these fluctuations are futile and do not lead to

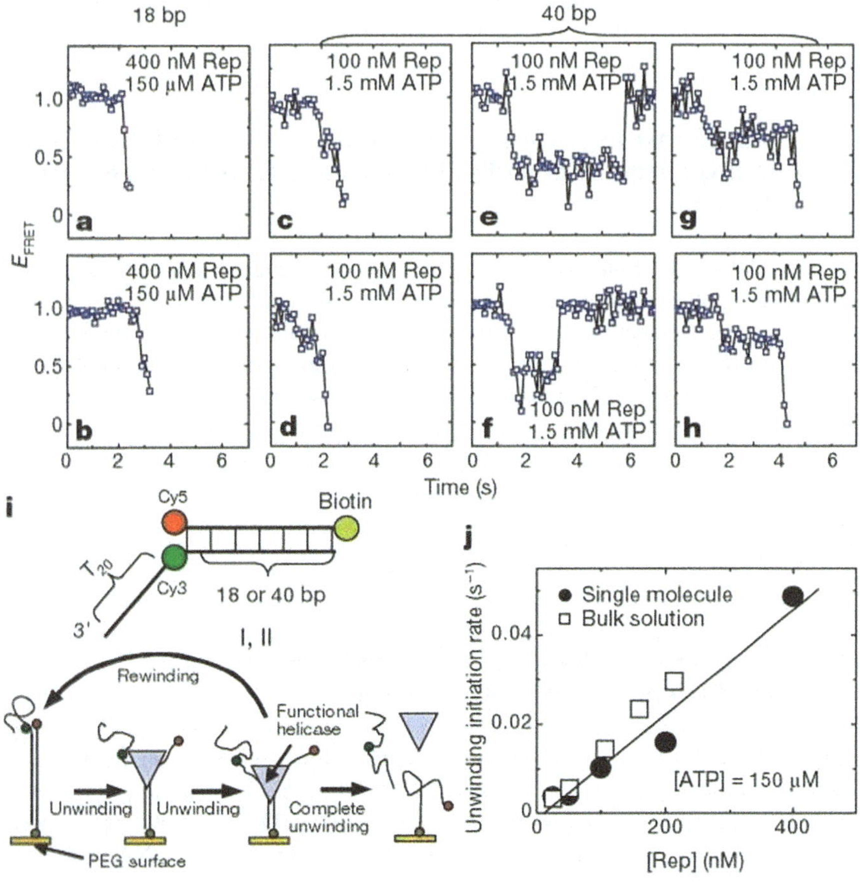

Fig. 6.9 DNA unzipping catalyzed by helicase. (**i**) Proposed model activity and schematic representation illustrating single-molecule unwinding experiments *E. coli* Rep helicase. (**j**) The unwinding initiation rate with respect to Rep helicase concentration. (**a–h**) Single-molecule FRET records of duplex DNA unwinding having length 18 bp and 40 bp, 40 bp DNA unwinding showed frequent stalling that probably leading to DNA rewinding or unwinding restart. Image courtesy: https://doi.org/10.1038/nature01083

DNA unwinding. It has been suggested that the unwinding initiation required the action of more than a helicase monomer and the unwinding restart required free helicase molecules in solution. Therefore, they proposed that the unwinding of replication fork stalls when the active helicase complex dissociates, leaving a monomer on the DNA, while the rewinding is observed if the remaining monomer dissociates; however, the unwinding can resume only if a functional helicase reassembles at the stall site (Fig. 6.9).

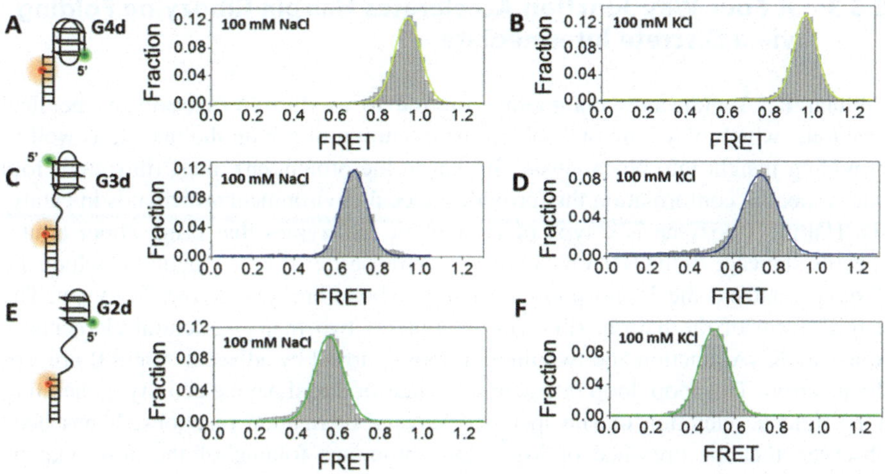

Fig. 6.10 Formation of intermediate transient structures during the proposed folding mechanism of G-ss quadruplexes, like G-triplex and G-hairpin at the 5′ end of a duplex DNA in 100 mM NaCl or KCl. (**a**) smFRET histograms of G4d in 100 mM NaCl have 1 peak at E0.94. (**b**) smFRET histograms of G4d structures in 100 mM KCl showing 1 peak at E0.96. (**c**) smFRET histograms of G3d structures in 100 mM NaCl having 1 peak at E0.69. (**d**) smFRET histograms of G3d structures in 100 mM KCl having 1 peak at E0.73. (**e**) smFRET histograms of G2d structures in 100 mM NaCl having 1 peak at E0.56. (**f**) smFRET histograms of G-hairpin structures in 100 mM KCl showing 1 peak at E0.50. Image courtesy: https://doi.org/10.1093/nar/gkx766

6.6.2 Involvement of G-triplex and G-hairpin in the Multipathway Folding of Human Telomeric G-quadruplex

Eukaryotic chromosomes are terminated by telomeres which primarily protect chromosomes from recombination and degradation. The single-stranded segment repeating nucleotide sequence TTAGGG at the human telomeres, which overhangs at the end of the double-stranded DNA helix, revealed the formation of a secondary DNA structure known as G quadruplex through high-resolution structural studies. It consists of four hydrogen-bonded guanines that are hydrogen-bonded and are originated by the folding of repeats of G-rich sequence motifs that are held together by Hoogsteen hydrogen bonds in a horizontal planar arrangement, and three such layers are stacked on top of each other.

In 2017, Hou et al. employed single-molecule FRET to characterize the conformational dynamics of human telomeric G4 in NaCl and observed four folding states that are initially assigned as anti-parallel G4, G-triplex, G-hairpin, and unfolded ssDNA. Using this technique, they proved intrastrand G-triplex/G-hairpin can indeed be formed and undergo dynamic transitions between different states. Besides, the effects of cation type and concentration on G4, G-triplex, and G-hairpin folding were carefully characterized. These results were further corroborated by MD simulations which also confirmed the presence of the middle FRET states. Thus, smFRET could successfully prove the multipathway folding mechanism for human telomeric G4 under different salt concentrations (Fig. 6.10).

6.6.3 A Four-Way Junction Accelerates Hairpin Ribozyme Folding via a Discrete Intermediate

Cellular RNA molecules commonly referred to as ribozymes catalyze chemical reactions, which play a pivotal role in the evolution of life on the planet, as well as providing insight into biocatalysis. Similar to the protein enzymes, ribozymes fold into a specific conformation that provides a local environment which aids in catalysis. Hairpin ribozyme is a type of nucleolytic ribozymes that bring about a site-specific cleavage of the RNA backbone via a transesterification reaction in which the 2′ oxygen attacks the 3′ phosphorus, leading to bond breakage to the 5′ oxygen. The natural form of the hairpin ribozyme comprises two major structural elements: a four-way RNA junction and two internal loops carried by adjacent A and B arms of the junction. This loop–loop interaction is vital for the ribozyme activity, generating a 10^5-fold acceleration of site-specific cleavage or ligation reactions. It has been observed that the presence of Mg^{2+} ions promotes folding of the arms via the docking of the two loops, and cleavage and ligation reactions of a specific bond within the active site.

In 2003, Tan et al. monitored the ribozyme folding by labeling fluorophores at the ends of A and B arms with the donor and acceptor which due to folding would bring the fluorophores in close proximity, yielding high FRET. The time-dependent fluorescence intensities of the donor and acceptor entities within a single ribozyme molecule are investigated by immobilizing it to a glass surface via a biotin–streptavidin interaction. The folding dynamics are monitored starting from the unfolded low FRET state (high donor signal and low acceptor signal) in the absence of Mg^{2+}. However, in presence of a buffer containing 10 mM Mg^{2+}, the molecule folded immediately as marked by an abrupt increase in FRET (donor signal decrease and acceptor signal increase). The molecule remained stably folded for 180 s until it started showing rapid fluctuations due to the intrinsic dynamics of the junction, indicating the release of the cleavage product (Fig. 6.11).

6.6.4 Monitoring of Structural Dynamics of a Holliday Junction [28]

The four-way DNA junction, commonly known as the Holliday junction, is an important intermediate structure in several crucial pathways involving DNA such as recombination and repair which can exist in three structural conformations—open planar, stacked isoform 1, and isoform 2 (Fig. 6.12a, d). Although NMR and time-resolved fluorescence resonance energy transfer studies predicted the coexistence of conformational states, it was smFRET (single-molecule fluorescence resonance energy transfer) that deciphered the transition rates, kinetics, and population distribution of conformational states as a function of nucleotide sequence, presence of different types of ions, and temperature. smFRET experiments and data analysis of the time traces revealed the following: junction 7 shows almost equal proportions of the two FRET states (high and low), whereas junction 3 displays a strong inclination toward the high FRET state (Fig. 6.13a). The rate of transition from

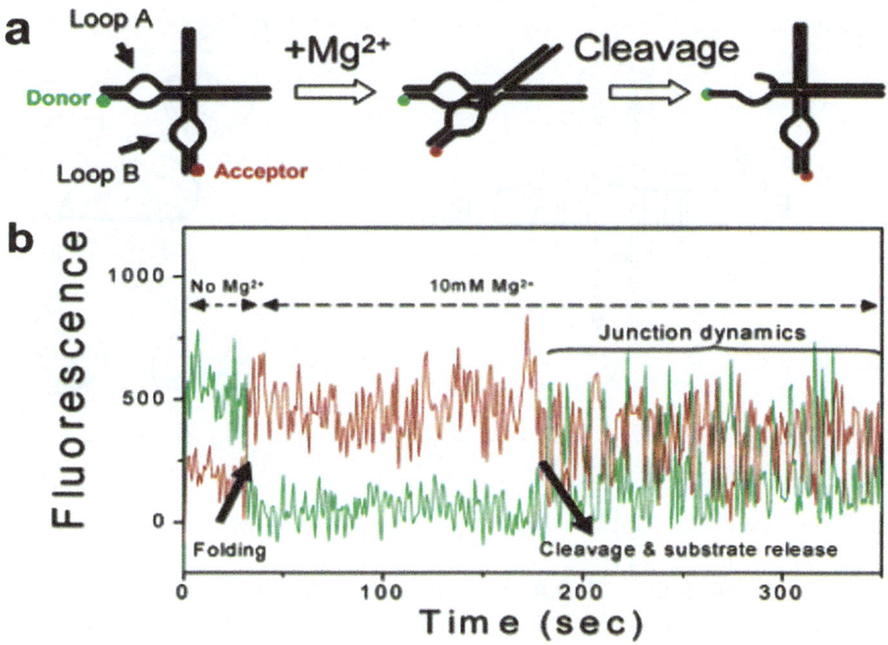

Fig. 6.11 Folding dynamics of hairpin ribozyme and their subsequent cleavage. (**a**) Schematic view of stepwise folding and catalysis of the hairpin ribozyme via FRET in which the two helices carrying the essential loops A and B are labeled by the donor and acceptor fluorophores. (**b**) Time traces of the donor (green) and acceptor (red) when hairpin ribozyme folds on the addition of Mg^{2+} are followed by the cleavage and release after ~180 s of showcasing the four-way junction dynamics. Image courtesy: https://doi.org/10.1073/pnas.1233536100

isoI to isoII and from isoII back to isoI at 50 mM $MgCl_2$ for junction 3 was calculated to be 12 s^{-1} and 3.5 s^{-1}, respectively. For junction 7, the values were 5.7 s^{-1} and 6.1 s^{-1} (Fig. 6.13b) [28].

It was found that the transition rate between the two isoforms was different in the presence and absence of 50 mM NaCl (Fig. 6.14). Besides, the decrease in transition rate between the two conformers with increasing concentration of Mg^{2+} indicated that Mg^{2+} alters the activation energy rather than stabilizing individual conformation. It also implied the importance of the open planar conformation in absence of Mg^{2+} as a transient structure in conformer switching. This also explained how electrostatic interactions between cations (Mg^{2+}) and negatively charged phosphates are screened by a monovalent ion like Na^+ [28].

The energy landscape model proposed based on the above observations attempts to explain the effect of Mg^{2+} concentration on conformer transition and branch migration. The results from single-molecule studies suggest that the rate of branch migration is maximum at 0.5 mM Mg^{2+}; however, the rate of conformer transition changes steeply till about 50 mM Mg^{2+} (Fig. 6.14). It indicates the influence of Mg^{2+} on the energy barrier in the conformer transition.

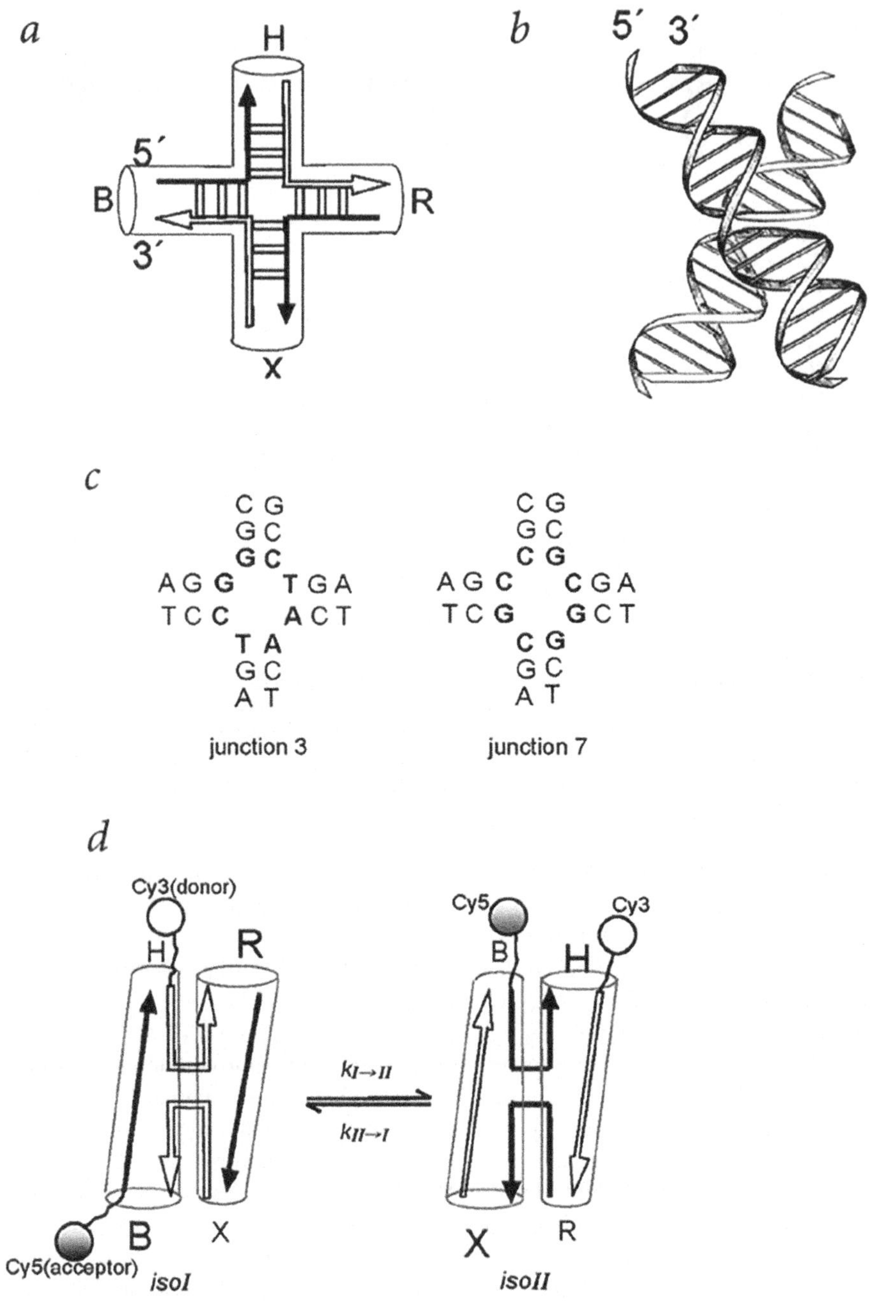

Fig. 6.12 Different conformations of Holliday junction. (**a**) open planar conformation with four helical arms named sequentially as B, H, R, and X, which is found in absence of any metal ion. (**b**) stacked isoform found in presence of divalent cations that causes coaxial stacking of adjacent arms. (**c**) two junctions with different central sequences. (**d**) the two stacked isoforms (depending on the combination of stacking partners), BX-HR (where the donor fluorophore Cy3 and acceptor Cy5 are farthest), and BH-XR (Cy3 and Cy5 are close). Isoform 1 yields low fret values, whereas isoform 2 yields high fret values. https://doi.org/10.1038/nsb883

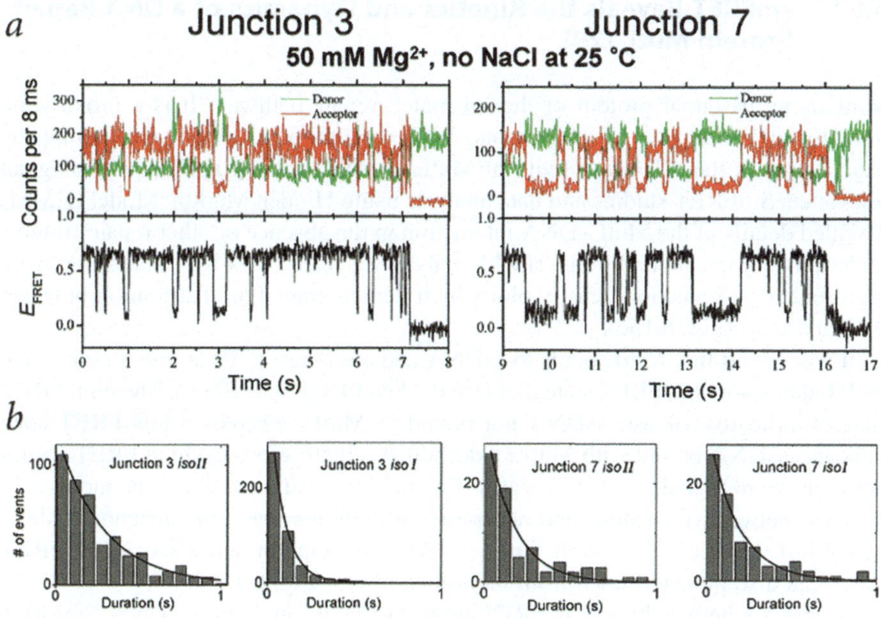

Fig. 6.13 Single-molecule time traces and data analysis. (**a**) time traces of donor (I_D) and acceptor intensities (I_A) in green and red, respectively, reflecting transitions between the two stacked isoforms of individual molecules of two different junctions (Junctions 3 and 7). Graphs at the bottom represent FRET efficiencies (E_{FRET}). (**b**) Histograms represent the lifetime of each isoform for junctions 3 and 7. https://doi.org/10.1038/nsb883

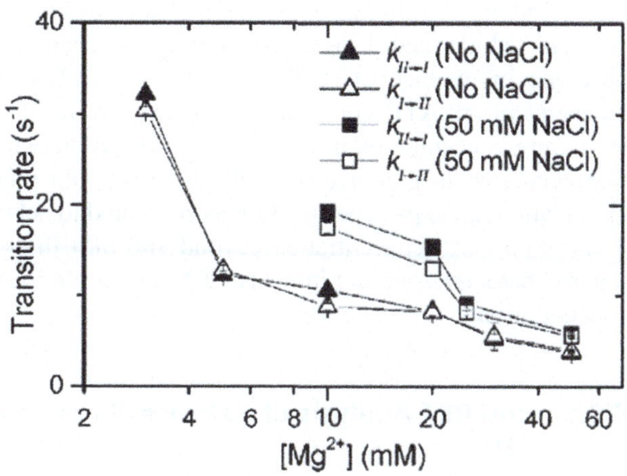

Fig. 6.14 Effect of Na^{+} and Mg^{2+} on conformer transition rate. Conformer transition rates from isoI to isoII (open symbols) and isoII to isoI (filled symbols) are given for Mg^{2+} with 50 mM NaCl (square) and without NaCl (triangle). https://doi.org/10.1038/nsb883

6.6.5 smFRET Reveals the Kinetics and Dynamics of a DNA Repair Protein MutL [29]

MutL is an essential protein of the mismatch repair pathway. It is a prokaryotic homolog of eukaryotic MLH proteins. While controversies were surrounding the exact status of the ATPase activity of MutL in DNA binding and its physiological relevance, SmFRET studies and data analysis using Hidden Markov Model (HMM) revealed details of the MutL–DNA interaction in the absence of other repair proteins under physiological conditions. HMM analysis detected two FRET states that were represented by transition density plot which depicts transition distribution between 0.44 and 0.26 (Fig. 6.15c).

It was found that MutL binds to ssDNA and stretches it. Time traces reveal two FRET states—a high FRET state (0.44) and a low FRET state (0.26). The high FRET state is indicative of free ssDNA not bound to MutL, whereas a low FRET state indicates ssDNA bound with MutL. Additionally, there was a cycle in FRET values between the high and low FRET states (Fig. 6.15d). It suggested that the increase in distance between the donor and acceptor fluorophores was time-dependent along with MutL binding. The dwell times of MutL association and dissociation FRET states can also be determined from individual traces (Fig. 6.15e).

To check whether changes in FRET are due to MutL binding to ssDNA, SmFRET was performed with a mutant protein (MutL R266E). A FRET efficiency of 0.4 ± 0.16 suggested results identical to MutL unbound ssDNA (Fig. 6.16a). MutL–ssDNA binding was also assessed with different lengths of ssDNA that were blocked at one end with digoxigenin. It suggested that longer ssDNA increases the increased rate of MutL association, and free ssDNA ends are not required for MutL binding (Fig. 6.16) [29].

ATP hydrolysis by MutL is essential for its activity and interaction with other proteins (MutH and UvrD). No significant changes in FRET efficiency were observed in absence of ATP. The rate of MutL dissociation was found to be independent of ATP concentration. However, MutL association increased with increasing concentration of ATP (Fig. 6.17b). Also, while the rate of MutL association decreased in presence of ADP, there was not much difference in the rate of dissociation in the presence of ADP as compared to ATP (Fig. 6.17c). No FRET was found with NaCl concentration beyond 100 mM that is in agreement with earlier reports that suggested the inability of MutL to bind ssDNA at physiological ionic concentrations.

6.6.6 smFRET Analysis of Helicases Involved in DNA Replication [30]

Helicases are important proteins in DNA replication and transcription. Its mechanism of function has been mostly studied by structural methods and ensemble techniques. However, smFRET allows the mechanistic study of helicase action in detail.

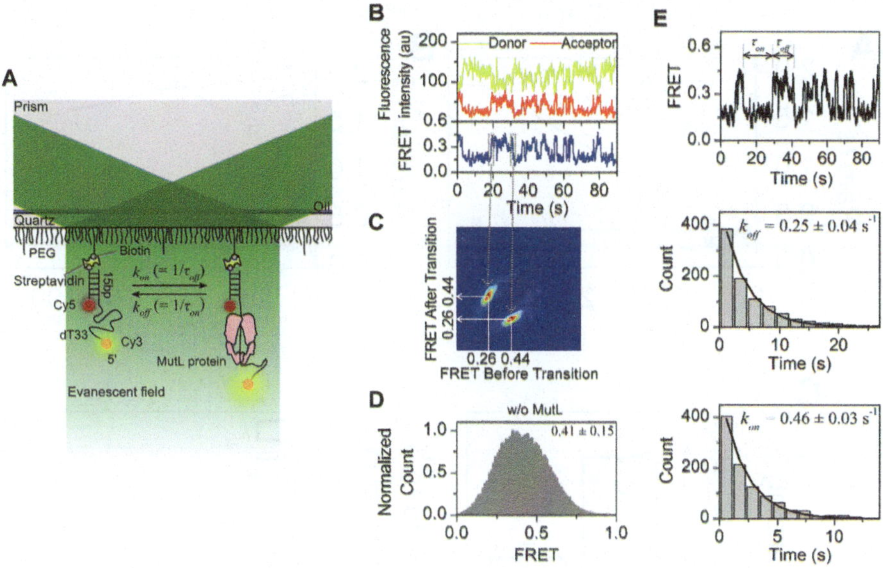

Fig. 6.15 smFRET analysis of MutL-ssDNA interaction. (**a**) Pictorial representation of SmFRET setup. (**b**) time trace representing Cy3 and Cy5 intensities at 50 nM MutL. (**c**) transition density plot depicting two FRET states (0.44 and 0.26). (**d**) Histogram suggesting FRET value of 0.41 ± 0.15 for MutL unbound ssDNA. (**e**) association and dissociation time of MutL. Determination of rate constants from the histogram of dwell times. https://doi.org/10.1371/journal.pone.0015496 [29]

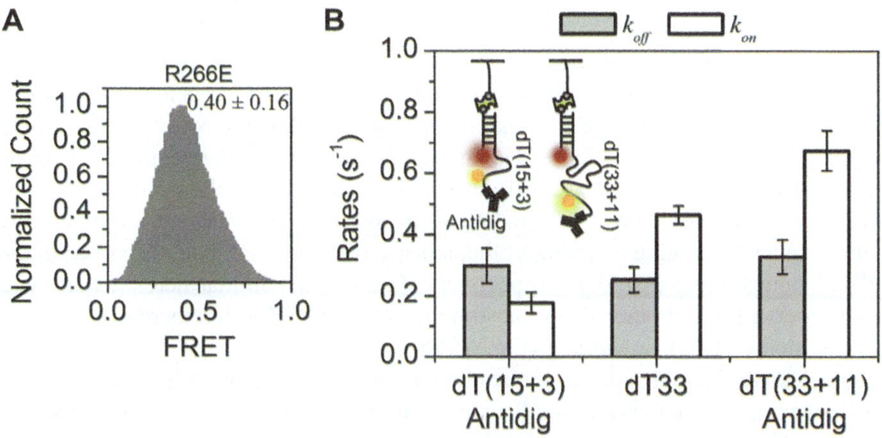

Fig. 6.16 (**a**) Histogram of SmFRET experiments with MutL R266E shows a FRET efficiency of 0.4 ± 0.16. (**b**) Pictorial representation of different lengths of end blocked ssDNA depicting the rate of association and dissociation in each case. https://doi.org/10.1371/journal.pone.0015496

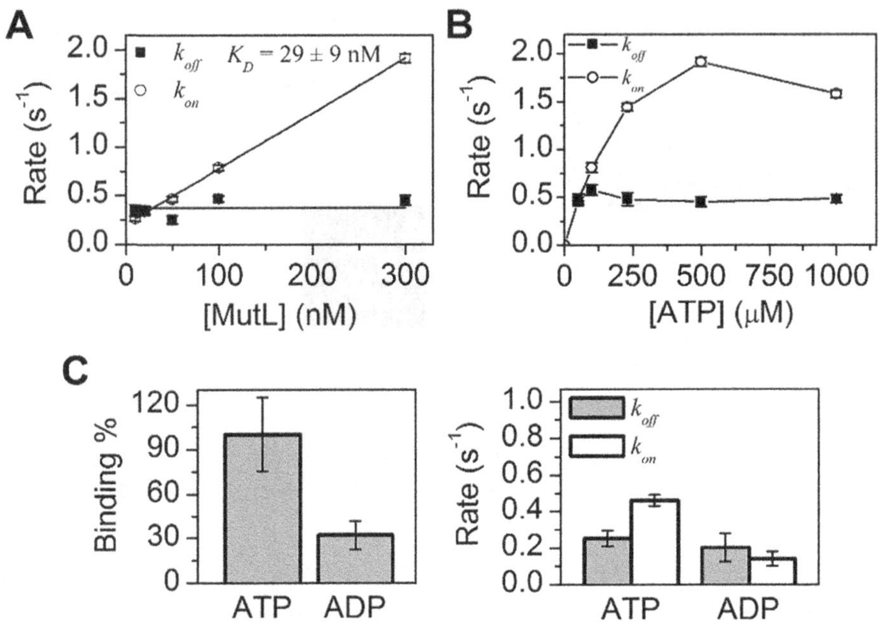

Fig. 6.17 Kinetics of MutL and ssDNA interaction. (**a**) The rate of MutL association and dissociation is dependent on MutL concentration. (**b**) dependence of association and dissociation rate on ATP concentration. (**c**) percentage of binding events and kinetic rate in presence of ATP and ADP. https://doi.org/10.1371/journal.pone.0015496 [29]

As the E1 helicase unwinds the replication fork, the donor and acceptor come close to each other giving a high fluorescence intensity and hence a high FRET value (Fig. 6.18a, b). However, subpopulations are seen within DNA substrates of both types. About 11% of molecules displayed monotonicity (decrease in FRET value steadily) in which the helicase did not slip from the DNA while unwinding. The majority of the molecules (46% and 43%) of the traces were found to be partially repetitive and repetitive (slipping of the helicase once or more than once resulting in FRET values larger than 0.4), respectively (Fig. 6.19a). Similar heterogeneity was also observed in the other DNA substrate in which FRET value increased with the progression of DNA unwinding (Fig. 6.19b).

The above examples depict the usefulness and intricacy of smFRET to monitor and detect minute details and small changes in cellular machinery that often result in major effects in a specific biochemical pathway or complex biophysical processes [30].

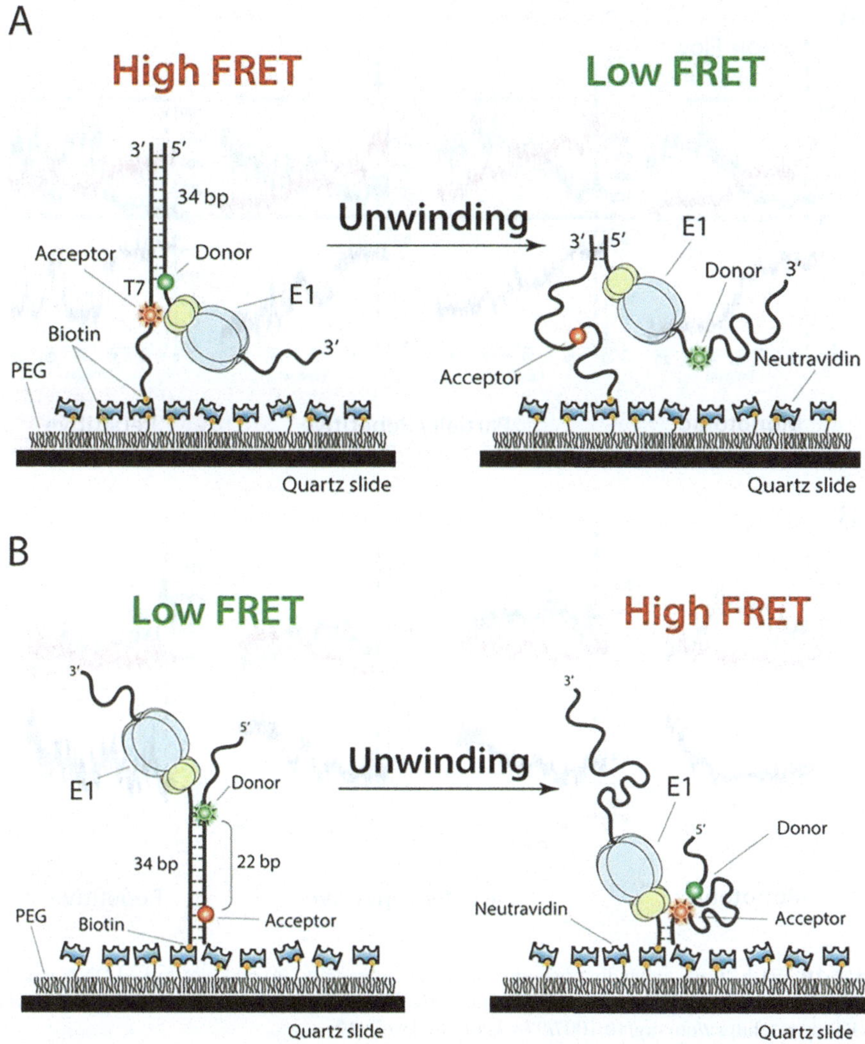

Fig. 6.18 A smFRET assay to study unwinding action of helicases on Replication Fork DNA. (**a**) DNA has been designed in such a way that as the helicase unwinds the DNA fork, the FRET signal decreases (high to low). (**b**) FRET signal increases (low to high). https://doi.org/10.1007/978-1-4939-8556-2_12 [30]

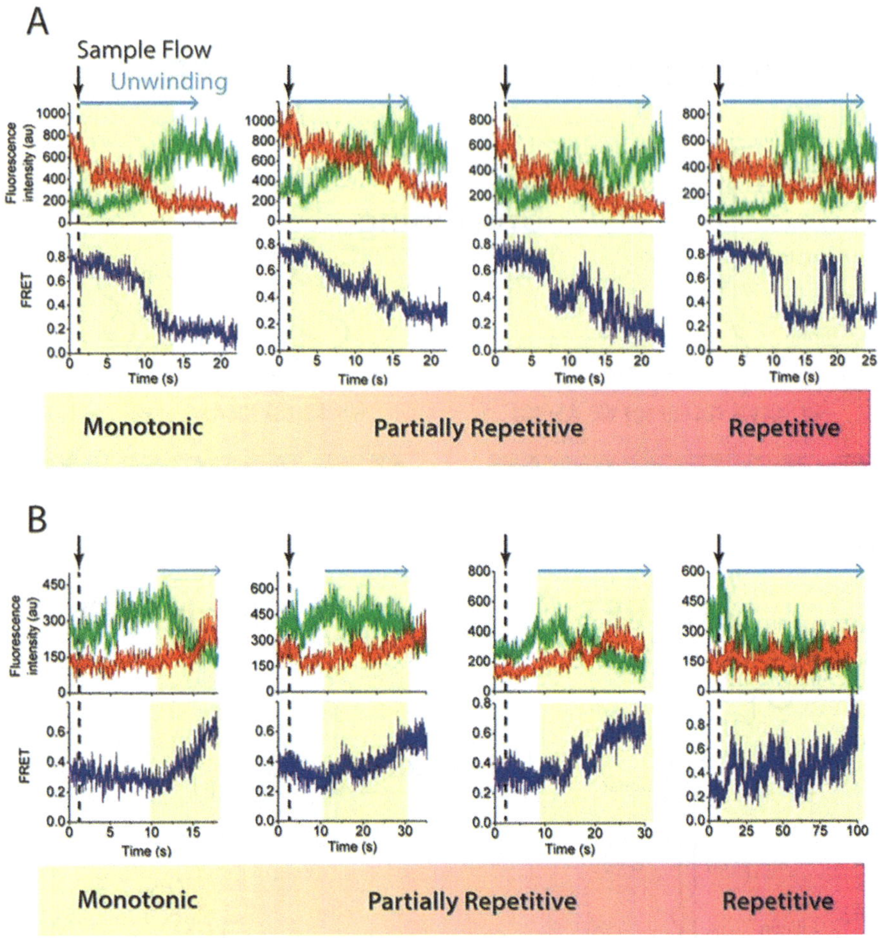

Fig. 6.19 E1 helicase unwinding pattern exhibits heterogeneity. (**a**) traces of forked DNA whose unwinding results in a decrease in FRET signal. (**b**) DNA substrates whose unwinding increases FRET signal. https://doi.org/10.1007/978-1-4939-8556-2_12

6.7 Summary

In summary, smFRET studies of surface-immobilized biomolecules offer distinct opportunities for studying and characterizing complex biomolecular reactions over extended time scales with minute detail as it eliminates ensemble averaging and the need to synchronize reactions. So, it is a powerful imaging approach that provides great insights in studying and determining folding kinetics, conformational dynamics, and interactions of biologically important molecules. But, there are certain limitations of this technique that have to be taken into consideration: (1) As it requires attachment of at least two extrinsic dyes to the molecule(s) of interest, it is relatively insensitive to incomplete labeling like if the donor is missing, the molecule could not be observed, or if the acceptor is missing, the donor-only species shows up as a zero FRET population giving inaccurate information; (2) As detection is achieved via spatial separation, weakly interacting fluorescent species becomes difficult to study; (3) As FRET is a distance-dependent phenomenon, it is insensitive to distance changes outside the 2–8 nm interdye distance range for $R_0 = 5$ nm. (4) To achieve an optimum signal-to-noise ratio, ~100 total photons need to be detected because more than 10^5 photons can be collected from single-dye molecules before photobleaching, yielding more than 10^3 data points; (5) Lastly, absolute distance estimation is challenging as the fluorescence properties and energy transfer greatly depend on the environment and orientation of the dyes. Therefore, the application of smFRET is often integrated with other single-molecule techniques to study biological events, as, through the combination of one or more techniques, it will be possible to monitor several parameters of different properties simultaneously providing a wide understanding of the interaction events. Often, smFRET studies combine with AFM, optical tweezer, and a magnetic tweezer to gain a better understanding and thus providing very promising results.

References

1. Stokes GG (1852) XXX. On the change of refrangibility of light. Philos Trans R Soc Lond 142: 463–562
2. Lakowicz JR (1983) Principles of fluorescence spectroscopy. Plenum Press, New York. 496 pp
3. Jabłoński A (1935) Über den mechanismus der Photolumineszenz von Farbstoffphosphoren. Z Phys 94(1–2):38–46
4. Lakowicz JR (2013) Principles of fluorescence spectroscopy. Springer
5. Sahoo H (2011) Förster resonance energy transfer–A spectroscopic nanoruler: principle and applications. J Photochem Photobiol C: Photochem Rev 12(1):20–30
6. Paul T, Bera SC, Mishra PP (2017) Direct observation of breathing dynamics at the mismatch induced DNA bubble with nanometre accuracy: a smFRET study. Nanoscale 9(18):5835–5842
7. Ha T et al (1999) Polarization spectroscopy of single fluorescent molecules. ACS Publications
8. Weiss VH et al (2000) The structure and oligomerization of the yeast arginine methyltransferase, Hmt1. Nat Struct Mol Biol 7(12):1165–1171
9. Ha T (2001) Single-molecule fluorescence methods for the study of nucleic acids. Curr Opin Struct Biol 11(3):287–292
10. Lakowicz JR, Masters BR (2008) Principles of fluorescence spectroscopy. J Biomed Opt 13(2): 029901
11. Selvin PR (2000) The renaissance of fluorescence resonance energy transfer. Nat Struct Mol Biol 7(9):730
12. Ha T (2001) Single-molecule fluorescence resonance energy transfer. Methods 25(1):78–86
13. Clegg RM (1992) [18] Fluorescence resonance energy transfer and nucleic acids. Methods Enzymol 211:353–388
14. Weiss S (1999) Fluorescence spectroscopy of single biomolecules. Science 283(5408): 1676–1683

15. Bandyopadhyay D, Mishra PP (2020) Real-time monitoring of the multistate conformational dynamics of polypurine reverse Hoogsteen hairpin to capture their triplex-affinity for gene silencing by smFRET microspectroscopy. J Phys Chem B 124(38):8230–8239
16. Mondal S, Mishra PP (2021) Direct observation of effect of crowding induced macromolecular hydration on molecular breathing in the stem of Fork-DNA by single-molecule FRET microspectroscopy. Int J Biol Macromol 167:559–569
17. Roy R, Hohng S, Ha T (2008) A practical guide to single-molecule FRET. Nat Methods 5(6): 507–516
18. Sisamakis E et al (2010) Accurate single-molecule FRET studies using multiparameter fluorescence detection. Methods Enzymol 475:455–514
19. Patonay G et al (2004) Noncovalent labeling of biomolecules with red and near-infrared dyes. Molecules 9(3):40–49
20. Ha T, Selvin PR (2008) Single-molecule techniques: a laboratory manual. Cold Spring Harbor Laboratory Press
21. Ishikawa-Ankerhold HC, Ankerhold R, Drummen GP (2012) Advanced fluorescence microscopy techniques—Frap, Flip, Flap. Fret and flim Molecules 17(4):4047–4132
22. Walter NG et al (2008) Do-it-yourself guide: how to use the modern single-molecule toolkit. Nat Methods 5(6):475–489
23. Sasmal DK et al (2016) Single-molecule fluorescence resonance energy transfer in molecular biology. Nanoscale 8(48):19928–19944
24. Cosa G et al (2001) Photophysical properties of fluorescent DNA-dyes bound to single-and double-stranded DNA in aqueous buffered solution¶. Photochem Photobiol 73(6):585–599
25. Duzdevich D, Redding S, Greene EC (2014) DNA dynamics and single-molecule biology. Chem Rev 114(6):3072–3086
26. Lamichhane R et al (2010) Single-molecule FRET of protein–nucleic acid and protein–protein complexes: surface passivation and immobilization. Methods 52(2):192–200
27. Ha T et al (2002) Initiation and re-initiation of DNA unwinding by the Escherichia coli Rep helicase. Nature 419(6907):638–641
28. McKinney SA et al (2003) Structural dynamics of individual Holliday junctions. Nat Struct Biol 10(2):93–97
29. Park J et al (2010) Single-molecule analysis reveals the kinetics and physiological relevance of MutL-ssDNA binding. PLoS One 5(11):e15496
30. Lee S-J, Syed S, Ha T (2018) Single-molecule FRET analysis of replicative helicases. In: Molecular motors. Springer, pp 233–250

Nanosecond Time-Resolved Fluorescence Assays

7

Yan-Cen Liu and Andreas Hennig

Abstract

Fluorescence-based assays are widely used in clinical diagnosis, drug development, environmental pollution detection, and chemical biology research due to their high sensitivity and rapid read-out times. Undesirable background fluorescence from complex matrices such as biological samples can be easily suppressed by time-resolved fluorescence (TRF) detection.Therein, long-lived probes are used and a delay time, after which all short-live background fluorescence has decayed, is applied to selectively detect the long-lived emission. In addition, to the established lanthanide-based probes with lifetimes in the millisecond time range, probes with much shorter lifetimes (>10 ns) can be used as well, for which the name Nano-TRF assays has been coined. This book chapter reviews the current state-of-the-art of Nano-TRF assays including probes based on Ru (II) complexes, pyrenes, and fluorazophores, as well as their applications.

Y.-C. Liu
Department of Life Science and Chemistry, Jacobs University Bremen, Bremen, Germany

A. Hennig (✉)
Institute of Chemistry of New Materials, Universität Osnabrück, Osnabrück, Germany

Department of Life Science and Chemistry, Jacobs University Bremen, Bremen, Germany
e-mail: andreas.hennig@uni-osnabrueck.de

© The Author(s), under exclusive license to Springer Nature Singapore Pte Ltd. 2022

H. Sahoo (ed.), *Optical Spectroscopic and Microscopic Techniques*,
https://doi.org/10.1007/978-981-16-4550-1_7

7.1 Introduction

7.1.1 Assay Development

The definition of an assay by the IUPAC is "a set of operations having the object of determining the value of a quantity" [1]. Within this definition, essentially any analytical procedure to measure a concentration is an assay, but the meaning of "quantity" in an assay procedure is usually much broader, in particular in the life sciences. It does not only refer to the amount or concentration of a specific target analyte, but the result of a bioassay could essentially refer to any functional activity. For example, the biological "activity" of a certain substance (e.g. vitamins, hormones, or antibiotics) is determined in an assay, which measures its effect on an organism or tissue compared to a standard preparation, such as a placebo in clinical studies [1]. As another example, the toxicity of a substance is usually assessed by an assay, in which the "quantity" to be determined is the probability whether the tested organism is alive or dead after administering a particular concentration of the substance [2].

Nowadays, assays are widely used in clinical diagnosis, drug development [3], environmental pollution detection [4], and chemical biology research [5–7]. For many routine analytical procedures in life science laboratories, assay kits have become commercially available, which include standardized reagents and other necessary materials all packaged together. These assay kits have been thoroughly validated and commonly provide the desired value of a specified quantity with impressive reliability, when used according to the instructions [1]. Moreover, many standardized assay procedures can be automated by the use of pipetting robots including automated sample preparation and data analysis [8].

Automated assay procedures have also become very useful for drug discovery in the pharmaceutical industry. In high-throughput screening (HTS) [9–13], large libraries of synthetic compounds are screened for their potential to serve as a drug lead structure against a specific biological target. With the advancement of assays, laboratory infrastructure, and computer software, the number of samples that can be processed increased from several hundred samples per week in the 1980s to nowadays more than 100,000 samples per day [14]. Microplates are the standard tool for HTS, whose working volume ranges from 30 to 250 μL for the 96-well microplate and from 15 to 80 μL for 384-well microplate. Further miniaturization, termed ultrahigh-throughput screening (uHTS), involves 1536-well and 3456-well microplates with total sample volumes down to 1 μL.

In contrast to the thousands of conceivable read-out parameters in macroscale, bench-top assays, the small sample sizes of HTS require read-out methods with high sensitivity. Efforts are undertaken to enable HTS with electrophysiological [15], mass spectrometric [16], and microcalorimetric methods [17], but the most competitive assays in terms of throughput are all either based on radioactivity, e.g., the scintillation proximity assay (SPA) [18], or involve photon absorption or photon emission (Fig. 7.1) [9, 19]. Among these three read-out methods, assays based on photon emission are by far the most popular, because of their excellent sensitivity compared to absorption-based assays as well as their sustainability compared to

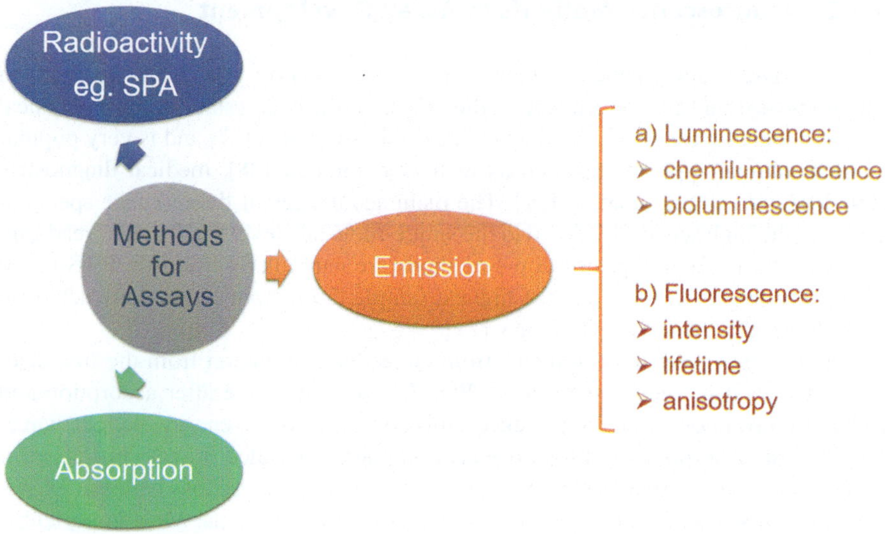

Fig. 7.1 Read-out methods in assays for high-throughput screening (HTS)

radioactive materials. Photon emission can be achieved through chemiluminescence and bioluminescence (which is a special form of chemiluminescence) [19], as well as by fluorescence, which is the most applied methodology in HTS due to its sensitivity, robustness, miniaturizability, and biocompatibility [20, 21]. Fluorescence-based read-out methods that have been used in HTS include, for example, steady-state fluorescence, lifetime-related [22, 23], and fluorescence polarization measurements [24], as well as flow cytometry and fluorescence correlation spectroscopy [25].

Herein, we are going to review fluorescence assays, which are based on probes with fluorescence (or luminescence) lifetimes that are unusually long (>10 ns) and thus enable nanosecond time-resolved fluorescence (Nano-TRF) detection. We will start with a brief, general introduction into fluorescence methods in assay development as well as with the photophysical basics of fluorescence and the fluorescence lifetime. Then, we will explain the principle of time-resolved fluorescence (TRF) detection, which has been widely explored with lanthanide-based probes having millisecond lifetimes. The largest part of this review will then focus on fluorescent probes with lifetimes between 10 ns and ca. 1000 ns and provide an overview of the assay methods that have been developed with them. These probes are comparably stable, offer unusual photophysical properties, or have an attractive biocompatibility and thus complement lanthanide-based probes. Most importantly, their long lifetime, in combination with fast-switching excitation sources and detection methods, enabled TRF detection in the nanosecond time regime, which can dramatically increase the signal-to-background ratio and, thus, enhance the reliability of assay results. This is highly desirable in the presence of an unknown and varying fluorescence background such as from autofluorescence in complex, biological environments as well as from hundred thousands of different compounds with unknown spectroscopic properties in HTS.

7.1.2 Fluorescence Methods in Assay Development

Besides assay development, fluorescence spectroscopy has been applied in photophysical and photochemical studies [26], and it is an established biophysical research method, for example, to analyze protein structures [27] and is very popular in bioanalytical applications such as in flow cytometry [28], medical diagnostics [29, 30], and cellular imaging [31]. The main advantages of fluorescence spectroscopy are the high sensitivity down to the single-molecule level and its rapid read-out. Actually, fluorescence spectroscopic methods are particularly useful in HTS-based assays for drug discovery, and the highest sample throughput can be only achieved with fluorescence-based HTS assays [21].

Fluorescence typically originates from emission of a photon from the first electronically excited, singlet state (S_1 in Fig. 7.2) of a molecule after absorption and internal conversion. The fluorescence emission is at lower energy (longer wavelength) than absorption, and the difference in energy is called Stokes shift after Sir G. G. Stokes, who observed the phenomenon first. The energy loss between excitation and emission for fluorescent molecules has various reasons. Most important is that fluorescence emission is accompanied by a vertical transition, which proceeds exclusively from the lowest vibrational level of the S_1 state to higher vibrational levels of the ground state as postulated by the Franck–Condon principle (note in Fig. 7.2 that the red arrows for absorption are longer than the blue arrows for emission). In addition, solvent relaxation effects, excited-state reactions, complex formation, and/or energy transfer can contribute to the Stokes shift of a fluorescent molecule. With respect to assay development, a large Stokes shift is usually considered an advantage, because the emission light can be easily separated from the excitation light, which reduces the background from scattered excitation light [32].

The fluorescence lifetime, τ, and the fluorescence quantum yield, Q, are two important characteristics of any fluorescent molecule. To understand these two concepts, we need to consider that the excited state of a fluorophore is depopulated via an emissive (fluorescence or, more general, luminescence) and a nonemissive decay pathway with the respective radiative, Γ, and nonradiative decay rate constant, k_{nr}. The average time that the molecule spends in the excited state before it returns to the ground state is the excited-state lifetime, which is defined as the inverse of the sum of the depopulation rates, so, the fluorescence lifetime, τ, is $\tau = 1/(\Gamma + k_{nr})$. This is best understood, when considering that in the absence of nonradiative pathways, the fluorescence lifetime equals the radiative lifetime, i.e., $\tau = 1/\Gamma$. The fluorescence quantum yield, which is defined by the ratio of the number of emitted photons to the number of absorbed photons, can also be expressed as the fraction of the fluorescent decay pathway over the sum of all pathways, i.e., $Q = \Gamma/(\Gamma + k_{nr})$.

Fluorescence lifetime measurements are very informative for mechanistic studies, for example, to dissect static from dynamic quenching, or to distinguish two same fluorophores located in different environment, such as two tryptophan residues in a protein. Two different experimental methods have been established for the measurement of fluorescence lifetimes, namely time-domain and frequency-domain fluorescence lifetime spectroscopy [33]. In time-domain measurements, a short light pulse,

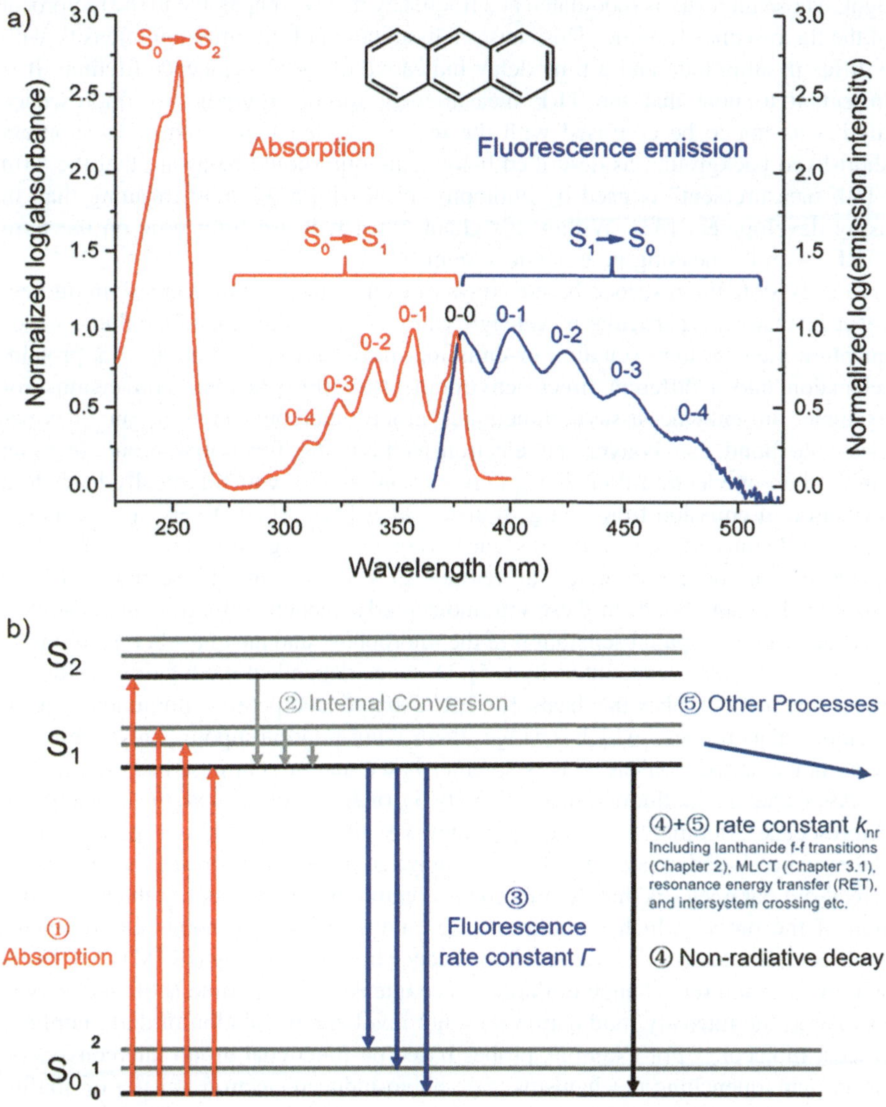

Fig. 7.2 (**a**) Absorption spectrum and fluorescence emission spectrum of anthracene in cyclohexane; (**b**) Jablonski diagram relating absorption and emission band shapes and position to transitions between energy levels

typically from a laser, is used to excite the sample, and photons are collected depending on the arrival time at the detector after the excitation pulse, e.g., by time-correlated single-photon counting (TCSPC). The method is, therefore, also called pulse fluorometry or time-resolved fluorescence (TRF) spectroscopy. In frequency-domain fluorescence lifetime measurements, the sample is excited with light, whose intensity is modulated at a frequency that is comparable to the reciprocal of the fluorescence lifetime. This causes a modulation of the emission intensity with a different amplitude and a time delay indicative of the fluorescence lifetime. It is important to note that the TRF measurement for determining the fluorescence lifetime is not to be confused with the use of a gated time window to suppress short-lived background as described below, and one should be aware that the term "TRF measurement" is used by photophysicists with a different meaning than in assay development [33]. Within this chapter, we will use from now on the term "TRF" with its meaning in assay development.

Steady-state fluorescence-based assays rely on an increase or decrease in fluorescence intensity at a particular wavelength caused by a modulation of the fluorescence quantum yield or the excitation or emission maximum upon binding to a protein, relocation into a different (micro)environment, or (bio)chemical conversion. For example, in enzyme assays, functional group interconversion at an enzyme-cleavable bond can convert an electron-withdrawing functional group (e.g. an amide) into an electron-donating group (e.g. an amine), which typically leads to a switch-on of emission from charge transfer states [34]. Alternatively, a fluorophore and a collision-induced or fluorescence resonance energy transfer (FRET)-based quencher can be incorporated at remote sites in a substrate separated by an enzyme-cleavable bond. In these intramolecularly quenched fluorescent substrates, cleavage causes a spatial separation of the fluorophore and the quencher, resulting in an increased fluorescence intensity [35]. Distance-dependent modulation of fluorescence intensity is also the basis for FRET-based competitive immunoassays or biomolecular probes, which change their conformation upon analyte binding (e.g. molecular beacons as well as aptamer-based and protein-based sensors) [36].

As a change in fluorescence intensity is often accompanied by a change in fluorescence lifetime, the latter provides an alternative read-out parameter in fluorescence-based assays. The key advantage of using the fluorescence lifetime as a read-out parameter is that the lifetime is commonly independent of the concentration of the probe, which eliminates uncertainties in heterogeneous environments such as in cellular assays based on fluorescence lifetime imaging (FLIM) [22]. Noteworthy, a concurrent change in fluorescence intensity and lifetime is often observed as long as the intensity modulation does not involve a population of dark, nonfluorescent molecules. For example, probes based on functional group interconversion or on static quenching mechanisms, e.g., photo-induced electron transfer or nonfluorescent aggregate formation, typically do not involve a change in the fluorescence lifetime [37–39].

Another very useful read-out parameter in fluorescence-based assays is fluorescence anisotropy or fluorescence polarization (which is used synonymously) [33, 40]. Anisotropy-based measurements rely on the fact that emission occurs on a much longer timescale (typically nanoseconds) than absorption (femtoseconds). When the fluorescent probe is excited with polarized light, only a subpopulation of the fluorophores are excited, namely those fluorophores, whose transition dipole moments are aligned parallel to the electric field vector of the incident photons, and when the orientation of the fluorophores remains fixed during its excited-state lifetime, all emitted light will be polarized and can thus be detected behind an appropriately aligned polarizer. Since rotational diffusion proceeds within the same time range as fluorescence emission, it is, however, more common that the excited fluorophores have rotated out of their ideal orientation with respect to the emission polarizer and, thus, less light will be detected. When the excited state had sufficient time to completely randomize its orientation by rotational diffusion, the emitted light will be fully anisotropic. Assays based on fluorescence anisotropy have been widely used in biochemistry, in particular when a small and, thus, rapidly rotating probe binds to a large macromolecule, which severely decelerates the probe's rotation, e.g., in antigen–antibody binding, protein–ligand associations, or probing the fluidity of biomembranes [24, 41–43].

7.2 Time-Resolved Fluorescence (TRF) Assays

Since fluorescence is a very sensitive analytical method, it is easily influenced by background and impurity. Even in controlled conditions, fluorescence spectroscopic measurements are prone to artifacts, and it is usually considered to be very easy to detect fluorescence, but difficult to see nothing else. The situation becomes even more severe in biological environments, such as in cell culture media or in cellular imaging. A typical biological environment has a very high fluorescence background from proteins in the UV region, but also in the visible range of the spectrum, a significant background fluorescence is caused by the presence of NADH, flavins, chlorophyll, and pyridoxal derivatives. The typical approach to address this problem is the development of fluorescent probes with long emission wavelengths and a large Stokes shift to reduce background from the fluorescence of biomolecules and scattered light in tissues [32]. Background emission is also an issue in HTS-based assays for drug discovery, in which hundred thousands of compounds with unknown optical properties are screened for their potential as drug lead structures in fluorescence-based assays.

An elegant method to suppress background emission has been devised by exploiting probes with long-lived emission lifetimes as illustrated in Fig. 7.3. Since typical fluorescence lifetimes are less than 10 ns, the start of the recording of the probe fluorescence can be delayed be a certain time, t_{delay}, during which the short-lived background fluorescence (blue line in Fig. 7.3) has already significantly decayed, whereas the long-lived emission from the probe (red line in Fig. 7.3) still shows enough intensity to be detected.

The strategy, which is now commonly referred to as time-resolved fluorescence (TRF) assays, has been first proposed at around 1980 under the name time-filtered detection [44, 45]. Meanwhile, TRF assays are well established and mainly based on chelates of the lanthanides europium (Eu), terbium (Tb), samarium (Sm), and dysprosium (Dy) [45–49]. Emission from these chelates proceeds via inner-shell 4f-4f transitions of the lanthanide cation, which are well shielded and thus sharp and not very sensitive to their environment. Moreover, these transitions are parity-forbidden leading to very long luminescence lifetimes in the millisecond range and

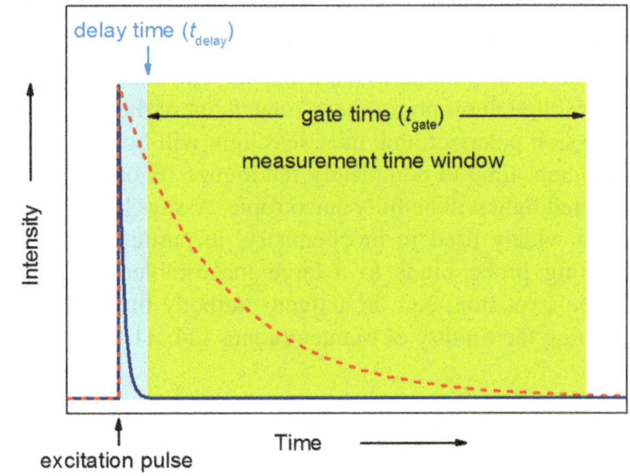

Fig. 7.3 Principle of time-resolved fluorescence (TRF) assays. The blue solid line represents the rapid decay of short-lived background fluorescence, and the red dotted line represents the decay of a long-lived probe. Efficient background suppression can be achieved by starting the measurement after a delay time, t_{delay}, during which the short-lived background has completely decayed and a sufficiently long gate time, t_{gate}, to record enough signal from the probe

a low oscillator strength with molar absorption coefficients $\varepsilon < 1\ M^{-1}\ cm^{-1}$. Efficient excitation of these complexes thus requires a sensitizer (also referred to as antenna) for photon absorption, intersystem crossing to the antenna triplet state, and energy transfer to the lanthanide ion [50]. Overall, this leads to sharp emission peaks, large Stokes shifts, and, most important, ultralong luminescence lifetimes in the millisecond range, which can be temporally separated from the short-lived background fluorescence.

Initially, lanthanide-based assay formats such as the dissociation-enhanced lanthanide fluorescent immunoassay (DELFIA) were heterogeneous and involved several washing and incubation steps, in which addition of the DELFIA inducer dissociates lanthanide ions from an antibody labeled with a chelating ligand and transferred it into a micellar environment, in which it could be efficiently sensitized [51]. Enhanced, homogeneous immunoassay formats relied on luminescence resonance energy transfer (LRET) [52, 53] and led to the development of lanthanide chelate energy transfer (LANCE) assays [54]. Lanthanide-based probes and assay formats have been subsequently developed for various types of enzymes [55, 56], as PET-based protease probes [57], for detection of oligonucleotides [58], as nucleotide sensors [59], and were converted into genetically encoded sensors [60]. Meanwhile, lanthanide-based probes have also been combined with quantum dots as RET acceptors and developed into multiplexed sensors [61, 62].

7.3 TRF Probes with Nanosecond Lifetimes

While lanthanide-based TRF assays are well established, TRF measurements based on probes with nanosecond fluorescence lifetimes are a more recent development, which has been referred to as Nano-TRF. These probes are an attractive alternative to lanthanide-based probes, which can be more stable, have an unusual high biocompatibility, or offer unusual photophysical properties to enable innovative assay principles. As a noteworthy asset, Nano-TRF probes can be excited by pulsed light sources with a higher repetition rate than probes with millisecond lifetimes, which allows to collect more photons within the same measuring time thereby increasing sensitivity.

7.3.1 Ru(II)-Based Nano-TRF Probes

Ru(II) complexes with aromatic, nitrogen-containing heterocycles such as $[Ru(bpy)_3]^{2+}$ (tris(bipyridine)ruthenium(II)) show exceedingly long luminescence lifetimes of up to several microseconds, which can be ascribed to emission from the long-lived metal-to-ligand charge transfer (MLCT) excited state. These complexes are typically stable in acidic or basic media and have relatively high molar absorption extinction coefficients (e.g. $[Ru(bpy)_3]^{2+}$ in aqueous solution at room temperature, $\varepsilon = 14{,}800\ M^{-1}\ cm^{-1}$ at 452 nm) [63, 64] comparable to lanthanide sensitizers [65]. They were first explored for their potential as Nano-TRF probes by Bannwarth and coworkers in 1988, who labeled oligonucleotides with Ru(II) bathophenanthroline complexes [66]. The resulting bioconjugates showed luminescence lifetimes longer than 1000 ns in the absence and presence of complementary DNA. Although DNA hybridization did not affect the emission from these probes, the more than 100 times longer lifetime than the normal fluorescence background rendered them ideal for TRF detection.

7.3.1.1 Hydrolase Assays with Ru(II) Complexes

Subsequently, the Bannwarth group developed two FRET pairs involving Ru (II) complexes (Fig. 7.4), which were all suitable for TRF detection [67–70]. For example, peptides could be covalently labeled at the N-terminus with a Ru (II) complex and a carbostyril derivative of phenylalanine (carbostyril-Phe) during solid-phase peptide synthesis, in which the emission spectrum of the carbostyril dye showed an efficient spectral overlap with the absorption of the Ru(II) complex. The carbostyril dye could thus serve as an efficient FRET donor for the Ru(II) complex. Specifically, two oligopeptides were designed as substrates for the protease thrombin [67, 68]. When the peptides were excited at around 340 nm, a clear emission signal at the emission of the Ru(II) complex around 615 nm was observed indicative of efficient FRET. Moreover, the lifetime of the emission from the Ru(II) complex was 530 ns, while that of the donor dye was around 0.6 ns in presence as well as in absence of the Ru(II) FRET acceptor. Proteolytic cleavage of the peptide chain by thrombin led to a spatial separation of FRET donor and acceptor and, therefore, inefficient FRET, such that emission from the Ru(II) acceptor decreased significantly.

Hydrolysis could also be monitored with oligonucleotides labeled with carbostyril as FRET donor and Ru(II) as acceptor [69]. A uridine unit was incorporated into the oligonucleotide, which could be specifically cleaved under basic conditions. The uncleaved oligonucleotide showed efficient FRET, whereas after addition of base, the fluorescence at around 615 nm decreased to a level, which

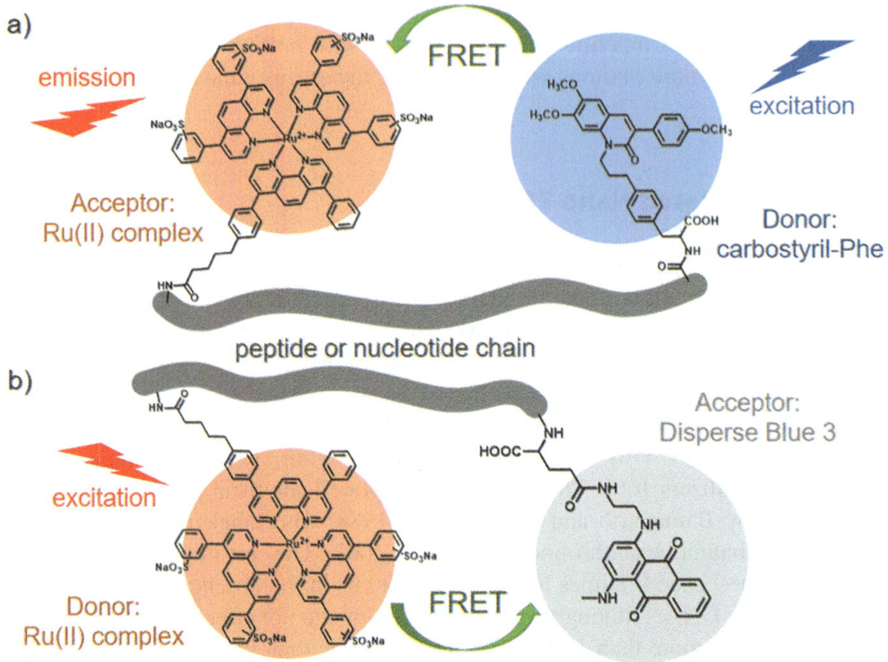

Fig. 7.4 FRET pairs of (**a**) Ru (II) complex and carbostyril-Phe, (**b**) Ru (II) complex and Disperse Blue 3

was in the same range as that of the independently synthesized fragments. The fluorescence lifetime of the donor dye was ca. 0.52 ns, while that of the Ru (II) acceptor was 400 ns, enabling the detection of the FRET-sensitized emission by TRF measurements. Within the same report, the authors also introduced a molecular beacon based on the carbostyril/Ru(II) FRET pair, in which the oligonucleotide formed a hairpin structure bringing donor and acceptor into close spatial proximity. Addition of a complementary DNA strand and hybridization led to an increase in the donor/acceptor distance, such that FRET was no longer efficient and the intensity of the Ru(II)-based FRET acceptor decreased.

To invert the direction of FRET (Fig. 7.4b), a FRET pair was also established, in which the Ru(II) complex served as a FRET donor [70]. The corresponding acceptor dye is a nonfluorescent quencher called Disperse Blue, which has an absorption peak in the range of 500–700 nm, which overlaps very well with the emission from the Ru (II) complex. Donor and acceptor were attached to a peptide sequence susceptible to the protease thrombin, which showed that the fluorescence of the Ru(II) complex is quenched by Disperse Blue with a quenching efficiency of more than 60%, while successful cleavage by thrombin increased the fluorescence intensity of the Ru (II) complex by a factor of ca. 3 to its original value in absence of the RET acceptor. Noteworthy, the lifetime of the Ru(II)-labeled peptide fragment after cleavage was 2150 ns, which is significantly longer than the lifetime of the Ru(II) complex when it

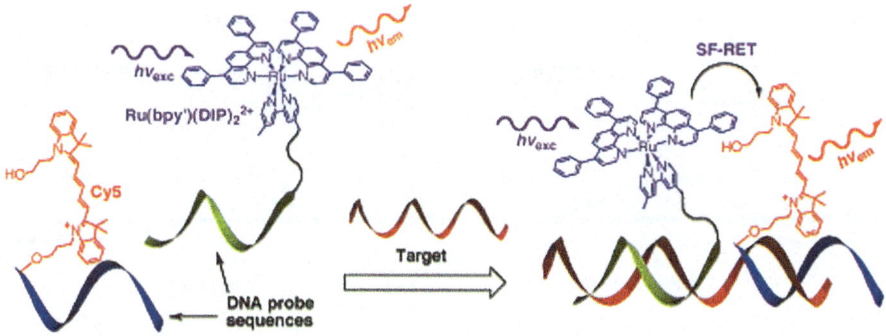

Fig. 7.5 Binary probes as sensors for DNA based on spin-forbidden resonance energy transfer (SF-RET) between two strands labeled with either $Ru(bpy')(DIP)_2^{2+}$ as SF-RET donor and Cy5 as acceptor, which are brought in proximity by hybridization with the complementary DNA strand. (Reprinted from Ref. [71], Copyright © 2007 American Chemical Society)

is not covalently bound to a peptide indicating a potentially beneficial shielding effect on the photophysical properties of the Ru(II) complex.

7.3.1.2 Ru(II)-Based Binary Probes for DNA Detection

Turro and coworkers explored binary probes for DNA detection based on Ru (II) complexes (Fig. 7.5) [71]. Binary probes are generally based on two fluorescently labeled oligonucleotides, which bind selectively to different regions of the target oligonucleotide strand. If the two complementary oligonucleotide probes bind correctly to the target DNA, the fluorescence signal changes by bringing the fluorescent dyes into close spatial proximity [72, 73]. The authors used oligonucleotides labeled with $Ru(bpy)(DIP)_2^{2+}$ (bpy = bipyridine, DIP = 4,7-diphenyl-1,10-phenanthroline) as energy donor and the dye Cy5 as energy acceptor and pointed out that energy transfer from the MLCT state to the singlet state of the organic fluorophore Cy5 is actually spin-forbidden [71]. Although spin-forbidden RET should proceed at a slower rate than FRET and, thus, be less efficient, a sufficiently high RET efficiency to enable an assay was observed, which can be ascribed to the long lifetime of the Ru(II)-based $Ru(bpy)(DIP)_2^{2+}$ probe (1.8 μs).

To probe the distance dependence of spin-forbidden RET, three oligonucleotide strands with different lengths were labeled with $Ru(bpy)(DIP)_2^{2+}$ at the 5′-end, and one strand was modified with Cy5 at the 3′-end. Combination of the Cy5-labeled oligonucleotide with three different Ru(II)-labeled oligonucleotides and hybridization with the complementary strand positioned $Ru(bpy)(DIP)_2^{2+}$ and Cy5 at a distance of four, five, and eight nucleotides. These three different combinations gave Cy5 fluorescence lifetimes of 45, 46, and 69 ns in the presence of target and with indirect excitation by RET from the Ru(II) complex, whereas the fluorescence lifetime of directly excited Cy5 is only ca. 2 ns. Such a delayed fluorescence lifetime is the result of the comparably slow energy transfer rate as confirmed by the dependence of the delayed lifetime on the distance between Ru

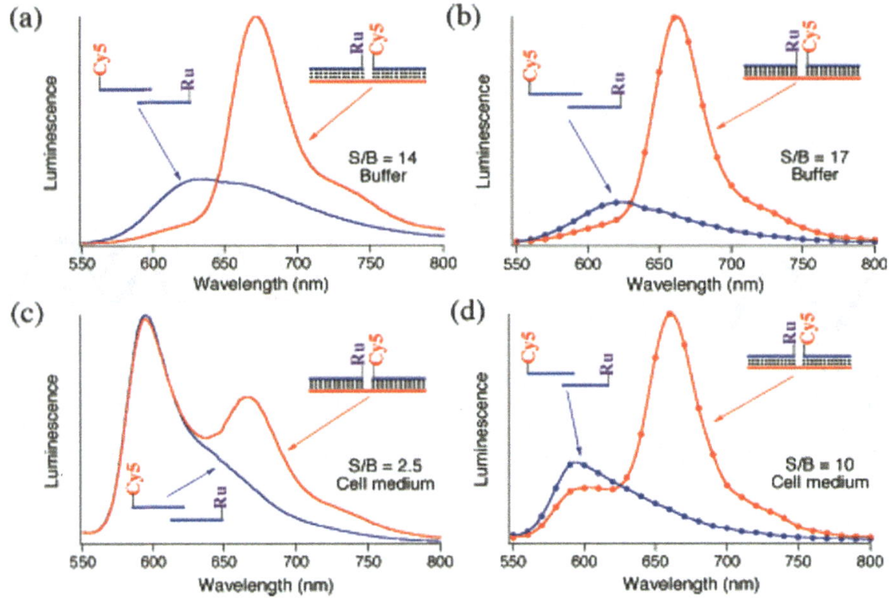

Fig. 7.6 (**a**, **c**) Steady-state and (**b**, **d**) time-resolved emission spectra (recorded 59–77 ns after the excitation pulse) of Ru probe and Cy5 probe in (**a**, **b**) buffer or (**c**, **d**) cell medium containing the fluorophore RedX as background in absence (red) and presence (blue) of the target DNA. (Reprinted from Ref. [71], Copyright © 2007 American Chemical Society)

$(bpy)(DIP)_2^{2+}$ donor and Cy5 acceptor. As a control, a spin-allowed energy transfer pair, Alexa and Cy5, was also tested, and no significant increase of the acceptor's fluorescence lifetime was observed [71, 74].

To illustrate the effect of TRF detection, time-resolved emission spectra recorded 59–77 ns after the excitation pulse were compared to steady-state emission spectra (Fig. 7.6). This demonstrated that the target DNA could be detected in buffer and in cell medium with an improved signal-to-background ratio. Whereas in buffer, the signal-to-background ratio increased only about 20%, an impressive suppression of the background fluorescence from the fluorophore RedX was achieved in cell medium leading to a 400% improved signal-to-background ratio. For the control oligonucleotides labeled with Alexa Fluor 488 and Cy5, TRF detection did not give an enhancement compared to steady-state detection.

7.3.1.3 Ru(II)-Based Immunoassays

Meanwhile, an assay kit for cAMP detection based on Ru(II) complexes as an RET donor and an Alexa Fluor 700 RET acceptor has been commercialized by Roche Diagnostics, Mannheim, Germany. The assay principle (Fig. 7.7) is based on an anti-cAMP IgG antibody labeled with Alexa Fluor 700, which efficiently binds a Ru (II)-labeled antigen leading to RET from the Ru(II) complex to the Alexa Fluor 700 dye. In the competitive immunoassay, cAMP competes with the Ru(II)-labeled

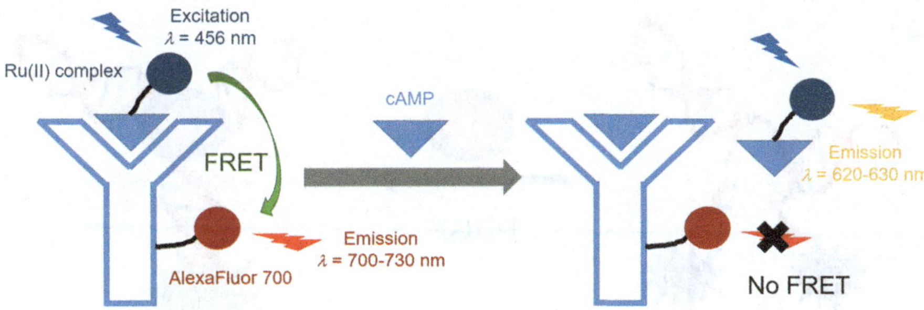

Fig. 7.7 Working principle of a competitive immunoassay for cAMP using Nano-TRF detection

antigen for the antibody-binding site and displaces the antigen. As a consequence of the presence of cAMP, RET no longer applies leading to an increased luminescence from the Ru(II) complex and a decreased long-lived emission from the Alexa Fluor 700 dye. The method is efficient, convenient, and sufficiently robust to serve as an assay method in HTS-based applications such as the search for lead structures for regulating the activity of cAMP-dependent kinases [75].

7.3.2 Nano-TRF Probes Based on Pyrene

Although assays exploiting the long-lived emission from lanthanide chelates or Ru (II) complexes are termed TRF assays, their emission is strictly not fluorescence, because it does not originate from a singlet excited state. Long-lived probes with lifetimes exceeding a few nanoseconds that emit genuine fluorescence are actually very rare with pyrene (up to 400 ns in degassed organic solvents) and coronene (up to 200 ns) being the prototypical examples [33].

In 2005, a method for quantification of the protein platelet-derived growth factor (PDGF) in cells was reported by Tan, Turro, and coworkers [76], which is based on a PDGF aptamer labeled with two pyrene units at the 3′- and 5′-ends (Fig. 7.8). The nucleotide sequence at the terminal ends of the aptamer was optimized such that they do not hybridize in absence of the target analyte, whereas the conformational rearrangement during PDGF binding facilitates hybridization and thereby brings both pyrene units into close spatial proximity. The PDGF-induced conformational changes of the pyrene-labeled aptamer probe led to an increased fluorescence intensity around the emission wavelength of the excimer at 485 nm [77, 78] and enabled the selective detection of PDFG at low nanomolar concentrations in buffer by ratiometric steady-state fluorescence measurements. However, when the method was used in cell media, the fluorescence increase was much smaller due to the interfering background fluorescence from the cell medium (Fig. 7.8b). Exploiting the long-lived fluorescence from pyrene monomer and excimer of up to 100 ns, the short-lived background fluorescence (<5 ns) was successfully suppressed by applying a 40 ns delay time (Fig. 7.8c). In the respective time-resolved emission spectrum,

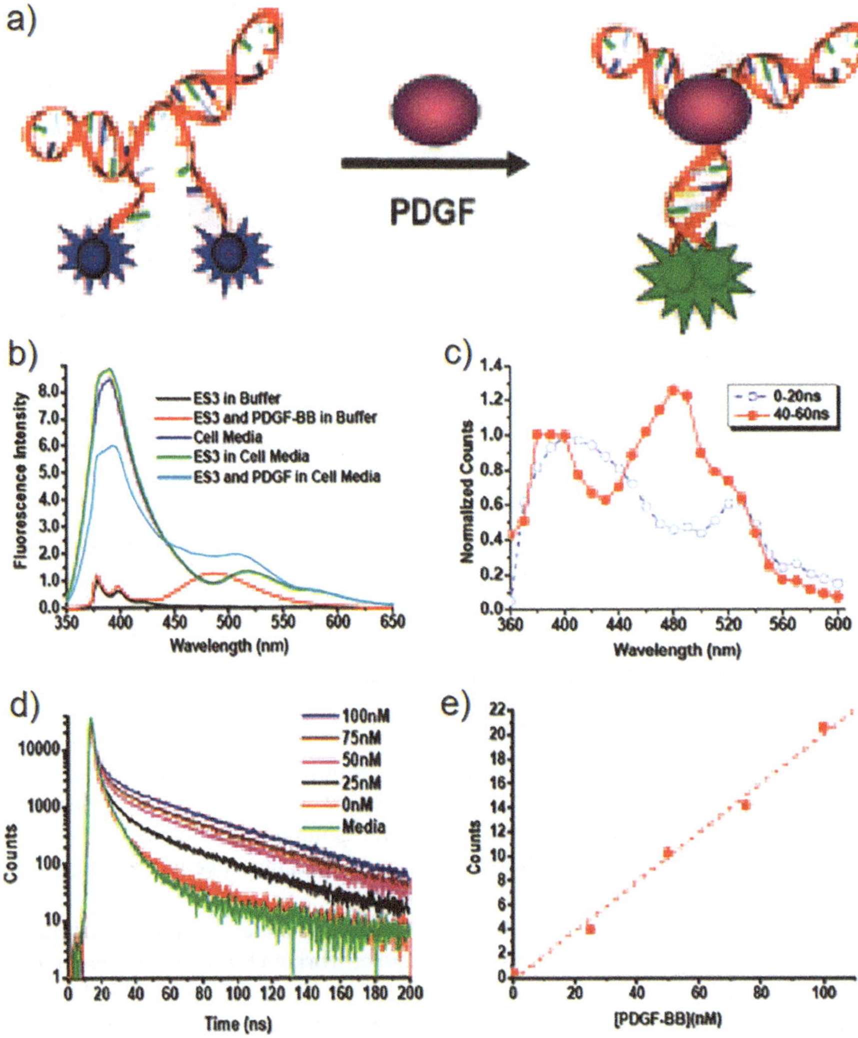

Fig. 7.8 Aptamer-based sensing of platelet-derived growth factor (PDGF) in cell media. (**a**) Working principle of the sensor. (**b**) Steady-state fluorescence spectra of the two-pyrene-modified aptamer sensor in buffer (red and black) and in cell media (cyan and green). (**c**) Time-resolved fluorescence emission spectra of the sensor in presence of PDGF in cell media recorded with 0–20 ns (blue) and 40–60 ns (red) time windows after the excitation pulse. (**d**) Fluorescence decays of the probe in cell media with varying concentrations of PDGF. (**e**) Response of fluorescence intensity (time window 40–100 ns, emission wavelength 480 nm) to the change of PDGF protein concentration. (Reprint from Ref. [76], Copyright (2005) National Academy of Sciences)

the fluorescence intensity at 485 nm was well resolved resulting in a linear relationship with PDGF concentration (Fig. 7.8e). Furthermore, the method also allows quantitative protein analysis in situ, which was proved by detecting PDGF secrete from PC3 cells. Since aptamers are principally available for any target of choice such as drugs, proteins, or other inorganic or organic molecules, the established working principle could be applied to other analytical targets as well [79–81].

In a similar approach, Tan, Wang, and coworkers designed a hairpin DNA strand, which binds to an anti-lysozyme aptamer, and modified it with two pyrene groups at both ends (Fig. 7.9) [82]. The principle is based on molecular beacons [78, 83, 84], which are single-stranded oligonucleotide hybridization probes that form a stem-and-loop structure. In the absence of lysozyme, the anti-lysozyme aptamer binds with the molecular beacon probe resulting in a spatial separation of the two pyrene groups, whereas in the presence of lysozyme, the molecular beacon is released from

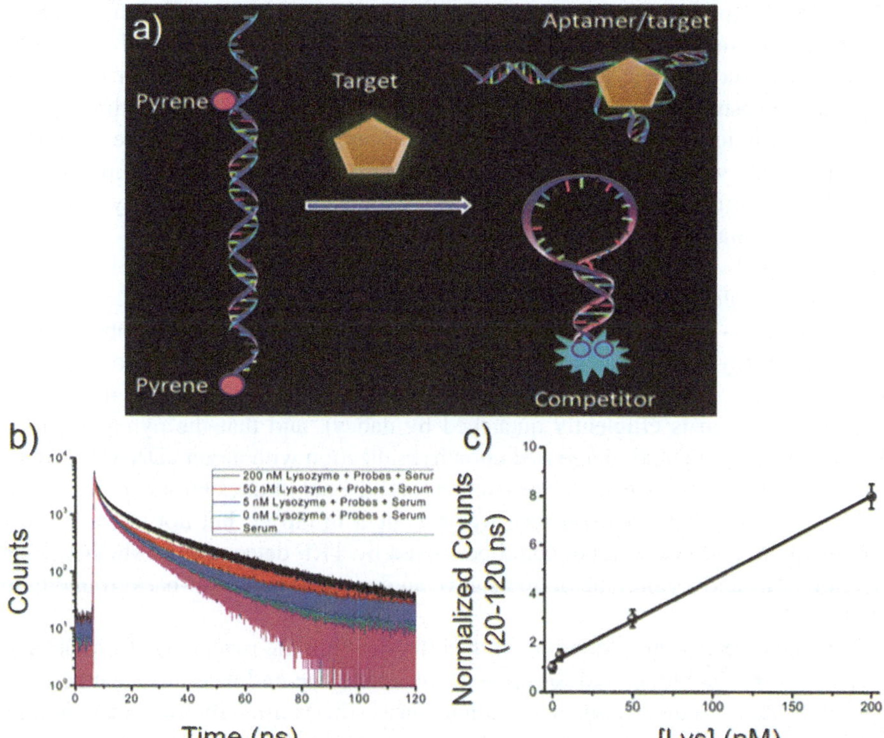

Fig. 7.9 (**a**) Working principle of pyrene-modified aptamer (competitor) as a lysozyme probe; (**b**) Fluorescence decays (at 495 nm) of 200 nM probes in serum with various concentrations of lysozyme at excitation of 355 nm. (**c**) The response of normalized counts between 20 and 120 ns in (**a**) to each concentration of lysozyme in human serum. (Reprint from Ref. [82], Copyright © 2010 American Chemical Society)

the anti-lysozyme aptamer and returns to its hairpin conformation, which brings the two pyrene groups at both ends together leading to the formation of a pyrene excimer. As a result, the fluorescence emission maximum shifts from ca. 400–495 nm enabling the detection of lysozyme by ratiometric fluorescence measurements.

The design required optimization of the loop length, since it influences the affinity to the anti-lysozyme aptamer. If the loop of the hairpin DNA strand is too short, hybridization with anti-lysozyme may be inefficient because binding to the aptamer is weaker than the self-hybridization of the stem ends. On the other hand, if the loop is too long, its binding affinity to anti-lysozyme is too strong to compete with the target lysozyme. The optimal compromise was found for a loop of eight nucleotides, i.e., 5′-pyrene-$(CH_2)_6$-**ATCAGGGC**ACTCTTTA**GCCCTGAT**-$(CH_2)_6$-pyrene-3′ (stem region in bold), which showed maximal signal change among four molecular beacons with different loop lengths. The sensitivity and selectivity with this probe using steady-state fluorescence measurements was sufficient to detect lysozyme down to 200 pM in buffer, as well as in presence of typical proteins such as thrombin, bovine serum albumin, and human IgG, but in human blood serum, the background fluorescence significantly interfered with detection of the probe signal. Lifetime measurements, however, showed that the fluorescence lifetime of the excimer emission increased significantly with the addition of lysozyme, and TRF measurements with a delay time of 20 ns and a gate time of 100 ns improved the signal-to-background ratio significantly and enabled quantification of lysozyme in serum at low nanomolar concentrations (Fig. 7.9c).

A modified version of pyrene-based TRF probes has been suggested, in which at one end of a oligonucleotide, multiple pyrene groups were attached by a dendritic linker, whereas the other end was labeled with 4-(dimethylaminoazo)benzene-4-carboxyl (dabcyl) as a quencher (Fig. 7.10) [85, 86]. The molecular beacon with two pyrene groups was investigated in more detail, which showed that the pyrene excimer emission is efficiently quenched by dabcyl, and that the pyrene excimer fluorescence significantly increased upon hybridization with target cDNA due to the increased distance between pyrene excimer and dabcyl in the stem-open conformation. Again, the probe performance was excellent in buffer, but not in Dulbecco's cell medium, and the problem could be solved by TRF detection, in which a delay time of 60 ns and a gate time of 50 ns gave an optimized signal-to-background ratio (Fig. 7.10c).

To gain a deeper understanding of the effect of multiple pyrene labels, molecular beacons with one, three, and four pyrene groups at one end were also synthesized. This indicated that the excimer emission intensity, the excimer fluorescence lifetime, as well as the quantum yield increased with increasing number of pyrene groups in the presence of excess cDNA, whereas the quenching efficiency in the absence of cDNA decreased with the number of pyrene units. As a result, the molecular beacon containing two pyrene groups had the best signal-to-background ratio among these four molecular beacons. The authors suggested that tuning the length of linkers or varying the number of quenching units could help to overcome the drawbacks of the reduced quenching efficiency with more than two pyrene units.

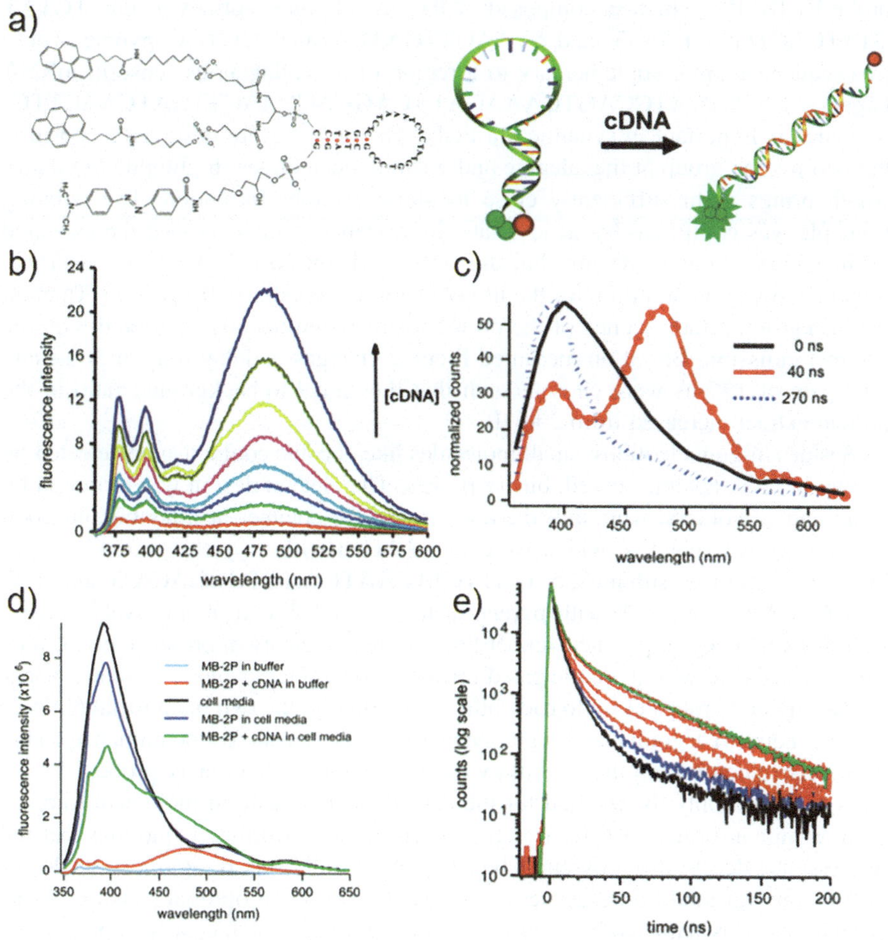

Fig. 7.10 (**a**) Structural representation and signaling principle of the molecular beacon probe modified with two pyrene groups at one end and a dabcyl quencher at the other end. (**b**) Emission spectra of 1 μM probe with increasing concentrations of the complementary target DNA (cDNA) in buffer solution. (**c**) Comparison of steady-state fluorescence spectra of the molecular beacon probe in cell media (blue and green) and buffer (red and cyan) in absence (blue and cyan) and presence (red and green) of cDNA. The background emission of the cell medium is also shown (black). (**d**) Time-resolved emission spectra of the molecular beacon in the presence of cDNA in cell growth media recorded with 0 ns (black), 40 ns (red), and 270 ns (blue) delay times. (**e**) Fluorescence decay traces at 480 nm of the molecular beacon (blue) and with increasing concentrations of cDNA (red) in cell growth media. The autofluorescence decay trace from the cell media is shown in black. (Reprinted from Ref. [85], Copyright © 2008 American Chemical Society)

Pyrene has been previously established as suitable fluorophore label for binary DNA probes [87] and was subsequently also applied for TRF detection of complementary nucleotide sequences, for example for the detection of sensorin mRNA [88] and cocaine [89]. In 2006, the Turro group [88] established the pyrene-based binary probe Py-BP-R2, which is composed of the two oligonucleotides pyrene-TGTAT GTTTCACTGGATGA-3′ and 5′-AAGTTGATCAAGTTGGTGC-pyrene. These two sequences are complementary to a region of an endogenous sensorin mRNA target, i.e., 5′-GTCATCCAGTGAAACATAC**AG**CACCAACTTGATCAACTTG-3′, which is important in synaptic plasticity. The probes were designed to position the two-pyrene group at the adenine and guanine nucleotides (highlighted in bold), which brings them sufficiently close to allow excimer formation. The working principle was confirmed by steady-state fluorescence, which showed the expected excimer peak around 485 nm, but did only work moderately well in a neuronal protein extract, i.e., a mimic for the in vivo neuronal condition (Fig. 7.11). Therein, the background fluorescence of the protein extract significantly interfered with the excimer emission. Based on measured lifetime changes, a delay time of 30 ns and time gate of 120 ns were chosen, such that the signal-to-background ratio in the protein extract increased from 2 to 10.

Besides oligonucleotides, small molecules like cocaine could also be detected by pyrene-labeled, aptamer-based, binary probes. The elegant design was proposed by Yang and coworkers, who split the known cocaine aptamer at a nonbinding loop region into two subunits, which were each labeled with a pyrene unit (Fig. 7.12) [89]. The optimized subunits, 5′-GGTTCTTCAATGAAGTGGGACA-3′ and 5′-G GGAGTCAAGAACC-3′ with pyrene at the 5′- and 3′-end, respectively, gave a significant increase of pyrene excimer fluorescence intensity in presence of cocaine. This is consistent with the anticipated formation of a heterotrimeric structure bringing two pyrene groups close to each other. Importantly, the two main metabolites of cocaine, benzoyl ecgonine and ecgonine methyl ester, had almost no influence on the fluorescence confirming the selectivity of this aptamer-based binary probe.

In order to apply the method for the detection of cocaine in biological samples such as human urine, TRF measurements were also performed exploiting that the fluorescence decay of the pyrene-labeled aptamer fragments is significantly slower than the background fluorescence in urine (Fig. 7.12c). An optimal signal-to-background ratio was achieved by using a gate time of 40 ns, which clearly improved the signal-to-background ratio (Fig. 7.12d).

Moreover, a small-molecule probe for pH sensing using pyrene-related TRF detection was designed by Zhang and coworkers (Fig. 7.13) [90]. Therefore, two pyrene groups and fluorescein isothiocyanate were covalently linked by *N*,*N*-bis (2-aminoethyl)ethane-1,2-diamine. In the probe design, the pyrene excimer was supposed to serve as FRET donor and fluorescein as the acceptor as the absorption spectrum of the open form of fluorescein overlaps better with the emission of the pyrene excimer than the closed, lactone form. As a result, the emission intensity ratio of fluorescein and pyrene excimer becomes pH-dependent due to the increasing FRET efficiency with increasing pH. The reversibility was also demonstrated and the method allowed naked eye observation of pH changes in buffer. Similar like before, TRF detection with a delay time of 20 ns clearly reduced the interference from the intense background fluorescence in a cell medium in acidic and basic condition leading to an improved signal-to-background ratio.

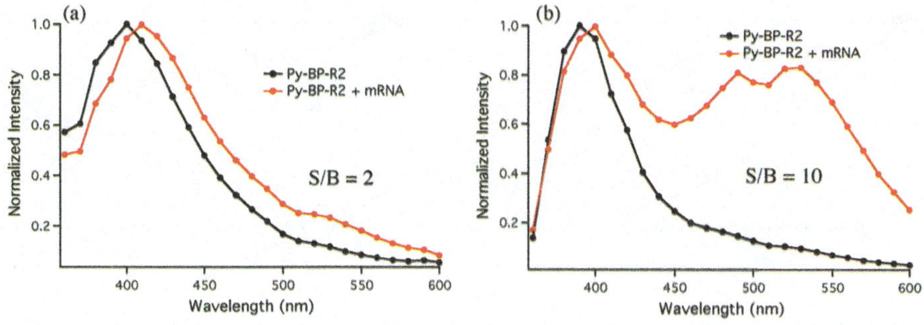

Fig. 7.11 Time-resolved fluorescence spectra of Py-BP-R2 with and without target in neuronal protein extract, gating (**a**) from 0 to 300 ns and (**b**) from 30 to 150 ns. (Reprinted from Ref. [88], Copyright © 2006, Oxford University Press)

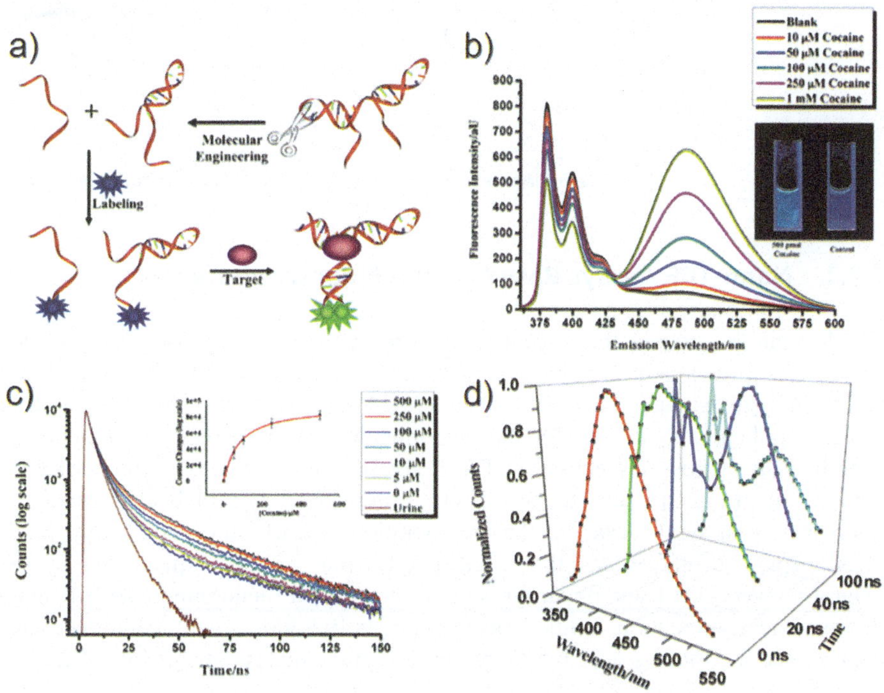

Fig. 7.12 (**a**) Working principle of target-induced complementary aptamer fragment binding based on pyrene excimer formation. (**b**) Detection of cocaine in buffer by steady-state fluorescence spectroscopy. (**c**) Fluorescence decays of aptamer fragments in urine with varying concentrations of cocaine and respective change in fluorescence intensity (inset). (**d**) Time-resolved emission spectra of the aptamer fragments and 500 μM cocaine in urine recorded with delay time of 0 ns (red), 20 ns (green), 40 ns (blue), and 100 ns (cyan). (Reprinted from Ref. [89], Copyright © 2010 Elsevier B.V)

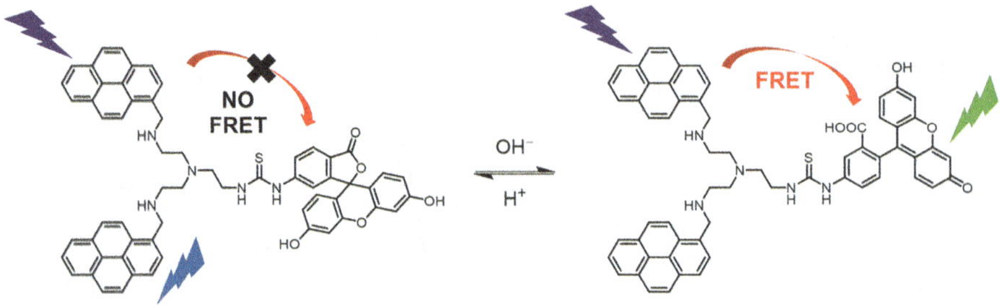

Fig. 7.13 Structure and working principle of a pH sensor based on pyrene excimers. (Adapted from Ref. [90], Copyright © 2014 American Chemical Society)

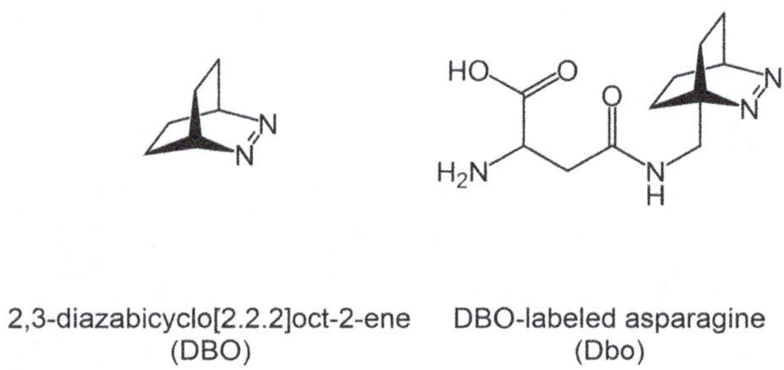

Fig. 7.14 Chemical structures of DBO and Dbo

7.3.3 Nano-TRF Assays Based on the Fluorazophore DBO

Another purely organic fluorescent probe with an exceedingly long fluorescence lifetime of up to 1 μs is 2,3-diazabycyclo[2.2.2]oct-2-ene (DBO, Fig. 7.14). DBO and its derivatives, which have also been termed fluorazophores, have been used in various applications such as for the determination of the association kinetics of supramolecular host molecules and for investigating diffusional properties of antioxidants in membrane models [91]. Moreover, a DBO-labeled asparagine (Dbo) has been synthesized for the incorporation of DBO into peptides by solid-phase peptide synthesis and the efficient, diffusion-controlled quenching by the natural amino acids tryptophan, tyrosine, cysteine, and methionine enabled short-distance measurements in single-labeled peptides [92–95] as well as the measurement of peptide folding rates by fluorescence lifetime measurements [96, 97].

7.3.3.1 Nano-TRF Protease Assays with DBO

After its establishment in peptide folding, Dbo was also incorporated into various enzyme substrates to establish single-label, DBO-based enzyme assays and explore the possibility for Nano-TRF detection. This included the proteases carboxypeptidase A (CPA) [98], trypsin, pepsin, leucine aminopeptidase, and chymotrypsin [99], as well as tyrosine kinases and phosphatases [100].

As a first proof of concept, a Dbo-based CPA protease assay was developed (Fig. 7.15) [98]. Therefore, three short peptide substrates, i.e., Ac-Dbo-Trp-OH,

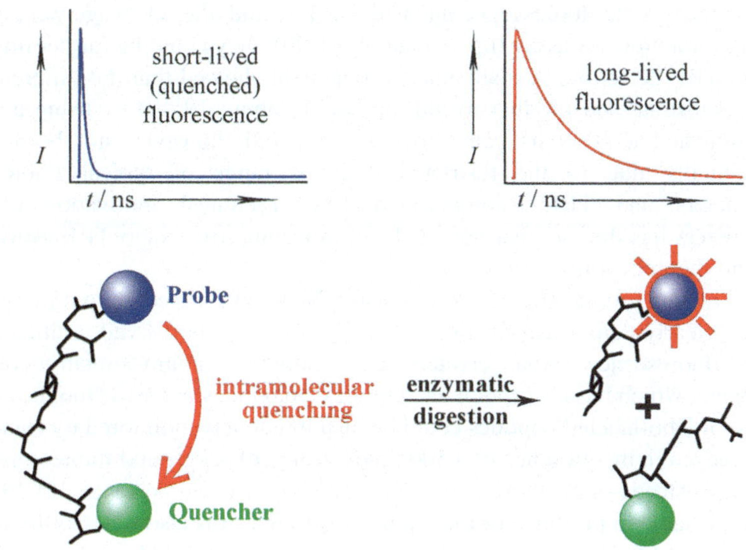

Fig. 7.15 Principle of a fluorescent protease or, more generally, hydrolase assay based on intramolecularly quenched substrates. (Reprint from Ref. [98], Copyright © 2006 Wiley-VCH Verlag GmbH & Co. KGaA, Weinheim)

H-Dbo-Gly-Trp-OH, and H-Dbo-Gly-Gly-Trp-OH, were designed for CPA, which is known to cleave aromatic amino acids from the C-terminus of a peptide chain. In the uncleaved form, Dbo is efficiently quenched by Trp through a collision-induced, aborted electron transfer mechanism, which reduced the fluorescence intensity as well as the fluorescence lifetime by a factor of >10. Removal of the C-terminal Trp residue upon cleavage by CPA allowed Dbo and quencher to diffuse apart, such that Dbo is no longer quenched by Trp. This gave a pronounced increase in steady-state fluorescence intensity and fluorescence lifetime. Interestingly, the enzyme kinetic parameters were compared with a very similar series of peptides, in which Dbo was replaced by a dansyl fluorophore as a FRET acceptor for Trp. This indicated that the catalytic turnover number, k_{cat}, of CPA increased significantly when the dansyl group was at the P3 position of the cleavage site, which is in accordance with a synergetic binding of aromatic residues to the hydrophobic pocket of CPA, whereas labeling with Dbo had only a minor influence on the enzyme kinetic parameters. This clearly demonstrated that the small and hydrophilic, fluorescent DBO probe can be considered as more biocompatible than the typical large aromatic fluorescent dyes. Substrate labeling with the latter is well-known for its potentially detrimental influence on biomolecular recognition [101, 102].

The fluorescence lifetimes before and after enzymatic digestion were around 10 ns and 350 ns, respectively, and the monoexponential fluorescence decay after enzymatic digestion confirmed that the hydrolysis of the peptides was complete. The difference between the fluorescence lifetimes before and after enzymatic digestion of 1–2 orders of magnitude clearly called for TRF detection. While the experimental ratios of steady-state fluorescence intensities before and after cleavage were already excellent for a protease assay (up to a factor of 100), they could be further improved by Nano-TRF detection. A mathematical treatment showed that the differentiation between substrate and product would increase to about 10^{18}(!) by using a 150 ns delay time and a 2000 ns gate time, which could, however, not been shown experimentally due to the restricted dynamic range of typical fluorescence spectrometers and TRF microplate readers. Technical innovation of TRF spectrometers has thus a great potential to further improve Nano-TRF assays with almost no theoretical limitation.

Efficient background suppression by using Nano-TRF detection with DBO was, however, clearly demonstrated (Fig. 7.16). Therefore, a short-lived, artificial background fluorescence was created by adding 7-amino-4-methyl-coumarin ($\tau = 4.9$ ns), which has absorption and emission maxima similar to Dbo. Proteolytic cleavage of Dbo-labeled peptides could be unambiguously monitored by Nano-TRF detection even in the presence of a 500 times stronger background fluorescence. For comparison, steady-state fluorescence spectra in absence and presence of DBO were indistinguishable in presence of this high background. It is also noteworthy that the increased signal fluctuation from Dbo in presence of a 500-fold background fluorescence measured by Nano-TRF (blue triangles in Fig. 7.16a) is mainly due to a saturation of the photomultiplier by the very strong background fluorescence. It is technically very challenging to switch on the high voltage of the photomultiplier tube within the short delay time, such that the instrument used in this report continuously reads out the fluorescence intensity, but only records the fluorescence after a certain delay time by using a boxcar integrator.

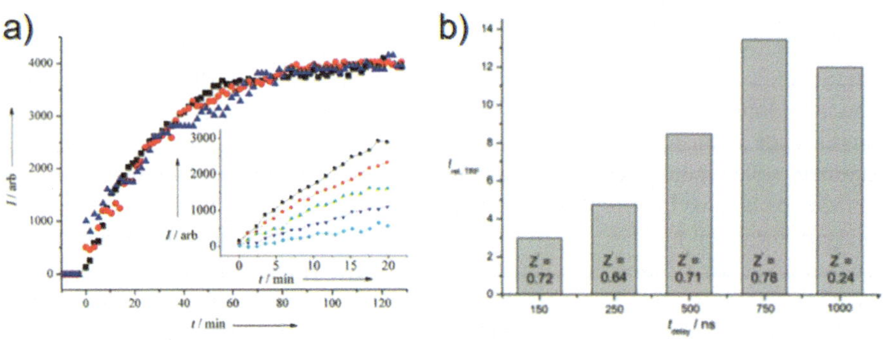

Fig. 7.16 (**a**) Evolution of Nano-TRF fluorescence intensity (delay time 150 ns, gate time 2 μs) during cleavage of the peptide H-Dbo-Gly-Trp-OH by CPA without background (black square) and with a 50 (red circle) or 500 times (blue triangle) stronger steady-state fluorescence background produced by addition of 7-amino-4-methylcoumarin. The inset shows the initial linear portions of fluorescence increase in a 384-well microplate assay during cleavage of the peptide H-Dbo-Gly-Gly-Trp-OH (different concentrations) by CPA. (**b**) Optimization of substrate–product fluorescence differentiation for the peptide H-Tyr-Gln-Ile-Phe-Val-Lys-Dbo-NH_2 cleaved by trypsin using varying delay times in Nano-TRF mode and respective Z'-factors as a measure of assay performance. (Reprinted from Ref. [98], Copyright © 2006 WILEY-VCH Verlag GmbH & Co. KGaA, Weinheim. For panel **b** Reprinted from Ref. [99], Copyright © 2006 Elsevier Inc.)

The Dbo-based protease assay was subsequently evaluated with other proteases as well. Therefore, 11 more Dbo-modified peptides and four more enzymes were investigated, resulting in 28 examined combinations in total [99]. The peptides were designed based on reported enzyme cleavage sites, and the chosen enzymes include trypsin, chymotrypsin, pepsin, and leucine aminopeptidase (LAP), and the obtained results provided insight into the design criteria for Dbo-based protease assays. For example, the fluorescence lifetimes of all peptides after cleavage were above 300 ns, which is comparable to the lifetime of the unquenched, parent DBO, whereas the fluorescence lifetimes before cleavage were largely dependent on the peptide sequence. Shorter peptides clearly increased the probability for collision-induced quenching, e.g., the lifetime of H-Dbo-Gly-Trp-OH (5.9 ns) is shorter than that of H-Dbo-Gly-Gly-Trp-OH (7.6 ns), and more flexible peptide sequences also showed shorter fluorescence lifetimes, for example, H-Trp-Thr-Leu-Thr-Gly-Lys-Dbo-NH_2 (27 ns) contains more flexible amino acids [96] between probe (Dbo) and quencher (Trp) than H-Trp-Gln-Ile-Phe-Val-Lys-Dbo-NH_2 (90 ns). Moreover, Trp has been established as a more efficient quencher than Tyr by measuring the intermolecular quenching rate constants by Stern–Volmer plots, which is also reflected in the difference in the fluorescence lifetimes of peptides containing Trp or Tyr as quencher, e.g., H-Trp-Gln-Ile-Phe-Val-Lys-Dbo-NH_2 had a shorter fluorescence lifetime (90 ns) than H-Tyr-Gln-Ile-Phe-Val-Lys-Dbo-NH_2 (172 ns). Since deprotonation of tyrosine increases its quenching efficiency by a change of the quenching mechanism from aborted hydrogen transfer to electron transfer, it was suggested that a stopped enzyme assay by addition of base could provide a remedy for the decreased quenching efficiency of tyrosine at neutral pH.

Another way to improve the quality and robustness of the assay is the optimization of the TRF read-out parameters. As a standard delay time, 150 ns was used for Dbo-based assays, but increasing the delay time for longer-lived, quenched substrates could improve the differentiation between substrate and product. This was exemplified with the peptide H-Tyr-Gln-Ile-Phe-Val-Lys-Dbo-NH_2 (Fig. 7.16b), whose steady-state fluorescence intensity increases only by a factor of 1.8 after enzymatic cleavage. This could be improved to a factor of 3 using the standard 150 ns delay time, but with longer delay times, the differentiation was further enhanced to a factor of 13.5 for 750 ns, and the quality of the assay as assessed by the Z'-factor improved as well. Theoretically, the differentiation could be even higher by using even longer delay times, but this was not found experimentally. At long delay times, the signal from the cleaved probe became too small to be detected reliably with a Nano-TRF reader composed of a nitrogen laser and a regular photomultiplier tube, whereas another set-up using an Nd-YAG laser and an ICCD camera increases the differentiation to an impressive 2 orders of magnitude. Noteworthy, protease assays have also been established with peptides terminally labeled at both ends with pyrene, but the possibility for Nano-TRF detection has not been explored [103].

7.3.3.2 Nano-TRF Kinase Assays with DBO

As mentioned before, the quenching of Dbo by tyrosine proceeds via hydrogen abstraction from the phenolic hydrogen atom of tyrosine. Since phosphotyrosine (pTyr), the phosphorylated product of a tyrosine kinase, has no phenolic hydrogen atom, it causes no significant quenching, which called for the development of tyrosine kinase and phosphatase assays based on Dbo (Fig. 7.17) [100].

To illustrate the feasibility of Dbo-based, single-label phosphorylation assays, two kinases (EGFR kinase and $p^{60c\text{-}Src}$ Src kinase) and three phosphatases (the specific YOP protein tyrosine phosphatase, and the nonspecific alkaline and acid phosphatases) were selected, and 12 peptides were investigated in the respective enzyme assays. For all tested combinations, differentiation of substrate and product was achieved by steady-state fluorescence spectroscopy. Nano-TRF detection and the peculiar photophysical properties of Dbo provided the opportunity for performance optimization. For example, the differentiation of the dephosphorylated product from the substrate H-Dbo-EEEEpY-OH was readily improved from 2.8 for steady-state fluorescence spectroscopy to a factor of 3.8 by using Nano-TRF detection with a delay time of 500 ns (Fig. 7.17b). Addition of base at the end of the enzymatic reaction to increase the pH of the solution to 12 further increased

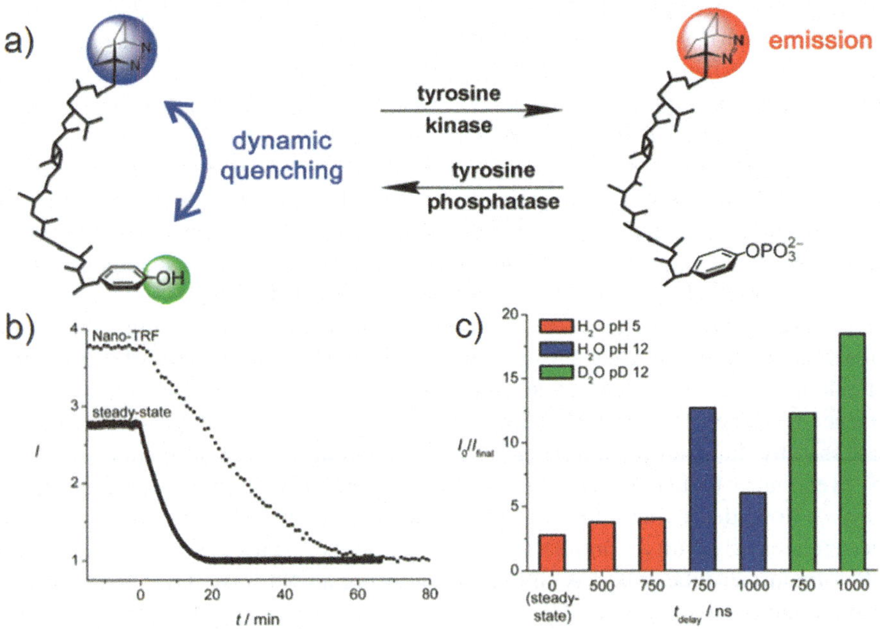

Fig. 7.17 (**a**) Principle of tyrosine kinase and phosphatase assays based on Dbo-labeled substrates. (**b**) Comparison of enhancement factor between state–state fluorescence intensity and Nano-TRF. For the experiments, H-Dbo-EEEEpY-OH was dephosphorylated by acid phosphatase at pH 5. Delay time 500 ns and gate time 2000 ns for Nano-TRF measurement. (**c**) Optimization of a phosphatase assay by Nano-TRF detection, by setting delay time, tuning pH, and using D_2O. (Reprinted from Ref. [100], Copyright © 2007, American Chemical Society)

the differentiation to a factor of ca. 12.5 by decreasing the fluorescence lifetime of the unphosphorylated product by a change in the quenching mechanism from aborted hydrogen atom transfer to aborted electron transfer. Furthermore, the assay was also performed in basic D_2O instead of the H_2O buffer. While this had no influence on the already efficiently quenched product, the lifetime of the phosphorylated substrate could be significantly increased leading to a further increase in the substrate–product differentiation. Under optimized conditions (D_2O, pH 12, 1000 ns delay time), the differentiation could thus be improved to a factor of nearly 20 (Fig. 7.17c).

7.4 Multiple-Pulse Pumping with Nano-TRF Probes

The previous chapters have shown the utility of long-lived probes combined with time-gated (i.e. TRF) detection to detect the probes more selectively by suppression of the short-lived background fluorescence. However, a long delay time may sacrifice a large fraction of the signal from the long-lived probe, because many photons have already been emitted during the delay time and before the signal is being recorded (see Figs. 7.16b and 7.17c). This led, for example, to the reduced substrate–product discrimination in phosphatase assays, when the delay time was increased from 750 to 1000 ns in basic, buffered H_2O (Fig. 7.17c). Moreover, only moderately long-lived probes with lifetimes around 20 ns are not well suited for TRF detection. As a remedy, Gryczynski and coworkers have recently proposed a novel method named multiple-pulse pumping [104].

In order to understand the principle behind multiple-pulse pumping, it is necessary to consider that contemporary pulsed laser diodes excite only a small fraction of the probe molecules. For example, the photon counting rate in TCSPC measurements is usually adjusted to only about 1% of the pulse rate to avoid the pile-up of photons, which would otherwise deteriorate the measured lifetime result. In multiple-pulse pumping, however, such a pile-up is highly desired (Fig. 7.18). When a long-lived probe is excited with a high repetition rate of the laser pulses such that the time between each pulse does not allow a complete decay of the long-lived fluorescence, the population of excited long-lived fluorescence probe accumulates from pulse to pulse, thereby increasing the signal intensity. In contrast, a shorter-lived probe will completely decay between each pulse, and no such signal enhancement due to the pile-up of excited fluorophores will occur. This clearly suggested that the signal from a long-lived probe could be selectively *enhanced* in presence of a short-lived probe; noteworthy, this contrasts time-gated TRF techniques, in which the long-lived probe is less *attenuated* than the short-lived probe.

To also proof this concept experimentally, solutions of commercially available tris(2,2′-bipyridyl)dichlororuthenium(II) hexahydrate ($Ru(bpy)_3$) and sulforhodamine B, whose fluorescence lifetime are around 380 ns and 1.7 ns, respectively, were prepared and excited with a varying number of laser pulses with a 80 Hz (12.5 ns) repetition rate (Fig. 7.19) [104]. The result clearly showed that the signal intensity of the long-lived $Ru(bpy)_3$ largely benefitted from the pumping

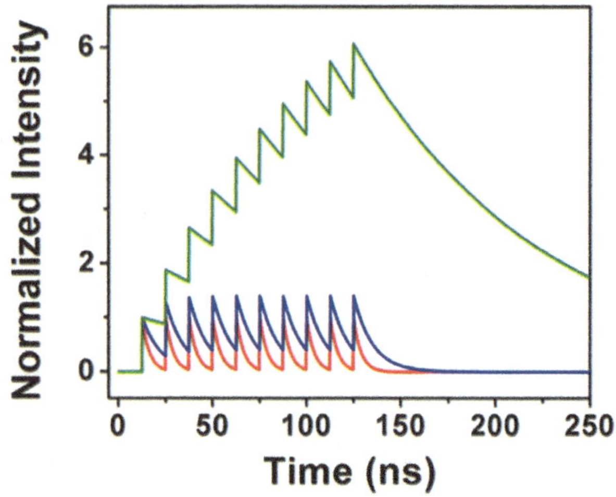

Fig. 7.18 Simulated results of fluorescence decay traces for probe lifetimes of 4 ns (red), 10 ns (blue), and 100 ns (green) after a burst of 10 excitation pulses at a repetition rate of 80 MHz (12.5 ns between each pulse). (Reprinted from Ref. [104], Copyright © 2013 Elsevier Inc.)

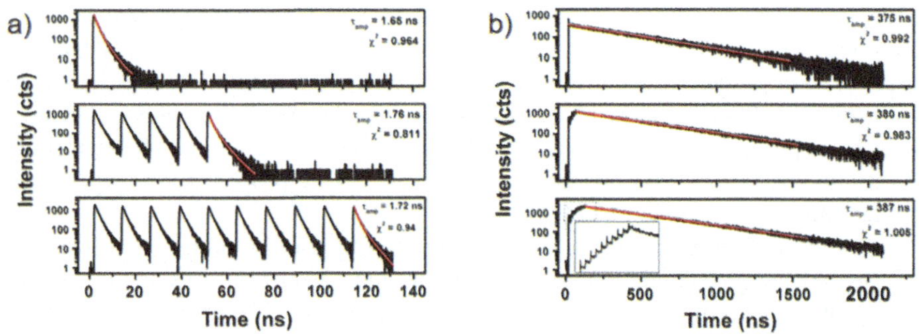

Fig. 7.19 Intensity decays of solutions containing (**a**) sulforhodamine B or (**b**) Ru(bpy)$_3$ with excitation bursts of one pulse (top), five pulses (middle), and ten pulses (bottom). (Reprinted from Ref. [104], Copyright © 2013 Elsevier Inc.)

effect and significantly increased with increasing number of pulses. On the contrary, an increasing number of excitation pulses had no influence on the peak intensity of sulforhodamine B, since its fluorescence decays completely between each pulse.

It was also shown that the method is of practical value in fluorescence imaging (Fig. 7.20). To illustrate, a Ru(bpy)$_3$-loaded bead was positioned on a surface that had been covered with sulforhodamine B to create an artificial background fluorescence. With increasing number of excitation pulses at the right repetition frequency, the signal-to-background ratio of the corresponding surface plot was significantly increased. Moreover, it was clearly shown that an increasing laser power (Fig. 7.20e, j) only gave an overall brighter image, in which the intensity of signal and the background was increased. Within the same report [104], the utility of multiple-pulse pumping was also demonstrated by detecting β-tubulin III in rat retinal tissue with a Ru(bpy)$_3$-labeled IgG antibody, which led to an enhanced signal-to-background ratio.

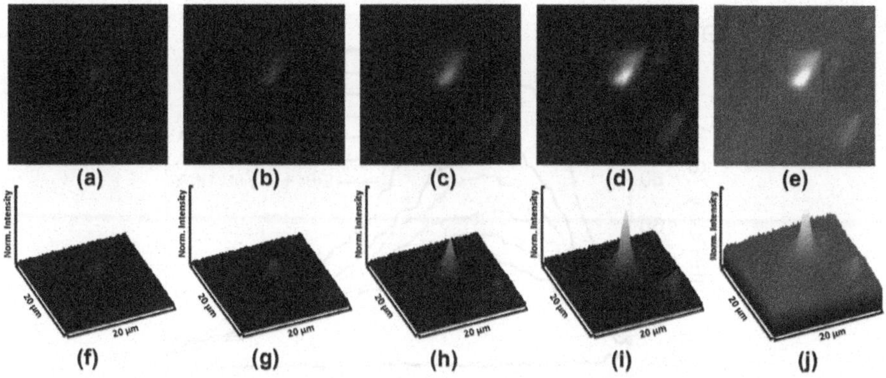

Fig. 7.20 Fluorescence imaging with multiple-pulse excitation. Images (**a–d**) depict a bead, which was labeled with long-lived Ru(bpy)$_3$ and placed on a surface labeled with short-lived sulforhodamine B. The intensity scales are normalized to the peak intensity of (**d**), and the different images were recorded after (**a**) 1-pulse, (**b**) 2-pulse, (**c**) 5-pulse, and (**d**) 10-pulse excitation. The image in (**e**) was recorded after 1-pulse excitation, but the average laser power has been increased to match the overall laser power of the pulse train in the 10-pulse laser burst in (**d**). The signal-to-background ratio in (**e**) differs by 3% from that of (**a**). Each surface plot, (**f**) through (**j**), corresponds to the image, (**a**) through (**e**) above it. (Reprinted from Ref. [104], Copyright © 2013 Elsevier Inc.)

Various other fluorescent probes have been subsequently explored with this multiple-pulse pumping strategy. For example, BSA-covered Au clusters emit in the red region with lifetimes exceeding 1 μs. They are, thus, principally suitable for biological imaging, but suffer from low quantum yields [105]. This limitation could be overcome by multiple-pulse pumping with a delay time of 50 ns, which increased the signal-to-background ratio to 30 and rendered the emission of the BSA Au clusters clearly distinguishable from the background.

Very recently, multiple-pulse pumping has been also applied to DNA detection [106]. Therefore, ethidium bromide (EtBr) was chosen as the probe, whose fluorescence intensity increases upon addition of calf thymus DNA, but which is also easily influenced by background fluorescence. The fluorescence lifetime of EtBr is around 1.5 ns when it is free in solution and around 22 ns when bound to DNA, which is incompatible with TRF detection, but appeared ideal for multiple-pulse pumping. In fact, the signal from EtBr increased almost 80 times in the presence of DNA when applying the three-pulse excitation and a delay time of 10 ns (Fig. 7.21), whereas the steady-state fluorescence intensity increased only by a factor of 2. The limit of detection for DNA could be lowered by multiple-pulse pumping to 20 nM DNA.

7.5 Conclusion

Fluorescence-based techniques are key in (biochemical) assay development due to their high sensitivity and rapid read-out times. Innumerable different assay formats have been developed, which all have their disadvantages and advantages in certain

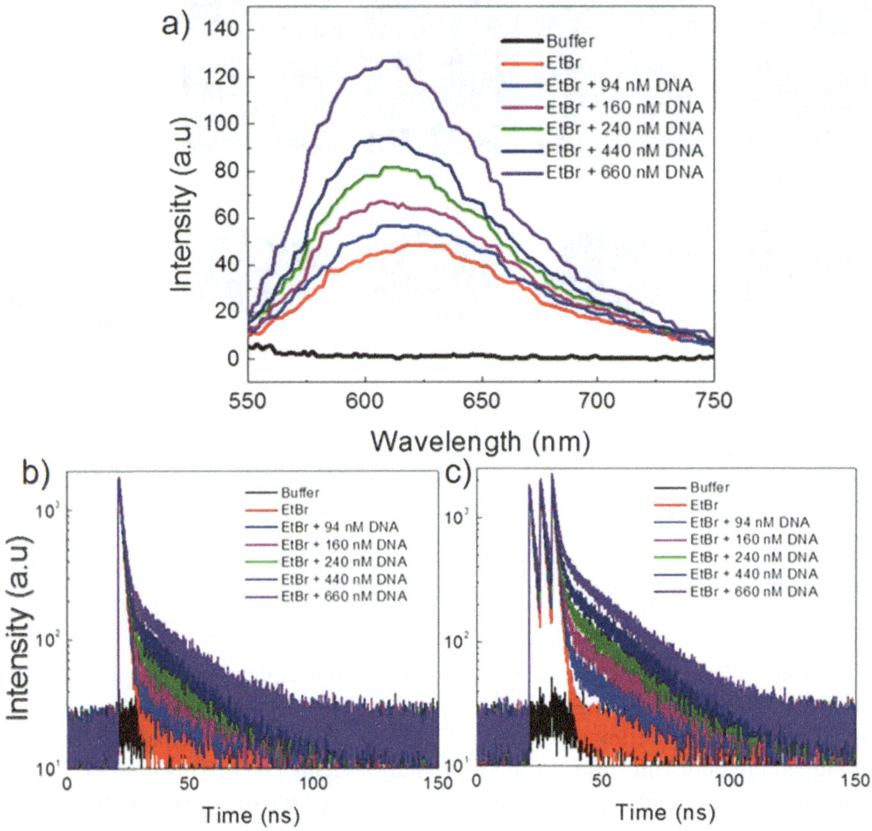

Fig. 7.21 (**a**) Emission spectra of ethidium bromide (EtBr) in absence and presence of different DNA concentrations. (**b**, **c**) Respective intensity decays using (**b**) one or (**c**) three excitation pulses. (Reproduced from Ref. [106], Copyright The Royal Society of Chemistry)

areas. To suppress background fluorescence, TRF detection involving long-lived probes allows detection after a delay time after which all short-lived background fluorescence has decayed and, thus, enables unprecedented improvements in signal-to-background ratio. The most advanced probes include lanthanide-based probes with lifetimes in the millisecond time range, but probes with much shorter lifetimes, yet exceeding ca. 10 ns, can be used as well. For the latter, the name Nano-TRF assays has been coined, and the novel assay formats have been developed. Nano-TRF probes feature advantages like enhanced stability, uncommon photophysics, and unusually high biocompatibility, and can, thus, be complementarily used to the existing lanthanide-based assay formats.

References

1. IUPAC (1997) Compendium of chemical terminology, 2nd edn. Blackwell Scientific Publications, Oxford. https://doi.org/10.1351/goldbook. Online version (2019) created by S. J. Chalk (the "Gold Book"). Compiled by A. D. McNaught and A. Wilkinson. ISBN 0-9678550-9-8
2. Repetto G, del Peso A, Zurita JL (2008) Neutral red uptake assay for the estimation of cell viability/cytotoxicity. Nat Protoc 3:1125–1131
3. Sittampalam GS, Grossman A, Brimacombe K et al (eds) (2004) Assay guidance manual. Eli Lilly & Company and the National Center for Advancing Translational Sciences, Bethesda, MD. https://www.ncbi.nlm.nih.gov/books/NBK53196/
4. Lee GF (1973) Chemical aspects of bioassay techniques for establishing water quality criteria. Water Res 7:1525–1546
5. Aydin S (2015) A short history, principles, and types of ELISA, and our laboratory experience with peptide/protein analyses using ELISA. Peptides 72:4–15
6. Nguyen BT, Anslyn EV (2006) Indicator–displacement assays. Coord Chem Rev 250:3118–3127
7. Reymond J-L (2006) Enzyme assays, 1st edn. Wiley-VCH, Weinheim
8. Hennig A, Nau WM (2019) Cucurbiturils and related macrocycles, 1st edn. RSC Press, Cambridge, pp 121–149
9. Inglese J, Johnson RL, Simeonov A, Xia M, Zheng W, Austin CP, Auld DS (2007) High-throughput screening assays for the identification of chemical probes. Nat Chem Biol 3:466–479
10. von Ahsen O, Bömer U (2005) High-throughput screening for kinase inhibitors. Chem Bio Chem 6:481–490
11. Hertzberg RP, Pope AJ (2000) High-throughput screening: new technology for the 21st century. Curr Opin Chem Biol 4:445–451
12. Hüser J (2006) High-throughput screening in drug discovery, vol 35, 1st edn. Wiley-VCH, Weinheim
13. Carter JM (2004) A guide to HTS assay development, 1st edn. D&MD, Westborough, MA
14. Entzeroth M, Flotow H, Condron P (2009) Overview of high-throughput screening. Curr Protoc Pharmacol 44:9.4.1–9.4.27
15. Dunlop J, Bowlby M, Peri R, Vasilyev D, Arias R (2008) High-throughput electrophysiology: an emerging paradigm for ion-channel screening and physiology. Nat Rev Drug Discov 7:358–368
16. Liesener A, Karst U (2005) Monitoring enzymatic conversions by mass spectrometry: a critical review. Anal Bioanal Chem 382:1451–1464
17. Torres FE, Kuhn P, De Bruyker D, Bell AG, Wolkin MV, Peeters E, Williamson JR, Anderson GB, Schmitz GP, Recht MI, Schweizer S, Scott LG, Ho JH, Elrod SA, Schultz PG, Lerner RA, Bruce RH (2004) Enthalpy arrays. Proc Natl Acad Sci U S A 101:9517–9522
18. Park Y-W, Cummings RT, Wu L, Zheng S, Cameron PM, Woods A, Zaller DM, Marcy AI, Hermes JD (1999) Homogeneous proximity tyrosine kinase assays: scintillation proximity assay versus homogeneous time-resolved fluorescence. Anal Biochem 269:94–104
19. Fan F, Wood KV (2007) Bioluminescent assays for high-throughput screening. Assay Drug Dev Tech 5:127–136
20. Eggeling C, Brand L, Ullmann D, Jäger S (2003) Highly sensitive fluorescence detection technology currently available for HTS. Drug Discov Today 8:632–641
21. Gribbon P, Sewing A (2003) Fluorescence readouts in HTS: no gain without pain? Drug Discov Today 8:1035–1043
22. Turconi S, Bingham RP, Haupts U, Pope AJ (2001) Developments in fluorescence lifetime-based analysis for ultra-HTS. Drug Discov Today 6:27–39
23. Moger J, Gribbon P, Sewing A, Winlove CP (2006) The application of fluorescence lifetime readouts in high-throughput screening. J Biomol Screen 11:765–772

24. Turek-Etienne TC, Small EC, Soh SC, Xin TA, Gaitonde PV, Barrabee EB, Hart RF, Bryant RW (2003) Evaluation of fluorescent compound interference in 4 fluorescence polarization assays: 2 kinases, 1 protease, and 1 phosphatase. J Biomol Screen 8:176–184
25. Neefjes J, Dantuma NP (2004) Fluorescent probes for proteolysis: tools for drug discovery. Nat Rev Drug Discov 3:58–69
26. Turro NJ, Ramamurthy V, Scaiano JC (2010) Modern molecular photochemistry of organic molecules, 1st edn. University Science Books, Herndon, VA
27. Eftink MR (1991) Methods of biochemical analysis. Wiley-VCH, Weinheim, pp 127–205
28. Nebe-von-Caron G, Stephens PJ, Hewitt CJ, Powell JR, Badley RA (2000) Analysis of bacterial function by multi-colour fluorescence flow cytometry and single cell sorting. J Microbiol Methods 42:97–114
29. Albani JR (2016) Encyclopedia of analytical chemistry. Wiley-VCH, Weinheim, pp 1–23
30. Hemmilä IA (1991) Applications of fluorescence in immunoassays, 1st edn
31. Forero-Shelton M (2019) Peering into cells at high resolution just got easier. Nat Methods 16:293–294
32. Ren T-B, Xu W, Zhang W, Zhang X-X, Wang Z-Y, Xiang Z, Yuan L, Zhang X-B (2018) A general method to increase stokes shift by introducing alternating vibronic structures. J Am Chem Soc 140:7716–7722
33. Lakowicz JR (2006) Principles of fluorescence spectroscopy, 3rd edn. Springer, New York, NY
34. Graham Knight C (1995) Methods in enzymology, vol 248. Academic Press, London, pp 18–34
35. Gershkovich AA, Kholodovych VV (1996) Fluorogenic substrates for proteases based on intramolecular fluorescence energy transfer (IFETS). J Biochem Biophys Methods 33:135–162
36. Tyagi S, Kramer FR (1996) Molecular beacons: probes that fluoresce upon hybridization. Nat Biotechnol 14:303–308
37. Marmé N, Knemeyer J-P, Wolfrum J, Sauer M (2004) Highly sensitive protease assay using fluorescence quenching of peptide probes based on photoinduced electron transfer. Angew Chem Int Ed 43:3798–3801
38. Packard BZ, Toptygin DD, Komoriya A, Brand L (1996) Profluorescent protease substrates: intramolecular dimers described by the exciton model. Proc Natl Acad Sci U S A 93:11640–11645
39. Johansson MK, Cook RM (2003) Intramolecular dimers: a new design strategy for fluorescence-quenched probes. Chem Eur J 9:3466–3471
40. Valeur B, Berberan-Santos MN (2012) Molecular fluorescence: principles and applications, 2nd edn. Wiley-VCH, Weinheim
41. Hall MD, Yasgar A, Peryea T, Braisted JC, Jadhav A, Simeonov A, Coussens NP (2016) Fluorescence polarization assays in highthroughput screening and drug discovery: a review. Methods Appl Fluores 4:022001
42. Coffin J, Latev M, Bi X, Nikiforov TT (2000) Detection of phosphopeptides by fluorescence polarization in the presence of cationic polyamino acids: application to kinase assays. Anal Biochem 278:206–212
43. Simeonov A, Bi X, Nikiforov TT (2002) Enzyme assays by fluorescence polarization in the presence of polyarginine: study of kinase, phosphatase, and protease reactions. Anal Biochem 304:193–199
44. Haugen GR, Lytle FE (1981) Quantitation of fluorophores in solution by pulsed laser excitation and time-filtered detection. Anal Chem 53:1554–1559
45. Soini E, Hemmilä I (1979) Fluoroimmunoassay: present status and key problems. Clin Chem 25:353–361
46. Halonen P, Meurman O, Lovgren T, Hemmila I, Soini E (1983) Detection of viral antigens by time-resolved fluoroimmunoassay. Curr Top Microbiol Immunol 104:133–146

47. Hemmilä I, Dakubu S, Mukkala V-M, Siitari H, Lövgren T (1984) Europium as a label in time-resolved immunofluorometric assays. Anal Biochem 137:335–343
48. Hemmilä I, Webb S (1997) Time-resolved fluorometry: an overview of the labels and core technologies for drug screening applications. Drug Discov Today 2:373–381
49. Hemmilá I, Mukkala V-M (2001) Time-resolution in fluorometry technologies, labels, and applications in bioanalytical assays. Coord Chem Rev 38:441–519
50. Bünzli J-CG (2010) Lanthanide luminescence for biomedical analyses and imaging. Chem Rev 110:2729–2755
51. Hemmilä I (2020) Fluoroimmunoassays and immunofluorometric assays. Clin Chem 31:359–370
52. Selvin PR, Rana TM, Hearst JE (1994) Luminescence resonance energy transfer. J Am Chem Soc 116:6029–6030
53. Selvin PR, Hearst JE (1994) Luminescence energy transfer using a terbium chelate: improvements on fluorescence energy transfer. Proc Natl Acad Sci U S A 91:10024–10028
54. Hemmilä I (1999) Lance™: homogeneous assay platform for HTS. J Biomol Screen 4:303–307
55. Hewitt SH, Butler SJ (2018) Application of lanthanide luminescence in probing enzyme activity. Chem Commun 54:6635–6647
56. Karvinen J, Laitala V, Mäkinen M-L, Mulari O, Tamminen J, Hermonen J, Hurskainen P, Hemmilä I (2004) Fluorescence quenchingbased assays for hydrolyzing enzymes. Application of time-resolved fluorometry in assays for caspase, helicase, and phosphatase. Anal Chem 76:1429–1436
57. Terai T, Kikuchi K, Iwasawa S-Y, Kawabe T, Hirata Y, Urano Y, Nagano T (2006) Modulation of luminescence intensity of lanthanide complexes by photoinduced electron transfer and its application to a long-lived protease probe. J Am Chem Soc 128:6938–6946
58. Johansson MK, Cook RM, Xu J, Raymond KN (2004) Time gating improves sensitivity in energy transfer assays with terbium chelate/dark quencher oligonucleotide probes. J Am Chem Soc 126:16451–16455
59. Wu M, Dul BE, Trevisan AJ, Fiedler D (2013) Synthesis and characterization of non-hydrolysable diphosphoinositol polyphosphate messengers. Chem Sci 4:405–410
60. Vuojola J, Syrjänpää M, Lamminmäki U, Soukka T (2013) Genetically encoded protease substrate based on lanthanide-binding peptide for time-gated fluorescence detection. Anal Chem 85:1367–1373
61. Charbonnière LJ, Hildebrandt N, Ziessel RF, Löhmannsröben H-G (2006) Lanthanides to quantum dots resonance energy transfer in time-resolved fluoro-immunoassays and luminescence microscopy. J Am Chem Soc 128:12800–12809
62. Geißler D, Charbonnière LJ, Ziessel RF, Butlin NG, Löhmannsröben H-G, Hildebrandt N (2010) Quantum dot biosensors for ultrasensitive multiplexed diagnostics. Angew Chem Int Ed 49:1396–1401
63. Kalyanasundaram K (1982) Photophysics, photochemistry and solar energy conversion with tris(bipyridyl)ruthenium(II) and its analogues. Coord Chem Rev 46:159–244
64. Zhou M, Robertson GP, Roovers J (2005) Comparative study of ruthenium(II) tris(bipyridine) derivatives for electrochemiluminescence application. Inorg Chem 44:8317–8325
65. Juris A, Balzani V, Barigelletti F, Campagna S, Belser P, von Zelewsky A (1988) Ru(II) polypyridine complexes: Photophysics, photochemistry, eletrochemistry, and chemiluminescence. Coord Chem Rev 84:85–277
66. Bannwarth W, Schmidt D, Stallard RL, Hornung C, Knorr R, Müller F (1988) Bathophenanthroline-ruthenium(II) complexes as nonradioactive labels for oligonucleotides which can be measured by time-resolved fluorescence techniques. Helv Chim Acta 71:2085–2099
67. Kainmüller EK, Ollé EP, Bannwarth W (2005) Synthesis of a new pair of fluorescence resonance energy transfer donor and acceptor dyes and its use in a protease assay. Chem Commun:5459–5461

68. Kainmüller EK, Bannwarth W (2006) A new robust and highly sensitive FRET donor–acceptor pair: synthesis, characterization, and application in a thrombin assay. Helv Chim Acta 89:3056–3070
69. Clima L, Hirtz-Haag C, Kienzler A, Bannwarth W (2007) Application of a highly robust and efficient fluorescence-resonance-energytransfer (FRET) system in DNA. Helv Chim Acta 90:1082–1098
70. Kramer RA, Kainmüller EK, Flehr R, Kumke MU, Bannwarth W (2008) Quenching of the long-lived Ru(II)bathophenanthroline luminescence for the detection of supramolecular interactions. Org Biomol Chem 6:2355–2360
71. Martí AA, Puckett CA, Dyer J, Stevens N, Jockusch S, Ju J, Barton JK, Turro NJ (2007b) Inorganic−organic hybrid luminescent binary probe for DNA detection based on spin-forbidden resonance energy transfer. J Am Chem Soc 129:8680–8681
72. Kolpashchikov DM (2010) Binary probes for nucleic acid analysis. Chem Rev 110:4709–4723
73. Guo J, Ju J, Turro NJ (2011) Fluorescent hybridization probes for nucleic acid detection. Anal Bioanal Chem 402:3115–3125
74. Martí AA, Li X, Jockusch S, Stevens N, Li Z, Raveendra B, Kalachikov S, Morozova I, Russo JJ, Akins DL, Ju J, Turro NJ (2007) Design and characterization of two-dye and three-dye binary fluorescent probes for mRNA detection. Tetrahedron 63:3591–3600
75. http://www.iom-berlin.de/html/eng/produkte/assay/2.04.2.1-assay.htm
76. Yang CJ, Jockusch S, Vicens M, Turro NJ, Tan W (2005) Light-switching excimer probes for rapid protein monitoring in complex biological fluids. Proc Natl Acad Sci U S A 102:17278–17283
77. Chunming W, Cuichen W, Yan C, Yanling S, Weihong T, Chaoyong James Y (2011) Pyrene excimer for DNA sensors. Curr Org Chem 15:465–476
78. Fujimoto K, Shimizu H, Inouye M (2004) Unambiguous detection of target DNAs by excimer−monomer switching molecular beacons. J Org Chem 69:3271–3275
79. Luzi E, Minunni M, Tombelli S, Mascini M (2003) New trends in affinity sensing: aptamers for ligand binding. TrAC Trend Anal Chem 22:810–818
80. Tombelli S, Minunni M, Mascini M (2005) Analytical applications of aptamers. Biosens Bioelectron 20:2424–2434
81. Hermann T, Patel DJ (2000) Adaptive recognition by nucleic acid aptamers. Science 287:820–825
82. Huang J, Zhu Z, Bamrungsap S, Zhu G, You M, He X, Wang K, Tan W (2010) Competition-mediated pyrene-switching aptasensor: probing lysozyme in human serum with a monomer-excimer fluorescence switch. Anal Chem 82:10158–10163
83. Zheng J, Yang R, Shi M, Wu C, Fang X, Li Y, Li J, Tan W (2015) Rationally designed molecular beacons for bioanalytical and biomedical applications. Chem Soc Rev 44:3036–3055
84. Wang K, Tang Z, Yang CJ, Kim Y, Fang X, Li W, Wu Y, Medley CD, Cao Z, Li J, Colon P, Lin H, Tan W (2009) Molecular engineering of DNA: molecular beacons. Angew Chem Int Ed 48:856–870
85. Conlon P, Yang CJ, Wu Y, Chen Y, Martinez K, Kim Y, Stevens N, Marti AA, Jockusch S, Turro NJ, Tan W (2008) Pyrene excimer signaling molecular beacons for probing nucleic acids. J Am Chem Soc 130:336–342
86. Matayoshi E, Wang G, Krafft G, Erickson J (1990) Novel fluorogenic substrates for assaying retroviral proteases by resonance energy transfer. Science 247:954–958
87. Masuko M, Ohtani H, Ebata K, Shimadzu A (1998) Optimization of excimer-forming two-probe nucleic acid hybridization method with pyrene as a fluorophore. Nucleic Acids Res 26:5409–5416
88. Martí AA, Li X, Jockusch S, Li Z, Raveendra B, Kalachikov S, Russo JJ, Morozova I, Puthanveettil SV, Ju J, Turro NJ (2006) Pyrene binary probes for unambiguous detection of mRNA using time-resolved fluorescence spectroscopy. Nucleic Acids Res 34:3161–3168

89. Wu C, Yan L, Wang C, Lin H, Wang C, Chen X, Yang CJ (2010) A general excimer signaling approach for aptamer sensors. Biosens Bioelectron 25:2232–2237
90. Wu Y-X, Zhang X-B, Li J-B, Zhang C-C, Liang H, Mao G-J, Zhou L-Y, Tan W, Yu R-Q (2014) Bispyrene–fluorescein hybrid based fret cassette: a convenient platform toward ratiometric time-resolved probe for bioanalytical applications. Anal Chem 86:10389–10396
91. Nau WM, Wang X (2002) Biomolecular and supramolecular kinetics in the submicrosecond time range: the fluorazophore approach. Chem Phys Chem 3:393–398
92. Sahoo H, Roccatano D, Hennig A, Nau WM (2007b) A 10-Å spectroscopic ruler applied to short polyprolines. J Am Chem Soc 129:9762–9772
93. Roccatano D, Sahoo H, Zacharias M, Nau WM (2007) Temperature dependence of looping rates in a short peptide. J Phys Chem B 111:2639–2646
94. Sahoo H, Nau WM (2007) Phosphorylation-induced conformational changes in short peptides probed by fluorescence resonance energy transfer in the 10 Å domain. Chem Bio Chem 8:567–573
95. Sahoo H, Roccatano D, Zacharias M, Nau WM (2006) Distance distributions of short polypeptides recovered by fluorescence resonance energy transfer in the 10 Å domain. J Am Chem Soc 128:8118–8119
96. Huang F, Nau WM (2003) A conformational flexibility scale for amino acids in peptides. Angew Chem Int Ed 42:2269–2272
97. Hudgins RR, Huang F, Gramlich G, Nau WM (2002) A fluorescence-based method for direct measurement of submicrosecond intramolecular contact formation in biopolymers: an exploratory study with polypeptides. J Am Chem Soc 124:556–564
98. Hennig A, Roth D, Enderle T, Nau WM (2006) Nanosecond time-resolved fluorescence protease assays. Chem Bio Chem 7:733–737
99. Hennig A, Florea M, Roth D, Enderle T, Nau WM (2007) Design of peptide substrates for nanosecond time-resolved fluorescence assays of proteases: 2,3-diazabicyclo[2.2.2]oct-2-ene as a noninvasive fluorophore. Anal Biochem 360:255–265
100. Sahoo H, Hennig A, Florea M, Roth D, Enderle T, Nau WM (2007) Single-label kinase and phosphatase assays for tyrosine phosphorylation using nanosecond time-resolved fluorescence detection. J Am Chem Soc 129:15927–15934
101. Sloniec J, Resch-Genger U, Hennig A (2013) Photophysics and release kinetics of enzyme-activatable optical probes based on Hdimerized fluorophores on self-immolative linkers. J Phys Chem B 117:14336–14344
102. Daugherty DL, Gellman SH (1999) A fluorescence assay for leucine zipper dimerization: avoiding unintended consequences of fluorophore attachment. J Am Chem Soc 121:4325–4333
103. Ahn T, Kim J-S, Choi H-I, Yun C-H (2002) Development of peptide substrates for trypsin based on monomer/excimer fluorescence of pyrene. Anal Biochem 306:247–251
104. Rich RM, Gryczynski I, Fudala R, Borejdo J, Stankowska DL, Krishnamoorthy RR, Raut S, Maliwal BP, Shumilov D, Doan H, Gryczynski Z (2014) Multiple-pulse pumping for enhanced fluorescence detection and molecular imaging in tissue. Methods 66:292–298
105. Raut SL, Fudala R, Rich R, Kokate RA, Chib R, Gryczynski Z, Gryczynski I (2014) Long lived BSA au clusters as a time gated intensity imaging probe. Nanoscale 6:2594–2597
106. Kimball JD, Maliwal B, Raut SL, Doan H, Nurekeyev Z, Gryczynski I, Gryczynski Z (2018) Enhanced DNA detection using a multiple pulse pumping scheme with time-gating (mpptg). Analyst 143:2819–2827

Fluorescence Correlation Spectroscopy: A Highly Sensitive Tool for Probing Intracellular Molecular Dynamics and Disease Diagnosis

8

Bikash Chandra Swain, Anand Kant Das, Janmejaya Rout, Shrutidhara Biswas, and Umakanta Tripathy

Abstract

Fluorescence correlation spectroscopy (FCS) is a very versatile and powerful technique that is based on time-averaging fluctuation analysis of fluorescence intensity generated in a tiny volume and can easily achieve single-molecule sensitivity. Briefly, in FCS, the fluctuations in fluorescence are recorded as a function of time and subsequently statistically analyzed by autocorrelation analysis. FCS helps to determine concentrations, diffusional dynamics, molecular interactions, intersystem crossing, and excited-state reactions of fluorescent species. Recent advances in related methods have pushed the frontiers such that FCS can now be applied to increasingly complex systems such as live cells and organisms to obtain quantitative data at physiological concentrations. In this chapter, we provide a brief overview of the basic principle of FCS, the experimental aspects, including the FCS data analysis, and emerging and efficient varieties of FCS that are currently being used to probe intracellular molecular dynamics and serve as a diagnostic tool for various disease conditions and characterization of biomedical samples. We intend to motivate the reader to appreciate the versatility of FCS as a tool being used in a plethora of disciplines ranging from photophysics to biophysical to biomedical sciences and across in vitro and in vivo systems.

B. C. Swain · J. Rout · U. Tripathy (✉)
Department of Physics, Indian Institute of Technology (Indian School of Mines), Dhanbad, Jharkhand, India
e-mail: utripathy@iitism.ac.in

A. K. Das · S. Biswas
Physics Program, New York University Abu Dhabi, Saadiyat Island, Abu Dhabi, UAE

© The Author(s), under exclusive license to Springer Nature Singapore Pte Ltd. 2022

H. Sahoo (ed.), *Optical Spectroscopic and Microscopic Techniques*,
https://doi.org/10.1007/978-981-16-4550-1_8

8.1 Introduction

The Fluorescence correlation spectroscopy (FCS) technique serves as a sensitive optical probe to determine the molecular dynamics and interactions kinetics [1]. In the early 1970s, Douglas Magde, Elliot Elson, and Watt W. Webb at Cornell University propounded FCS as a tool to measure diffusibility and chemical kinetics [2]. They developed this compelling method to probe the interaction of ethidium bromide with DNA double helix. Around that time, Mans Ehrenberg, Rudolf Rigler, and Manfield Eigen further developed the theory of FCS and built an instrument to record fluorescence fluctuation data [3, 4]. In brief, FCS deals with the correlative analysis of fluctuation of fluorescence intensity due to the Brownian motion of the particles. FCS can access all physical parameters that give rise to fluctuation in the fluorescence intensity.

In contrast to the long correlation time and large excitation volume that lead to damage of the sample during the early phase of FCS development, subsequent advancements increased sensitivity and decreased duration during FCS measurements leading to minimal photodestruction and phototoxicity. Since then, FCS has been widely used to measure properties such as concentrations, rotational diffusion coefficients, rate constants, photodynamics, intersystem crossing, and excited-state singlet-triplet dynamics [5–8]. Over the last several decades, serious advancements in computational methodologies and FCS-related techniques have expanded the application of FCS in diverse systems, ranging from aqueous solutions to cells to whole organisms [3, 9, 10].

The chapter is divided into four sections. The first section, "General Principles, Instrumentation, and Evaluation of FCS Data," presents a summary of the theoretical description of the FCS, the instrumental set-up, and various modalities used to analyze the FCS data. This is followed by "Developments in FCS and Related Techniques," where we discuss in detail the concepts behind several varieties and add-ons to the FCS technique and its utility. The third section deals with the application part where we highlight the "Intracellular Molecular Dynamics Measurements with FCS", which is followed by the emerging application of FCS in biomedical research and disease diagnosis described in Sect. 8.4 titled, "FCS as a Diagnostic Tool for Disease Conditions." The potential of FCS with a single-molecule detection limit can be applied to obtain quantitative parameters related to various aspects of cellular dynamics and across multiple disciplines.

8.2 General Principles, Instrumentation, and Evaluation of FCS Data

Introduced in the early 1970s, FCS is a time-averaging analysis of fluorescence fluctuation of the minute molecular ensemble, thereby combining high sensitivity and statistical confidence [11]. In this method, the biophysical parameters are extracted from the fluctuation in fluorescence intensity caused by fluorescently labeled molecules entering and leaving the probe region, a subfemtoliter detection volume. The detection volume is generated using a laser as an excitation source and a high-numerical-aperture objective. For detection of the fluorescence fluctuation, a highly sensitive detector called avalanche photodiodes (APD) is used, which collects the stream of single-photon arrival time. The parameters associated with the process responsible for fluctuations can be obtained from the analysis of the FCS data. But before that, the data need to undergo several rounds of processing.

The fluctuation in fluorescence intensity ($F(t)$) can be expressed as

$$\delta F(t) = F(t) - \langle F(t) \rangle \tag{8.1}$$

Where $F(t)$ is the fluorescence intensity at time t, and $\langle F(t) \rangle$ denotes the average fluorescence intensity over time t. $\langle F(t) \rangle$ can be expressed as

$$\langle F(t) \rangle = \frac{1}{T} \int_{1}^{T} F(t) \mathrm{d}t \tag{8.2}$$

The information on the parameters contributing to the fluorescence fluctuation is extracted from the FCS experiment by generating an autocorrelation curve that indicates the probability that the emanated signal at different times arises from the same molecule. Autocorrelation measures the self-similarity of the fluctuation with itself with a time lag τ, known as lag time. The mathematical expression for the autocorrelation function can be given as

$$G(\tau) = \frac{\langle \delta F(t) \rangle \langle \delta F(t+\tau) \rangle}{\langle F(t) \rangle^2} \tag{8.3}$$

Equation (8.3) is known as the generalized autocorrelation function where $G(\tau)$ is the autocorrelation function. The self-similarity between $\langle \delta F(t) \rangle$ and $\langle \delta F(t + \tau) \rangle$ is more at less lag time, and hence at zero lag time, the amplitude of autocorrelation function is highest. After a long lag time, the self-similarity of the signal becomes highly reduced.

Different parameters that contribute to the fluorescence intensity fluctuation can be extracted from the generalized autocorrelation function. For example, assuming the excitation profile to be a three-dimensional (3D) Gaussian illumination profile and the change in fluorescence intensity is only due to diffusion of fluorophores in and out of the observation volume, Eq. (8.3) can be derived as

$$G(\tau) = \frac{1}{N} \left(1 + \frac{\tau}{\tau_D}\right)^{-1} \left(1 + \frac{\tau}{K^2\, \tau_D}\right)^{-1/2} \tag{8.4}$$

Where N is the average number of fluorophores diffuses through the observation volume, and K ($= \omega_z/\omega_{xy}$) is the ratio of radial to the axial radius of the three-dimensional Gaussian volume, also known as structure factor. The effective observation volume can be expressed as

$$V_{\text{eff}} = \pi^{3/2} {\omega_{xy}}^2 \omega_z \tag{8.5}$$

The diffusion time τ_{D} is related to the size of the observation volume as

$$\tau_D = \frac{{\omega_{xy}}^2}{4D} \tag{8.6}$$

Where D is the diffusion coefficient. The hydrodynamic radius (R_{h}) of the sample can be calculated using the Stoke–Einstein relationship given by

$$D = \frac{k_{\text{B}}T}{6\pi\eta R_{\text{h}}} \tag{8.7}$$

Where k_{B} is the Boltzmann constant, and η is the coefficient of viscosity of the solution.

The molecular size obtained from the hydrodynamic radii is very useful as it allows the characterization of the aggregation state of the molecule as well as the binding between small fluorescent ligands with large complexes. The schematic diagram of a typical confocal FCS set-up is shown in Fig. 8.1. It consists of excitation, emission, collection, and data acquisition parts. The excitation part includes first the excitation source, which is usually a laser of a particular wavelength. The beam of the laser light is expanded by using a telescope made of two lenses to overfill the back aperture of the objective. A dichroic mirror reflects the expanded beam to the back aperture of the objective lens. The objective focuses the beam to a small observation volume, where the sample in solution is placed. The fluorescence signal is collected in the epi-direction of the objective lens and passed through the dichroic mirror and subsequently through an emission filter in order to remove the excitation light if any. Then the beam is fed to an achromatic lens to focus on the pinhole. The signal is then fed to a single-photon counting module. The output of the detector is connected to the correlator card for the autocorrelation purpose, and the card is connected to the computer. In some set-ups, the signal is divided into two parts using a beam splitter, and the two signals are fed to two different detectors. The signals from the two detectors were cross-correlated to get the final data. This technique helps to reduce the after pulsing artifact.

The quantitative FCS measurements rely heavily on the fidelity of the calibration of the size of the detection volume, which in turn is influenced by several factors. The data collected are usually in terms of the number of photon counts in a given time interval. The detected photons are then used to construct the fluctuation intensity trace, as shown in Fig. 8.1. The autocorrelation curves are subsequently calculated from the fluctuation traces, as shown in Fig. 8.1. The interpretation of the FCS data involves appropriate theoretical modeling of the fluorescence fluctuation by fitting the autocorrelation function to a solution based on an appropriate diffusion model using nonlinear curve fitting. The diffusion coefficient D is the molecular property and is independent of the instrumental parameters. For this reason, FCS is calibrated with fluorophores with known diffusion coefficients. Since the FCS

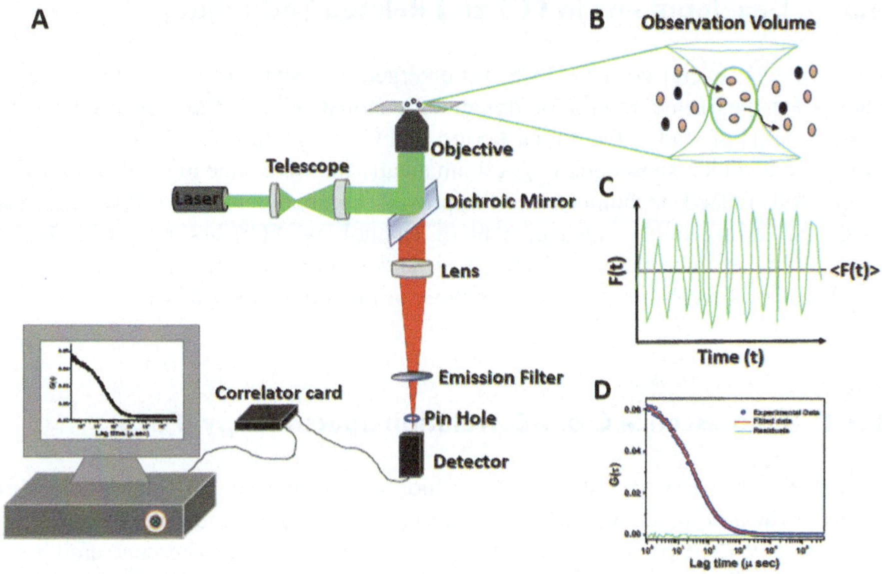

Fig. 8.1 Schematic presentation of the FCS set-up. (**a**) The fluorescence signal emitted by the fluorescing species (fluorescently labeled molecules/dyes) diffuses through the confocal volume element generated by a tightly focused laser spot and is transmitted to the detector. The number of fluorescent pulses sourced from the detected photons, which are recorded through a specific time interval, corresponds to the light intensity. Therefore, the fluctuation in fluorescence intensity over time is recorded. (**b**) The observation volume of the order of femtoliters. Fluorescent molecules diffuse in and out of this volume due to the Brownian motion. The green circle depicts the volume under observation at any given time point of the experiment. (**c**) The fluctuation in detected fluorescence intensity as a function of the time when molecules diffuse in the confocal volume is shown. (**d**) The electrical signal is collected at regular time intervals and is transferred to the signal correlation unit, and thus, the corresponding normalized autocorrelation curve $G(\tau)$ is calculated. The correlation function is plotted against the logarithm of delay time τ. The experimentally obtained autocorrelation curves are fitted with correlation function models to extract the parameters. The residual obtained from the fitting is shown in green

measurements involve the relative diffusion coefficient of two species, the result is useful even if the absolute diffusion coefficient values are inaccurate.

The dependence of the autocorrelation function on the diffusion coefficient has resulted in the widespread application of FCS. Deriving parameters from the FCS fit in the solution phase is relatively straightforward, but the complexity of the cellular environment poses serious challenges. There have been considerable improvements in FCS data analysis and the development of related techniques to expand the utility of FCS in cells and in vivo. A few of these techniques are described in the next section.

8.3 Developments in FCS and Related Techniques

Over the last several decades, FCS has emerged as a standard tool and method of choice for measuring molecular dynamics parameters and, therefore, utilized in several fields of study. The initial period of FCS was fraught with poor signal to noise ratios, which subsequently got dramatically improved due to the development of several related techniques and improved computational and data analysis methods. Owing to the advancement of technology, FCS has now widespread applications in microfluidic, biochemical, and cellular investigations. Out of the growing list of techniques, we discuss three of them in some detail.

8.3.1 Fluorescence Cross-Correlation Spectroscopy (FCCS)

While FCS measures the fluctuation in fluorescence intensity of single molecules diffusing in a confined volume under observation, fluorescence cross-correlation spectroscopy (FCCS) deals with the temporal analysis of fluorescent amplitude fluctuations of two distinct and differently labeled molecules under observation. If the fluctuations occur simultaneously in the two channels, then the inference drawn is that the two proteins are interacting and moving together in a complex. For a multicomponent system, FCCS is a preferred method of choice to measure molecular dynamics. The signals emanating from both the channels are cross-correlated and can measure rotational diffusion and association–dissociation kinetics [12, 13]. There are multiple variants both in terms of excitation and detection sources available these days for effective FCCS measurement and analysis. Some of the choices available for the laser sources could be the (a) dual-color FCCS with two visible lasers of distinct wavelengths. The schematics of the set-up are shown in Fig. 8.2. The sample containing two distinctly labeled fluorophores is subjected to dual laser excitation, each individually excited by the lasers. Then the fluorescence emission is separated with the help of dichroic mirrors and filters into two different detection channels. If the concentration of the reactants is constant, then the amplitude of the cross-correlation is directly proportional to the complex formed by the two components. The FCCS method is used effectively to quantify mRNA expression levels [14], monitor enzyme kinetics [15, 16], detect PCR complexes [17], chromatin dynamics [18], etc. (b) multiline laser FCS, (c) two-photon excitation FCCS with pulsed IR laser, (d) pulsed interleaved excitation FCCS, and (e) multiline laser FCCS with prism for selection. On the detection mode, there are several options, such as the dichroic with APD, PMT point detection, CCD detectors, or CMOS arrays, reviewed in [19].

For the spatial cross-correlation in two-color channels, the mathematical description is given below.

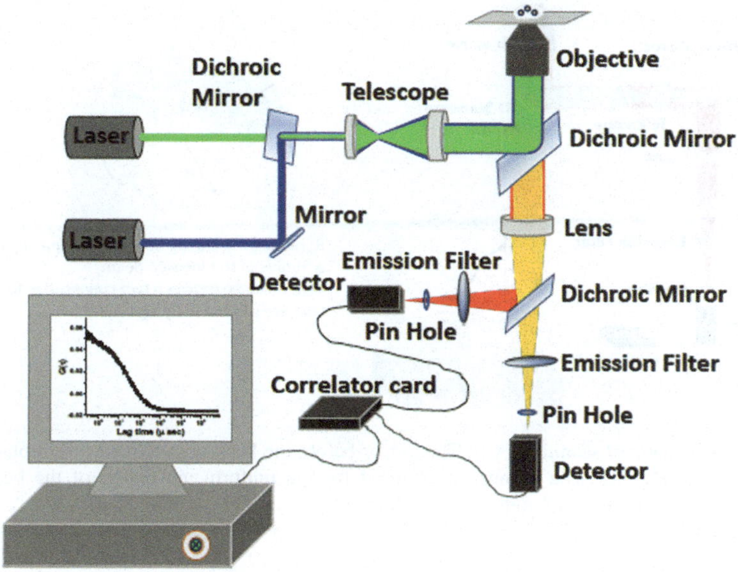

Fig. 8.2 Schematic of the dual-color fluorescence cross-correlation spectroscopy (FCCS) having two visible lasers of distinct wavelengths. The excitation and detection modes are depicted in the schematic

$$G_{cc}(\xi,\psi)=\left\langle\frac{I_1(x,y)I_2(x+\xi,y+\psi)\rangle}{\langle I_1(x,y)\rangle\langle I_2(x,y)\rangle}-1 \qquad (8.8)$$

Where indexes 1 and 2 indicate the two channels. Interestingly, the cross-correlation function is not affected by the detector shot noise because the noise in both the detector channels is not correlated. FCCS is a potent spectroscopic tool to probe molecular interactions with the limitation that it cannot detect higher-order oligomerization.

8.3.2 Scanning Fluorescence Correlation Spectroscopy (sFCS)

Scanning FCS (sFCS) is a common term used for the FCS type where the measurement volume moves relative to the sample or changes shape. The method is highly robust with short measurement times and a tiny probed area. In sFCS, the segments of the scanned path are continuously illuminated by the rotating beam generated due to the 2D scanning mirrors, as shown in Fig. 8.3. The calibration of the radius obtained by the scan is highly accurate and simple and helps in more robust measurements of the diffusion coefficients. The scanning mirror is placed in front of the objective without any relay lens, such that the displacement of the beam in the back focal aperture of the objective is minimal.

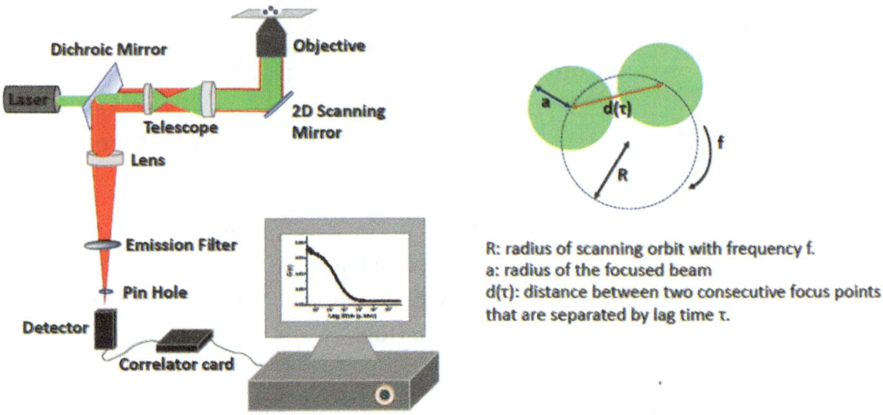

Fig. 8.3 Schematic of scanning FCS. The mirror before the back-objective entry is replaced by a two-axis piezo scanner. The scanner is required for the uniform movement of the beam with frequency "*f*," in a circle of radius "*R*," which in turn is comparable to the size "*a*" of the focused beam

sFCS can be efficiently used for robust measurements in complex cellular environments where prior knowledge of the measurement volume is not required, and the extra slow fluctuation due to sample movement can be corrected [20]. Scanning FCS substantially improves the accuracy of the measured signal, provides minimal phototoxicity and photobleaching effects, and can measure diffusion coefficients over a broad temporal regime [21–23]. The autocorrelation function of the sFCS is given as

$$S(\tau) = \exp\left(\frac{4A^2(1 - \cos(\omega\tau))}{\left(1 + \frac{4D_\tau}{w_0^2}\right)w_0^2}\right) \tag{8.9}$$

Where A is the radius of the orbit, and ω is the angular frequency of the circular scanning. The sFCS method, when coupled with photon counting histogram analysis, can serve as a powerful method to follow intracellular events such as probing yeast ribosomal structures or lipid domains in live cells [24]. sFCS can be used to be amenable to diffusion measurements in 3D using two-photon excitation. Taken together, sFCS can be used to probe quantitative dynamical interactions of molecules in membranes and complex environments.

8.3.3 Stimulated-Emission Depletion Microscopy–Fluorescence Correlation Spectroscopy (STED-FCS)

The far-field super-resolution stimulated-emission depletion microscopy (STED) is a suitable approach for subdiffraction scale imaging where an optically driven

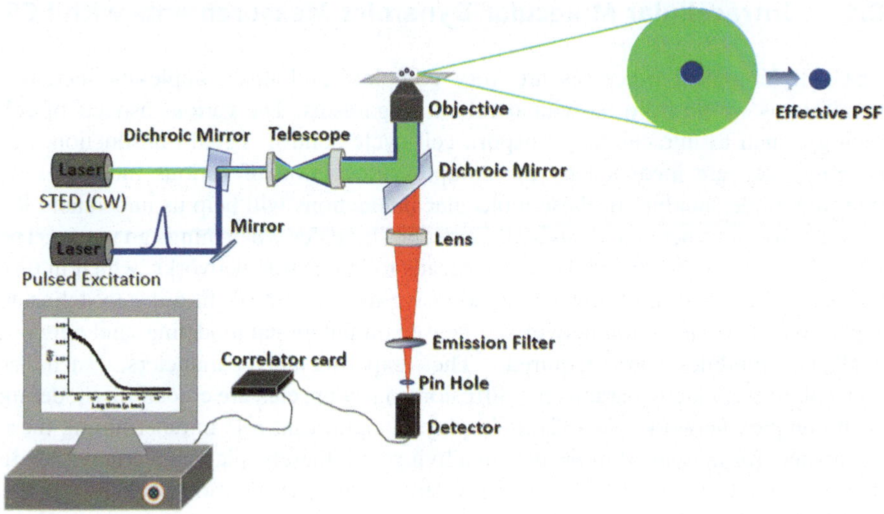

Fig. 8.4 Schematic representation of gated STED-FCS. The pulsed excitation beam and the continuous wave (CW) STED beams are superimposed using dichroic mirrors. The gated laser effective point spread function is also shown

reversible transfer takes place between states of different emission properties, thereby allowing modulation of fluorescence emission in space and time [25, 26]. STED is compatible with FCS and can be combined to provide the dual advantage of diffraction-unlimited spatial resolution from STED and fast molecular dynamics from FCS [27, 28]. The STED laser operating in continuous-wave (CW) mode generates a doughnut-shaped beam such that the central intensity is zero but the high intensity in the focal periphery, while the excitation laser is focused to a diffraction-limited spot as depicted in Fig. 8.4. The collection is done by a confocal pinhole imaged onto the single-photon counting avalanche photodiode.

The regular diffraction-limited focal spot arising from the excitation laser is used for fluorescence excitation, while the STED beam stimulates the excited molecules down to their ground state. One can suitably gate the STED-FCS by using a combination of the pulsed excitation beam and CW-STED, which ultimately leads to narrowing of the resulting observation spot owing to the reduction in peripheral fluorescence emission. The time-gated detection method can improve single-molecule detection and the effective observation spot. STED-FCS has been used to probe cell membrane architecture at the nanoscale [29] and to investigate lipid membrane dynamics [27].

8.4 Intracellular Molecular Dynamics Measurements with FCS

The cell biological processes are very complex, and this complexity increases manifold as we go from individual cells to organisms. The various aspects of cell biology, such as intracellular transport, cell-cycle control, signal transduction, cell motility, etc., are mediated by the complex interaction of various molecules. A thorough understanding of these molecular interactions will help us understand the cellular processes better and devise therapeutic strategies when things go wrong. The interaction of various molecular species leads to interaction networks, which in turn are part of more complex and more extensive networks [10]. In order to delineate such molecular interaction networks, intense computational modeling, and network analysis, modules are required. The experimental parameters, such as concentrations, kinetic parameters, diffusion constants, etc., are crucial for modeling such complex networks [30–32]. FCS plays a significant role in determining these parameters for individual molecules in a living cell, thereby playing a crucial role in delineating the complex networks prevalent in the living world. The two major challenges are how to detect the single molecule in a complex environment, and how to enhance the optical signal emitted per molecule for meaningful data analysis. Fluctuations in fluorescence occur when molecules diffuse in and out in an observation volume or due to conformational changes that induce a change in fluorescence brightness.

Although FCS is a highly versatile technique, the major drawback in its applications to more complex systems such as cells arises due to a lack of interpreting models that result in misleading data when measured in complex environments [33–35]. Major steps have been taken to overcome the lacunae and artifacts, leading to various extensions of the FCS method. In the section above, we have discussed in detail the technique of fluorescence cross-correlation spectroscopy (FCCS) that has been extensively applied to study molecular dynamics and interactions. There have been various up-gradation to this method, for example, the cross-talk-free dual-color FCCS technique that has been developed to measure enzyme activity during high-throughput screening with no spectral leakage in wrong color channels [36]. The information on fluorescence fluctuation on two different spectrally distinct color channels is collected and processed with a single-photon avalanche diode (SPAD) that prevents cross-emission, cross-excitation, and fluorescence-resonance energy transfer (FRET). Significant developments in areas of optics and detectors have reduced phototoxicity and allowed the usage of lower excitation irradiances [37]. An excellent step-by-step guide to the application of FCS in cells has been provided by Schwille and coworkers, which also includes methods for optimizing measuring conditions, reducing artifacts, and deriving meaningful parameters in the complex cell environment [6]. Notable additions to this by the same group have been the two-photon methods for FCS and FCCS [38, 39].

Interestingly, combining FCS with time-correlated single-photon counting (TCSPC), i.e., fluorescence lifetime correlation spectroscopy (FLCS), has led to a significant reduction in bleed-through and spectral cross-talk with the overall advantage of using one excitation source [40]. Another useful development in probing intracellular molecular dynamics is the combination of total internal reflection fluorescence (TIRF) with FCS, known as the TIR-FCS. The environment inside a cell is dense and chaotic. The minimal number of molecules required for effective FCS analysis remains a considerable challenge. The strategy then would be to limit the observation volume to the minimum. This is easily achieved by TIR illumination as the evanescent wave excites a very small volume of the exposed cell such that measurement in fluorescence fluctuation can easily be carried out in the evanescent field using TIR-FCS [41, 42]. Additionally, another useful combination is the use of two-photon FCS with one-photon single-particle tracking (SPT) [43]. This facilitates long-term tracking of the fluorescent species in the complex cellular environment. To track enhanced green fluorescent protein (eGFP) in the cytoplasm of *Escherichia Coli*, a dielectric or metallic mirror was used and set at the focal point of the excitation beam [44]. This significantly improves the fluorescence rate emitted per molecule as well as reduces the confocal analysis volume.

A novel addition to FCS has been the continuous fluorescence microphotolysis (CFM) using the 4Pi microscopy to significantly improve the spatial resolution [45] in living cells. 4Pi microscopy uses two opposite microscope objectives with high numerical apertures. This methodology takes the complicated 4Pi point spread function into account and utilizes the computational framework to improve the axial resolution of approximately 100 and 220 nm in the focal plane. Another exciting addition to FCS has been the near-field scanning optical microscopy (NSOM) [46, 47]. This method has led to the confinement of the detected volume to subdiffraction limited dimensions, thereby substantially improving the sensitivity of the system. This method has been successfully implemented to study proteins and their diffusion behavior through individual nuclear pore complexes of the nuclear envelope [46]. There have been substantial improvements in computational techniques in parallel to the photonic methods in order to facilitate data analysis in a complex cellular environment [9]. However, the computational strategies have not been covered in this book chapter.

In order to provide a flavor of the applicability of FCS and related techniques in tracking intracellular molecular dynamics and events on the plasma membrane, Table 8.1 has been provided.

FCS has played a crucial role in quantitative measurements in live cells and in vivo. Furthermore, new FCS modalities are continually being developed for improved data collection and analysis in complex systems. FCS has emerged as a standard tool for quantitative cell measurements. The next frontier is utilizing FCS in biomedical research, drug discovery, and disease diagnosis, which is described in the next section.

8.5 FCS as a Diagnostic Tool for Disease Conditions

The application of FCS and related techniques in disease diagnosis, drug discovery, and biomedical research is to be extremely promising [65]. FCS provides a unique advantage in terms of high sensitivity and efficiency, which is an important criterion for any effective diagnostic method. Apart from this, FCS has other advantages that

Table 8.1 Application of FCS in intracellular molecular dynamics studies

Active transport of GFP through the plastid tubule	[48]
Protein exchange through plant plastids	[49]
Molecular dynamics in live cells with two-photon excitation	[50]
Proinsulin C-peptide binding to the human cell membrane	[51]
Intranuclear diffusion of oligonucleotides in live cells	[52]
Diffusion of fluorescent probes in living cell nuclei	[53]
Diffusion of green fluorescent protein in the mitochondrial matrix	[54]
DNA mobility in cytoplasm and nucleus	[55]
Endocytic pathway of cholera toxin	[56]
Analysis of gene expression	[57]
Probing functional ribosomal complexes	[58]
Clustering of 5HT3 receptor in the plasma membrane	[59]
Tubulin transport in giant squid axons	[60]
Stability of drug-induced tubulin ring	[61]
Submicron cell membrane organization	[62]
Organization of lipid anchors in membrane	[63]
Association of unfolded protein with an ER stress sensor	[64]

make it suitable for a clinical laboratory, such as the requirement of a very small volume of the sample, single-molecule sensitivity, and no additional sample preparation or physical separation step. FCS reports the diffusion time, concentration, molecular association, and aggregation state serving as an important diagnostic tool for sparse molecules. FCS and its related techniques can be suitably used to track molecular markers of various diseases with high sensitivity and fidelity. There is a widespread application of FCS for the diagnosis of various diseases, which is indicated by the schematic in Fig. 8.5. A few of the examples illustrating the usage of FCS in various aspects of disease diagnosis, pharmacology, and biomedical research are listed below.

1. *Investigating GPCR pharmacology in membrane microdomains*: G protein-coupled receptors (GPCRs) are the largest family of proteins in the human genome with over 800 unique members and serve as important drug targets (nearly 40% of marketed drugs) for various disease conditions [66, 67]. GPCRs are critical players in signal transduction pathways, where extracellular signals initiate intracellular signaling cascades [68]. FCS and related methods such as photon counting histogram analysis have been extensively used to probe the properties and pharmacology of GPCRs present in membrane microdomains. For example, bradykinin B2 receptor, μ opioid receptor, $G\alpha_q$, tyrosine kinase EGF receptor, and $5HT_{2c}$ have been extensively studied using FCS [69–72].
2. *Detection of pathological prion protein aggregates*: Prion diseases are rapidly progressive, debilitating neurodegenerative diseases caused by the aggregation of misfolded infectious prion proteins [73]. The hallmark histological features of

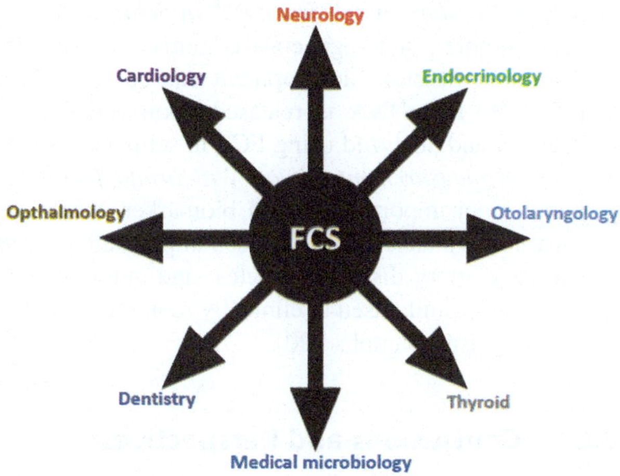

Fig. 8.5 FCS is an excellent diagnostic tool with high sensitivity and efficiency for various medical conditions

these diseases include deposits of aggregated prion proteins, spongiform encephalopathy, and neuronal cell death. The pathological prion protein (PrP^{sc}) is present in cerebrospinal fluid and has been successfully detected down to a concentration of 2 pM with confocal dual-color FCS by specific antibody probes tagged with fluorescent dyes [74] or by using FCS/FCCS [75].

3. *Detection of contactin-2 and amyloid-β in Alzheimer's disease patients*: Alzheimer's disease (AD) is the most common form of dementia that is characterized by abnormal protein aggregation of amyloid-β and tau in the brain, leading to neuronal loss and brain damage [76]. FCS has been used to probe protein aggregation of amyloid-β [77] and contactin, a biomarker of AD, in the CSF of AD patients [78].
4. *Monitoring blood nanocarriers in human blood*: Nanocarrier-based drug delivery system is a promising therapeutic strategy for various disease conditions. Nanometer-sized carriers known as nanocarriers protect the cargo (drug) from the environment during transport before it reaches its site of action [79]. FCS has been used to monitor the size, loading efficiency, and stability of drug nanocarriers in human blood [80], which is indeed a significant leap in quantitative aspects of nanocarrier-based therapeutics.
5. *Diffusion and reaction of bacteriophages inside biofilms*: Viruses that infect bacteria are called bacteriophages such that the phase development cycle, the infection pathway, and genetics have a major impact on basic and applied biology [81]. FCS has been used to monitor lactococcal c2 diffusion and reaction in bacterial biofilms [82].
6. *Polyglutamine protein oligomers in cells*: The aggregation of polyglutamine (poly(Q)) leads to a number of neurological disorders, including Alzheimer's disease, Parkinson's disease, and poly(Q) diseases [83, 84], collectively known as conformational diseases. FCS has been successfully utilized to detect poly (Q) oligomer formation in cells [82]

7. *Size distribution of cell-free DNA in Schizophrenia patients*: Schizophrenia is a very complex, heterogeneous cognitive disorder that seems to originate from disruption of brain development caused by genetic or environmental factors [85]. Cell-free DNA is released upon cell death and has been successfully detected and analyzed using FCS in schizophrenic patients [86].
8. *Clinical measurement of von Willebrand factor*: The protein von Willebrand factor is an important clinical biomarker, essentially functions as a tether and activator of platelets, and has been implicated in a variety of disease states such as coronary artery disease, diabetes, and autoimmune diseases [87–89]. FCS has been successfully used to clinically monitor the von Willebrand factor directly on human plasma samples [90].

8.6 Conclusions and Perspectives

From the time FCS was first devised in the 1970s to till date, it has been widely used by biophysicists and biochemists, including us, for quantitative analysis of mobility, local concentrations, interaction dynamics, and behavior of fluorescently labeled molecules in solution [2, 5, 11, 91–98]. This useful spectroscopic method is very dynamic and constantly being used with novel variations for wide applications. Over the last decade or so, the development of FCS-related modalities has opened a new era of biological research. With the continuous development of new models, computational power and advancement in simulations capabilities have led to the extraction of more useful parameters from the FCS data. The FCS and its related modalities are now being increasingly used by cell biologists and biomedical researchers. The utility of FCS in addressing the broad spectrum of biological questions is immensely promising. The "proliferative age" of the FCS-based techniques has now provided means to explore the previously uncharted territories in biomedicine, cell biology, and in vivo applications. The need to explore these frontiers will push the limits of the techniques associated with FCS for a better understanding of molecular behavior at various levels.

References

1. Chen H, Farkas ER, Webb WW (2008) Chapter 1: In vivo applications of fluorescence correlation spectroscopy. Methods Cell Biol 89:3–35
2. Magde D, Elson E, Webb WW (1972) Thermodynamic fluctuations in a reacting system measurement by fluorescence correlation spectroscopy. Phys Rev Lett 29(11):705
3. Vukojevic V, Pramanik A, Yakovleva T, Rigler R, Terenius L, Bakalkin G (2005) Study of molecular events in cells by fluorescence correlation spectroscopy. Cell Mol Life Sci 62(5): 535–550
4. Ehrenberg M, Rigler R (1976) Fluorescence correlation spectroscopy applied to rotational diffusion of macromolecules. Q Rev Biophys 9(1):69–81
5. Ries J, Schwille P (2012) Fluorescence correlation spectroscopy. BioEssays 34(5):361–368
6. Kim SA, Heinze KG, Schwille P (2007) Fluorescence correlation spectroscopy in living cells. Nat Methods 4(11):963–973

7. Digman MA, Gratton E (2011) Lessons in fluctuation correlation spectroscopy. Annu Rev Phys Chem 62:645–668
8. Dertinger T, Pacheco V, von der Hocht I, Hartmann R, Gregor I, Enderlein J (2007) Two-focus fluorescence correlation spectroscopy: a new tool for accurate and absolute diffusion measurements. ChemPhysChem 8(3):433–443
9. Machan R, Wohland T (2014) Recent applications of fluorescence correlation spectroscopy in live systems. FEBS Lett 588(19):3571–3584
10. Levin MK, Carson JH (2004) Fluorescence correlation spectroscopy and quantitative cell biology. Differentiation 72(1):1–10
11. Magde D, Elson EL, Webb WW (1974) Fluorescence correlation spectroscopy. II. An experimental realization. Biopolymers 13(1):29–61
12. Rigler R, Földes-Papp Z, Meyer-Almes F-J, Sammet C, Völcker M, Schnetz A (1998) Fluorescence cross-correlation: a new concept for polymerase chain reaction. J Biotechnol 63(2): 97–109
13. Rika J, Binkert T (1989) Direct measurement of a distinct correlation function by fluorescence cross correlation. Phys Rev A Gen Phys 39(5):2646–2652
14. Camacho A, Korn K, Damond M, Cajot JF, Litborn E, Liao B, Thyberg P, Winter H, Honegger A, Gardellin P, Rigler R (2004) Direct quantification of mRNA expression levels using single molecule detection. J Biotechnol 107(2):107–114
15. Rarbach M, Kettling U, Koltermann A, Eigen M (2001) Dual-color fluorescence cross-correlation spectroscopy for monitoring the kinetics of enzyme-catalyzed reactions. Methods 24(2):104–116
16. Kettling U, Koltermann A, Schwille P, Eigen M (1998) Real-time enzyme kinetics monitored by dual-color fluorescence cross-correlation spectroscopy. Proc Natl Acad Sci U S A 95(4): 1416–1420
17. Rigler R, Foldes-Papp Z, Meyer-Almes FJ, Sammet C, Volcker M, Schnetz A (1998) Fluorescence cross-correlation: a new concept for polymerase chain reaction. J Biotechnol 63(2): 97–109
18. Strohner R, Wachsmuth M, Dachauer K, Mazurkiewicz J, Hochstatter J, Rippe K, Langst G (2005) A 'loop recapture' mechanism for ACF-dependent nucleosome remodeling. Nat Struct Mol Biol 12(8):683–690
19. Hwang LC, Wohland T (2007) Recent advances in fluorescence cross-correlation spectroscopy. Cell Biochem Biophys 49(1):1–13
20. Petrasek Z, Schwille P (2008) Precise measurement of diffusion coefficients using scanning fluorescence correlation spectroscopy. Biophys J 94(4):1437–1448
21. Berland KM, So PT, Chen Y, Mantulin WW, Gratton E (1996) Scanning two-photon fluctuation correlation spectroscopy: particle counting measurements for detection of molecular aggregation. Biophys J 71(1):410–420
22. Petersen NO (1986) Scanning fluorescence correlation spectroscopy. I. Theory and simulation of aggregation measurements. Biophys J 49(4):809–815
23. Petrasek Z, Derenko S, Schwille P (2011) Circular scanning fluorescence correlation spectroscopy on membranes. Opt Express 19(25):25006–25021
24. Gunther G, Jameson DM, Aguilar J, Sanchez SA (2018) Scanning fluorescence correlation spectroscopy comes full circle. Methods 140–141:52–61
25. Hell SW (2003) Toward fluorescence nanoscopy. Nat Biotechnol 21(11):1347–1355
26. Hell SW (2009) Microscopy and its focal switch. Nat Methods 6(1):24–32
27. Clausen MP, Sezgin E, Bernardino de la Serna J, Waithe D, Lagerholm BC, Eggeling C (2015) A straightforward approach for gated STED-FCS to investigate lipid membrane dynamics. Methods 88:67–75
28. Kastrup L, Blom H, Eggeling C, Hell SW (2005) Fluorescence fluctuation spectroscopy in subdiffraction focal volumes. Phys Rev Lett 94(17):178104

29. Eggeling C, Ringemann C, Medda R, Schwarzmann G, Sandhoff K, Polyakova S, Belov VN, Hein B, von Middendorff C, Schonle A, Hell SW (2009) Direct observation of the nanoscale dynamics of membrane lipids in a living cell. Nature 457(7233):1159–1162
30. Chen Y, Muller JD, Berland KM, Gratton E (1999) Fluorescence fluctuation spectroscopy. Methods 19(2):234–252
31. Hess ST, Huang S, Heikal AA, Webb WW (2002) Biological and chemical applications of fluorescence correlation spectroscopy: a review. Biochemistry 41(3):697–705
32. Bacia K, Schwille P (2003) A dynamic view of cellular processes by in vivo fluorescence auto- and cross-correlation spectroscopy. Methods 29(1):74–85
33. Enderlein J, Gregor I, Patra D, Fitter J (2004) Art and artefacts of fluorescence correlation spectroscopy. Curr Pharm Biotechnol 5(2):155–161
34. Marrocco M (2004) Fluorescence correlation spectroscopy: incorporation of probe volume effects into the three-dimensional Gaussian approximation. Appl Opt 43(27):5251–5262
35. Nishimura G, Kinjo M (2004) Systematic error in fluorescence correlation measurements identified by a simple saturation model of fluorescence. Anal Chem 76(7):1963–1970
36. Lee W, Lee YI, Lee J, Davis LM, Deininger P, Soper SA (2010) Cross-talk-free dual-color fluorescence cross-correlation spectroscopy for the study of enzyme activity. Anal Chem 82(4): 1401–1410
37. Tian Y, Martinez MM, Pappas D (2011) Fluorescence correlation spectroscopy: a review of biochemical and microfluidic applications. Appl Spectrosc 65(4):115–124
38. Mutze J, Petrasek Z, Schwille P (2007) Independence of maximum single molecule fluorescence count rate on the temporal and spectral laser pulse width in two-photon FCS. J Fluoresc 17(6):805–810
39. Kim SA, Heinze KG, Bacia K, Waxham MN, Schwille P (2005) Two-photon cross-correlation analysis of intracellular reactions with variable stoichiometry. Biophys J 88(6):4319–4336
40. Chen J, Irudayaraj J (2010) Fluorescence lifetime cross correlation spectroscopy resolves EGFR and antagonist interaction in live cells. Anal Chem 82(15):6415–6421
41. Leutenegger M, Gosch M, Perentes A, Hoffmann P, Martin OJ, Lasser T (2006) Confining the sampling volume for Fluorescence Correlation Spectroscopy using a sub-wavelength sized aperture. Opt Express 14(2):956–969
42. Leutenegger M, Blom H, Widengren J, Eggeling C, Gosch M, Leitgeb RA, Lasser T (2006) Dual-color total internal reflection fluorescence cross-correlation spectroscopy. J Biomed Opt 11(4):040502
43. Hui YY, Zhang B, Chang YC, Chang CC, Chang HC, Hsu JH, Chang K, Chang FH (2010) Two-photon fluorescence correlation spectroscopy of lipid-encapsulated fluorescent nanodiamonds in living cells. Opt Express 18(6):5896–5905
44. Etienne E, Lenne PF, Sturgis JN, Rigneault H (2006) Confined diffusion in tubular structures analyzed by fluorescence correlation spectroscopy on a mirror. Appl Opt 45(18):4497–4507
45. Arkhipov A, Huve J, Kahms M, Peters R, Schulten K (2007) Continuous fluorescence microphotolysis and correlation spectroscopy using 4Pi microscopy. Biophys J 93(11): 4006–4017
46. Herrmann M, Neuberth N, Wissler J, Perez J, Gradl D, Naber A (2009) Near-field optical study of protein transport kinetics at a single nuclear pore. Nano Lett 9(9):3330–3336
47. Lu G, Lei FH, Angiboust JF, Manfait M (2010) Confined detection volume of fluorescence correlation spectroscopy by bare fiber probes. Eur Biophys J 39(5):855–860
48. Kohler RH, Schwille P, Webb WW, Hanson MR (2000) Active protein transport through plastid tubules: velocity quantified by fluorescence correlation spectroscopy. J Cell Sci 113 (Pt 22):3921–3930
49. Kohler RH, Cao J, Zipfel WR, Webb WW, Hanson MR (1997) Exchange of protein molecules through connections between higher plant plastids. Science 276(5321):2039–2042
50. Schwille P, Haupts U, Maiti S, Webb WW (1999) Molecular dynamics in living cells observed by fluorescence correlation spectroscopy with one- and two-photon excitation. Biophys J 77(4): 2251–2265

51. Rigler R, Pramanik A, Jonasson P, Kratz G, Jansson OT, Nygren P, Stahl S, Ekberg K, Johansson B, Uhlen S, Uhlen M, Jornvall H, Wahren J (1999) Specific binding of proinsulin C-peptide to human cell membranes. Proc Natl Acad Sci U S A 96(23):13318–13323
52. Politz JC, Browne ES, Wolf DE, Pederson T (1998) Intranuclear diffusion and hybridization state of oligonucleotides measured by fluorescence correlation spectroscopy in living cells. Proc Natl Acad Sci U S A 95(11):6043–6048
53. Wachsmuth M, Waldeck W, Langowski J (2000) Anomalous diffusion of fluorescent probes inside living cell nuclei investigated by spatially-resolved fluorescence correlation spectroscopy. J Mol Biol 298(4):677–689
54. Partikian A, Olveczky B, Swaminathan R, Li Y, Verkman AS (1998) Rapid diffusion of green fluorescent protein in the mitochondrial matrix. J Cell Biol 140(4):821–829
55. Lukacs GL, Haggie P, Seksek O, Lechardeur D, Freedman N, Verkman AS (2000) Size-dependent DNA mobility in cytoplasm and nucleus. J Biol Chem 275(3):1625–1629
56. Bacia K, Majoul IV, Schwille P (2002) Probing the endocytic pathway in live cells using dual-color fluorescence cross-correlation analysis. Biophys J 83(2):1184–1193
57. Korn K, Gardellin P, Liao B, Amacker M, Bergstrom A, Bjorkman H, Camacho A, Dorhofer S, Dorre K, Enstrom J, Ericson T, Favez T, Gosch M, Honegger A, Jaccoud S, Lapczyna M, Litborn E, Thyberg P, Winter H, Rigler R (2003) Gene expression analysis using single molecule detection. Nucleic Acids Res 31(16):e89
58. Jermutus L, Kolly R, Foldes-Papp Z, Hanes J, Rigler R, Pluckthun A (2002) Ligand binding of a ribosome-displayed protein detected in solution at the single molecule level by fluorescence correlation spectroscopy. Eur Biophys J 31(3):179–184
59. Pick H, Preuss AK, Mayer M, Wohland T, Hovius R, Vogel H (2003) Monitoring expression and clustering of the ionotropic 5HT3 receptor in plasma membranes of live biological cells. Biochemistry 42(4):877–884
60. Terada S, Kinjo M, Hirokawa N (2000) Oligomeric tubulin in large transporting complex is transported via kinesin in squid giant axons. Cell 103(1):141–155
61. Boukari H, Nossal R, Sackett DL (2003) Stability of drug-induced tubulin rings by fluorescence correlation spectroscopy. Biochemistry 42(5):1292–1300
62. Wawrezinieck L, Rigneault H, Marguet D, Lenne PF (2005) Fluorescence correlation spectroscopy diffusion laws to probe the submicron cell membrane organization. Biophys J 89(6): 4029–4042
63. Triffo SB, Huang HH, Smith AW, Chou ET, Groves JT (2012) Monitoring lipid anchor organization in cell membranes by PIE-FCCS. J Am Chem Soc 134(26):10833–10842
64. Oikawa D, Kitamura A, Kinjo M, Iwawaki T (2012) Direct association of unfolded proteins with mammalian ER stress sensor, IRE1beta. PLoS One 7(12):e51290
65. Shahzad A, Edetsberger M, Koehler G (2010) Fluorescence spectroscopy: an emerging excellent diagnostic tool in medical sciences. Appl Spectrosc Rev 45(1):1–11
66. Kilpatrick LE, Hill SJ (2016) The use of fluorescence correlation spectroscopy to characterize the molecular mobility of fluorescently labelled G protein-coupled receptors. Biochem Soc Trans 44(2):624–629
67. Briddon SJ, Kilpatrick LE, Hill SJ (2018) Studying GPCR pharmacology in membrane microdomains: fluorescence correlation spectroscopy comes of age. Trends Pharmacol Sci 39(2):158–174
68. Weis WI, Kobilka BK (2018) The molecular basis of G protein-coupled receptor activation. Annu Rev Biochem 87:897–919
69. Ward RJ, Pediani JD, Godin AG, Milligan G (2015) Regulation of oligomeric organization of the serotonin 5-hydroxytryptamine 2C (5-HT2C) receptor observed by spatial intensity distribution analysis. J Biol Chem 290(20):12844–12857
70. Calizo RC, Scarlata S (2013) Discrepancy between fluorescence correlation spectroscopy and fluorescence recovery after photobleaching diffusion measurements of G-protein-coupled receptors. Anal Biochem 440(1):40–48

71. Kluba M, Engelborghs Y, Hofkens J, Mizuno H (2015) Inhibition of receptor dimerization as a novel negative feedback mechanism of EGFR signaling. PLoS One 10(10):e0139971
72. Presman DM, Ganguly S, Schiltz RL, Johnson TA, Karpova TS, Hager GL (2016) DNA binding triggers tetramerization of the glucocorticoid receptor in live cells. Proc Natl Acad Sci U S A 113(29):8236–8241
73. Sigurdson CJ, Bartz JC, Glatzel M (2019) Cellular and molecular mechanisms of Prion disease. Annu Rev Pathol 14:497–516
74. Bieschke J, Giese A, Schulz-Schaeffer W, Zerr I, Poser S, Eigen M, Kretzschmar H (2000) Ultrasensitive detection of pathological prion protein aggregates by dual-color scanning for intensely fluorescent targets. Proc Natl Acad Sci U S A 97(10):5468–5473
75. Fujii F, Horiuchi M, Ueno M, Sakata H, Nagao I, Tamura M, Kinjo M (2007) Detection of prion protein immune complex for bovine spongiform encephalopathy diagnosis using fluorescence correlation spectroscopy and fluorescence cross-correlation spectroscopy. Anal Biochem 370(2):131–141
76. DeTure MA, Dickson DW (2019) The neuropathological diagnosis of Alzheimer's disease. Mol Neurodegener 14(1):32
77. Pitschke M, Prior R, Haupt M, Riesner D (1998) Detection of single amyloid beta-protein aggregates in the cerebrospinal fluid of Alzheimer's patients by fluorescence correlation spectroscopy. Nat Med 4(7):832–834
78. Chatterjee M, Noding B, Willemse EAJ, Koel-Simmelink MJA, van der Flier WM, Schild D, Teunissen CE (2017) Detection of contactin-2 in cerebrospinal fluid (CSF) of patients with Alzheimer's disease using Fluorescence Correlation Spectroscopy (FCS). Clin Biochem 50(18): 1061–1066
79. Lammers T, Kiessling F, Hennink WE, Storm G (2012) Drug targeting to tumors: principles, pitfalls and (pre-) clinical progress. J Control Release 161(2):175–187
80. Negwer I, Best A, Schinnerer M, Schafer O, Capeloa L, Wagner M, Schmidt M, Mailander V, Helm M, Barz M, Butt HJ, Koynov K (2018) Monitoring drug nanocarriers in human blood by near-infrared fluorescence correlation spectroscopy. Nat Commun 9(1):5306
81. Salmond GP, Fineran PC (2015) A century of the phage: past, present and future. Nat Rev Microbiol 13(12):777–786
82. Briandet R, Lacroix-Gueu P, Renault M, Lecart S, Meylheuc T, Bidnenko E, Steenkeste K, Bellon-Fontaine MN, Fontaine-Aupart MP (2008) Fluorescence correlation spectroscopy to study diffusion and reaction of bacteriophages inside biofilms. Appl Environ Microbiol 74(7): 2135–2143
83. Ross CA, Poirier MA (2005) Opinion: what is the role of protein aggregation in neurodegeneration? Nat Rev Mol Cell Biol 6(11):891–898
84. Carrell RW, Lomas DA (1997) Conformational disease. Lancet 350(9071):134–138
85. Owen MJ, Sawa A, Mortensen PB (2016) Schizophrenia. Lancet 388(10039):86–97
86. Jiang J, Chen X, Sun L, Qing Y, Yang X, Hu X, Yang C, Xu T, Wang J, Wang P, He L, Dong C, Wan C (2018) Analysis of the concentrations and size distributions of cell-free DNA in schizophrenia using fluorescence correlation spectroscopy. Transl Psychiatry 8(1):104
87. Ono T, Mimuro J, Madoiwa S, Soejima K, Kashiwakura Y, Ishiwata A, Takano K, Ohmori T, Sakata Y (2006) Severe secondary deficiency of von Willebrand factor-cleaving protease (ADAMTS13) in patients with sepsis-induced disseminated intravascular coagulation: its correlation with development of renal failure. Blood 107(2):528–534
88. Chion CK, Doggen CJ, Crawley JT, Lane DA, Rosendaal FR (2007) ADAMTS13 and von Willebrand factor and the risk of myocardial infarction in men. Blood 109(5):1998–2000
89. Nguyen TC, Carcillo JA (2007) Understanding the role of von Willebrand factor and its cleaving protease ADAM TS13 in the pathophysiology of critical illness. Pediatr Crit Care Med 8(2):187–189
90. Torres R, Genzen JR, Levene MJ (2012) Clinical measurement of von Willebrand factor by fluorescence correlation spectroscopy. Clin Chem 58(6):1010–1018

91. Koppel DE, Axelrod D, Schlessinger J, Elson EL, Webb WW (1976) Dynamics of fluorescence marker concentration as a probe of mobility. Biophys J 16(11):1315–1329
92. Webb WW (2001) Fluorescence correlation spectroscopy: inception, biophysical experimentations, and prospectus. Appl Opt 40(24):3969–3983
93. Das AK, Pandit R, Maiti S (2015) Effect of amyloids on the vesicular machinery: implications for somatic neurotransmission. Philos Trans R Soc Lond Ser B Biol Sci 370(1672):20140187
94. Das AK, Rawat A, Bhowmik D, Pandit R, Huster D, Maiti S (2015) An early folding contact between Phe19 and Leu34 is critical for amyloid-beta oligomer toxicity. ACS Chem Neurosci 6(8):1290–1295
95. Bera K, Das AK, Nag M, Basak S (2014) Development of a rhodamine-rhodanine-based fluorescent mercury sensor and its use to monitor real-time uptake and distribution of inorganic mercury in live zebrafish larvae. Anal Chem 86(5):2740–2746
96. Chandra B, Mithu VS, Bhowmik D, Das AK, Sahoo B, Maiti S, Madhu PK (2017) Curcumin dictates divergent fates for the central salt bridges in amyloid-beta40 and amyloid-beta42. Biophys J 112(8):1597–1608
97. Bhowmik D, Das AK, Maiti S (2015) Rapid, cell-free assay for membrane-active forms of amyloid-beta. Langmuir 31(14):4049–4053
98. Maity BK, Das AK, Dey S, Moorthi UK, Kaur A, Dey A, Surendran D, Pandit R, Kallianpur M, Chandra B, Chandrakesan M, Arumugam S, Maiti S (2019) Ordered and disordered segments of amyloid-beta drive sequential steps of the toxic pathway. ACS Chem Neurosci 10(5): 2498–2509

Principles and Applications of Fluorescence Microscopy

9

Bibhu Ranjan Sarangi

Abstract

Fluorescence spectroscopy and microscopy have been used extensively in diverse areas of both scientific research and industrial applications. Particularly, fluorescence microscopy is one of the most sought-after imaging techniques in biological research field. With the advent of many novel technologies, method of fluorescence imaging has grown manifolds. Our ability to visualize and resolve samples at different length scales using fluorescence imaging has improved significantly with a wide variety of microscopic techniques that are currently available. We have now access to various kinds of microscopes starting from basic epi-fluorescence microscope to highly advanced super resolution imaging techniques like photoactivated localization microscopy (PALM), stochastic optical reconstruction microscopy (STORM), etc. With such superresolution microscopic techniques, we now have the ability to detect and resolve the signals at the level of individual molecules. Overall fluorescence microscopic techniques have contributed immensely to the advancement of many biological research fields.

9.1 Introduction

Fluorescence spectroscopy and microscopy have been used extensively in diverse areas of both scientific research and industrial applications. Particularly, fluorescence microscopy is one of the most sought-after imaging techniques in biological

B. R. Sarangi (✉)
Department of Physics & Department of Biological Sciences and Engineering, IIT Palakkad, Palakkad, Kerala, India
e-mail: bibhu@iitpkd.ac.in

© The Author(s), under exclusive license to Springer Nature Singapore Pte Ltd. 2022

H. Sahoo (ed.), *Optical Spectroscopic and Microscopic Techniques*,
https://doi.org/10.1007/978-981-16-4550-1_9

research field. With the advent of many novel technologies, method of fluorescence imaging has grown many folds. Our ability to visualize and resolve samples at different length scales using fluorescence imaging has improved significantly with a wide variety of microscopic techniques that are currently available. We have now access to various kinds of microscopes starting from basic epi-fluorescence microscope to highly advanced super resolution imaging techniques like photoactivated localization microscopy (PALM), stochastic optical reconstruction microscopy (STORM), etc. With such superresolution microscopic techniques, we now have the ability to detect and resolve the signals at the level of individual molecules. Overall fluorescence microscopic techniques have contributed immensely to the advancement of many biological research fields.

This chapter aims to give a brief overview of a basic fluorescence microscope. First, a brief introduction to the phenomena of fluorescence is given, and major developments that have happened in this field have been highlighted. A brief account of various fluorescent molecules is discussed subsequently. The principle of fluorescence microscopy is also discussed, especially the epi-fluorescence microscopy. Finally, we give a brief overview of some advanced imaging techniques.

9.2 Phenomena of Fluorescence

Fluorescence is a manifestation of light–matter interaction. Transient electronic excitation due to the incident electromagnetic radiation leads to emission of photons of lower energy than that of the incident photon. This forms the basis of fluorescence. In particular, electrons can be excited from ground state to excited state with light of a specific wavelength. Typically, the excitation happens to a higher vibrational level of the electronic state. It quickly relaxes to the lowest vibrational level through a process known as internal conversion. Subsequently, it returns to the ground state with the emission of photon. Since the emission energy is lower than the excitation energy, the emission happens at a higher wavelength. Hence, for fluorescence emission, it is always the case that the excitation wavelength (λ_{ex}) < emission wavelength (λ_{em}). A more detailed description of the phenomena is explained in the following paragraph. Electronic excitation can also lead to other phenomena known as phosphorescence. Relaxation from a singlet excited state leads to fluorescence, whereas for some cases, the excited state can be a triplet state. Electrons in triplet state have the same spin orientation as that of the ground state. Since the transition from a triplet state to the ground state is forbidden, the rate of transition is slow ranging from milliseconds to seconds or even higher. A phosphorescence molecule upon excitation can glow for a long duration. This chapter mainly discusses the fluorescence phenomena.

The absorption and emission processes in fluorescent molecules are usually described with the help of Jablonski diagram named after Professor Alexander Jablonski.

A typical representation of Jablonski diagram is given in Fig. 9.1. The ground state, the first, and second excited states are shown as S_0, S_1, and S_2 respectively.

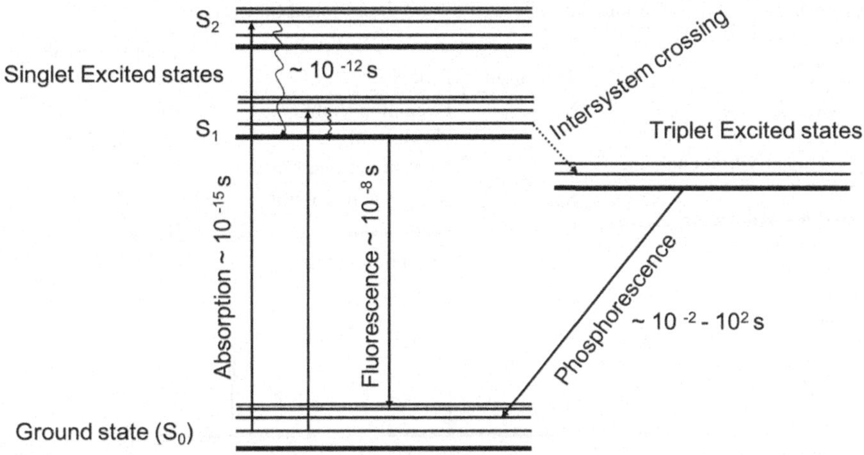

Fig. 9.1 Typical representation of Jablonski diagram showing the different electronic excitation states. Transition probabilities for fluorescence are very high, whereas for phosphorescence, transitions are forbidden, and hence, the rate is much slower

These electronic energy levels consist of several closely spaced vibrational levels. According to the Frank–Condon principle, the transitions between the electronic energy levels happen in a very short time, and therefore, the displacement of nucleus can be safely neglected. The typical transition time between the electronic energy levels is of the order of 10^{-15} s. Hence in Jablonski diagram, the transition between the levels is represented by vertical lines with arrows signifying the direction of transition. Upon the absorption of light of suitable wavelength, the molecule gets excited to a higher electronic energy level. Usually, the transitions happen to a higher vibrational level (S_n*) of S1 or S2. What follows this process is a rapid relaxation of the molecule to the lowest vibrational level of S1. This relaxation process (S_n*→ S_n) is known as internal conversion. The typical time scale of this relaxation is of the order of 10^{-12} s. Subsequently, the molecule returns to the ground state S_0 with the emission of a photon. One interesting feature is that the transition to ground state S_0 typically occurs to the higher vibrational level of the ground state. Subsequently, the molecule relaxes to the thermal equilibrium state with a similar time scale of 10^{-12} s. A summary of different radiative and nonradiative processes associated with fluorescence and phosphorescence along with their typical timescales is given in Table 9.1.

The important point to note here is that both the ground state (S_0) and the excited states (S_1 and S_2) are singlet states, that is, the spins are paired. Therefore, the transitions between these states are allowed. However, in certain cases, a spin conversion can happen in the S_1 state, which may lead to the molecule being in the first triplet state designated by T in the above diagram (Fig. 9.1). In triplet state, the spin orientations are parallel in both ground and excited state, and therefore, transition between such states (singlet→triplet or triplet→ singlet) is forbidden. This results in a very low transition rate or a relatively larger transition time scale of the

Table 9.1 Radiative and nonradiative processes and their typical time scales

States	Transition	Process	Typical time scales (in S)
S_0—Ground state S_n—Excited state S_n*—Higher vibrational state of S_n T—Triplet excited state	$S_0 \rightarrow S_n^*$	Excitation (light absorption)	~ 10^{-15}
	$S_n^* \rightarrow S_n$	Vibrational relaxation	~ 10^{-12}
	$S_n \rightarrow T$	Intersystem crossing	10^{-8}–10^{-3}
	$S_1 \rightarrow S_0$	Fluorescence	~ 10^{-8}
	$T \rightarrow S_0$	Phosphorescence	10^{-3}–10^{2}

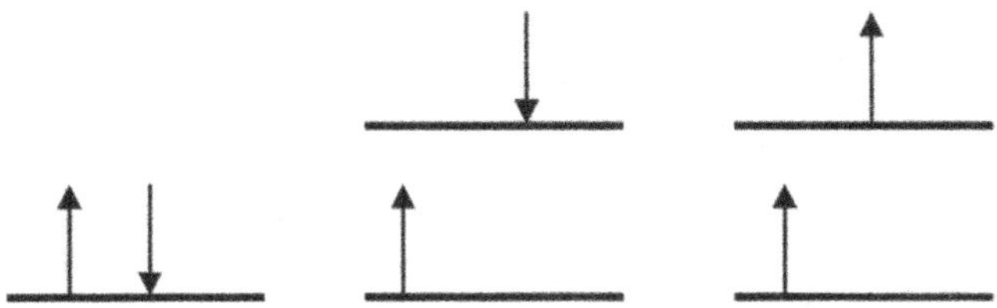

Fig. 9.2 Ground state and excited state spins in singlet and triplet configuration

order of milliseconds to seconds. The spin configurations in singlet and triplet states are shown in Fig. 9.2.

Out of all the processes, only fluorescence and phosphorescence result in the emission of a photon; hence, these processes are known as radiative processes. All other processes are nonradiative transitions.

Another interesting point regarding the transition involving fluorescence is that the emission spectrum is very similar to the absorption spectrum only differing in the peak positions. Closely spaced vibrational energy levels and thermal motion result in a spectrally broad absorption and emission band. A typical absorption and emission spectrum of fluorescein molecule is given in Fig. 9.3.

9.3 Major Developments

Although many observations related to the phenomena of fluorescence were reported before, the significant observation of fluorescence in a molecule was done by Sir John Fredrik William Herschel in 1845 [1]. He showed that quinine solution which is otherwise colorless and transparent exhibited a blue color when irradiated with sunlight. As we know now, the ultraviolet part of the sunlight excites the quinine molecule to emit blue light. Subsequently in 1852, Sir G. G. Stokes using a very simple experimental set-up like prism and stained glass as filter described Herschel's observation in details [2]. He showed that the ultraviolet light is only responsible for the fluorescence of quinine. This observation also led him to proclaim that

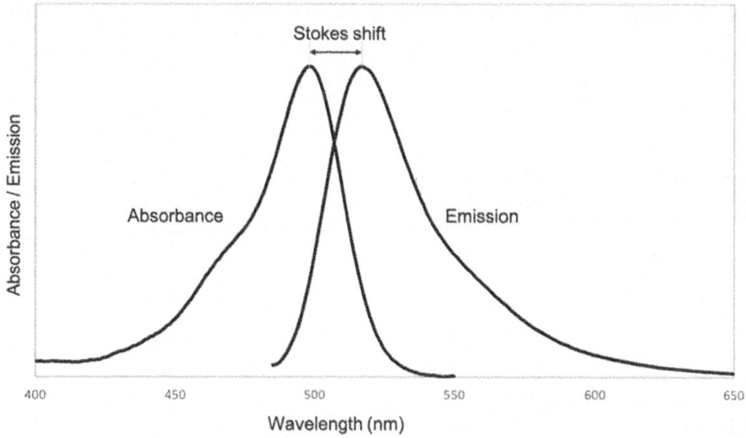

Fig. 9.3 The excitation and emission spectrum of Fluorescein molecule. The mean excitation peak is at 492 nm, whereas the mean emission peak lies at 520 nm. (Spectrum plotted with data collected from www.fpbase.org)

fluorescence emission has longer wavelength than that of the excitation wavelength. This shift is now known as the "Stokes shift". In the twentieth century, many important contributions to the field of fluorescence came from Nicolas and Merrit (first excitation spectrum of a dye molecule), Stern and Volmer (fluorescence quenching), Perrin (fluorescence polarization), Jablonski (Jablonski diagram), Forster (quantum theory of dipole–dipole interaction), and many others. In parallel, many compounds exhibiting fluorescence were also discovered. Mauveine (William Henry Perkin, 1856), fluorescein (Adolf Von Baeyer, 1871), and uranin (Paul Ehrlich, 1882) are some notable synthetic dye compounds. More detailed account of several research works in this regard can be found in many existing literature [3–7] (Table 9.2).

One of the most important discoveries happened in 1962, when Shimomura and his collogues gave evidence for green fluorescent proteins (GFP) [8, 9]. Subsequent investigation showed that GFP can be used as a fluorescent biomarker [10]. Subsequently, many other fluorescent proteins were developed with fluorescence emission spanning over entire visible spectrum. A concise list of some of the fluorescent proteins is presented in later sections (Table 9.3). In technological front, the first use of fluorescence dye in an optical microscope was demonstrated by the companies Carl Zeiss and Carl Reichert. After the development of protein labeling using green fluorescent proteins (GFP) and its variants, the fluorescence microscopy took a giant step toward the imaging of cellular molecules. With the advent of sophisticated microscopy techniques including PALM, STROM, etc., the fluorescence microscopy has grown leaps and bounds. We are now able to breach the optical resolution limit to the level of individual molecules using fluorescence techniques.

Table 9.2 Commonly used fluorescent probes and their excitation and emission wavelengths

Molecule name	Excitation wavelength (nm)	Emission wavelength (nm)
DAPI	345	460
Hoechst 33258	345	478
Alexa Fluor	345	442
Ethidium bromide	493	620
Alexa Fluor 488	394	517
FM-143	473	578
YOYO	491	509
Fluorescein FITC	495	518
Hex	535	556
Cy3	550	570
TRITC	547	572
Rhodamine red	560	580
Texas red	592	610
Cy5	650	670
TruRed	675	695
Cy7	743	770

Table 9.3 Commonly used fluorescent proteins and their properties

Name	Excitation wavelength (nm)	Emission wavelength (nm)
GFP (wild type)	475	509
EGFP	488	507
EBFP	383	445
ECFP	439	476
CyPet	435	477
EYFP	514	527
Topaz	514	527
Venus	525	528
mOrange	548	562
dTomato	551	581
TagRFP	555	584
DsRed	558	583
mCherry	587	610

9.4 Fluorescent Molecules

Molecules that exhibit fluorescence are known as fluorescent probes, fluorochromes, fluorophores, or fluorescent dyes. Fluorophores can be naturally occurring (intrinsic) or synthetically prepared (extrinsic). Intrinsic fluorescence molecules include mainly aromatic amino acids, flavins, chlorophyll, and derivative of pyridoxal. Proteins containing aromatic amino acids such as tryptophan, tyrosine, and phenylalanine are generally fluorescent. Enzyme cofactors such as NADH are also highly fluorescent. As is the case, generally many molecules of interest are nonfluorescent. In such cases, these molecules can be studied by labeling them with extrinsic fluorescent probes. For example, ethidium bromide can be used to label the DNA. Similarly, DAPI and Hoechst 33342 also bind to the DNA and make it fluorescently labeled. Several techniques are available to fluorescently label the biomolecules [11]. A huge number of such compounds are known and are available today. Here, we give a brief account of the general characteristics of these fluorescent molecules and their classifications.

9.4.1 Properties of Fluorescence Emission

The fluorescence excitation and emission spectrum displays some general characteristics across many fluorescent molecules. Although some exceptions to these characteristic features are also observed, the following features are frequently observed.

- Each fluorescent molecule exhibits unique absorption and emission spectra, depending on the molecular structure and sometimes also on their surroundings in the sample.
- Because of the rapid internal conversion to lowest vibrational energy level of the first excited state, the emission spectrum is independent of the excitation wavelength.
- The excitation and emission spectrum of fluorophore can overlap. As discussed previously, the difference of the energy between excitation and emission is called the Stokes shift. This shift can be measured as the difference between the excitation and emission maxima. Depending upon the fluorophore, the Stokes shift can range from a few nanometers to several hundred of nanometers. For example, DAPI (4′,6-diamidino-2-phenylindole) exhibits an excitation peak at 345 nm, whereas the emission peak is at 460 nm. The Stokes shift for DAPI is 115 nm. But for fluorescein, the Stokes shift comes out to be just 23 nm.
- Typically, the electron returns to higher vibrational energy levels in the ground state. This is because of the fact that the probability of an electron returning to a particular vibrational level of the ground state is similar to the probability of that electron in different vibrational levels before excitation. As a consequence, the emission spectrum is typically the mirror image of the excitation spectrum for a particular fluorescent probe. This is known as the “Mirror image rule”. It should be noted that exception to this mirror rules exists in many molecules.

9.4.2 Fluorescent Proteins

As discussed, many proteins are naturally fluorescent. Green fluorescent protein (GFP) is one of well-known names of similar proteins. Many variants of the GFP have been developed by introducing mutations into the amino acid sequences. Such mutants have different excitation/emission wavelengths as well as much better photostability. Several other types of similar protein have been discovered with excitation and emission spectrum spanning through most of the visible ranges. Yellow fluorescent proteins (YFPs), red fluorescent proteins (RFPs), etc. are some of the well-known examples. Some of the widely used fluorescent proteins along with their fluorescent properties are listed in Table 9.3. More details about the fluorescent proteins can be found out from existing literatures [12–17]. The fluorescent proteins have been widely used as biomarkers especially in live cell imaging applications [18].

9.5 Principles of Fluorescence Microscopy

Fluorescence microscopes enable the imaging of various fluorescence probes inside the samples. As discussed in the previous section depending upon the visualization requirement, a variety of fluorophores can be used to make the sample or a specific part of the sample (such as some organelles in living cells) fluorescent. Such samples can then be visualized with the help of specially designed optical microscope. The first practical realization of a fluorescence microscope was achieved by the companies Carl Zeiss and Carl Reichert at the beginning of the twentieth century. Since then, the technique has seen several advances that have led to the modern fluorescence microscopes that are being used today in many research labs. Modern fluorescence microscope improved both visualizations and quantifications of fluorescent-labeled samples. The following section discusses the principle and the components of a typical fluorescence microscope.

The basic requirement of a fluorescence microscope is an excitation source that is usually a mercury lamp/metal halide lamp/laser, an optical arrangement for selecting specific excitation and emission wavelengths, sample containing a fluorescent molecule, and detector for collecting the emitted fluorescence signal. The light from the source excites the sample containing the fluorescent probes. Subsequently, the fluorescent light emitted by the sample is collected using a detector (camera/eye). This process can be achieved essentially in two kinds of instrumental arrangements. One can use two separate objectives for excitation and emission. However, majority of biological fluorescence microscopy uses a method where the excitation of fluorescent sample and collection of emission signal are achieved using the same objective. A microscope with such an arrangement is called epi-fluorescence microscope.

A simplified diagram showing the light paths for excitation and emission in a typical epi-fluorescence microscope is shown in Fig. 9.4. Briefly, a narrow band of wavelength matching the excitation spectrum of the particular dye is selected from the light source using an excitation filter and is incident upon the fluorescently labeled sample with the help of a dichroic mirror. The light is absorbed by the molecule and the fluorescence emission from the sample transmitted through the dichroic and selected using an emission filter. This emitted intensity is recorded by a

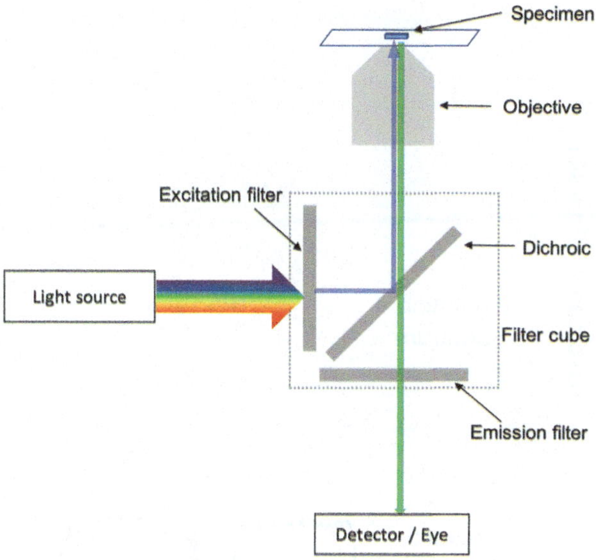

Fig. 9.4 Excitation and emission light path in an epi-fluorescence microscope. The excitation of fluorophore and the collection of emitted fluorescence signal

detector. The excitation filter, dichroic mirror, and the emission are assembled together inside a filter cube. A brief account of the main components of an epi-fluorescence microscope is discussed in the following section.

Light source

A bright light source is often used. It can be either a mercury or xenon lamp. Metal halide lamps which give an improved life time are also used in many microscopes. If the lamp housing is not directly attached to the microscope, the light is introduced to the microscope with the help of an optical fiber or a liquid guide. Many modern microscopes are now using the bright Light Emitting Diodes (LEDs) as an alternate light source. In case of LEDs, combinations of several LEDs are needed in order to have the access to wider selections of excitation wavelengths. Lasers are also used as the excitation source in many arrangements particularly for advanced fluorescence imaging techniques such as confocal microscopy.

Filter Set

Using a combination of filters and dichroic mirror, the filter cube selects the appropriate excitation wavelength from the light source and projects it on to the sample. Similarly, it also allows the emitted light from the sample to the detector for observation/recording. The filter cube consists of three components, i.e., the excitation filter, the dichroic mirror, and the emission filter. The excitation filter selectively transmits a narrow band of wavelengths corresponding to the excitation wavelength of the specific fluorophore present in the sample. The dichroic mirror is essentially a long pass filter designed specifically to reflect and transmit at specific wavelength

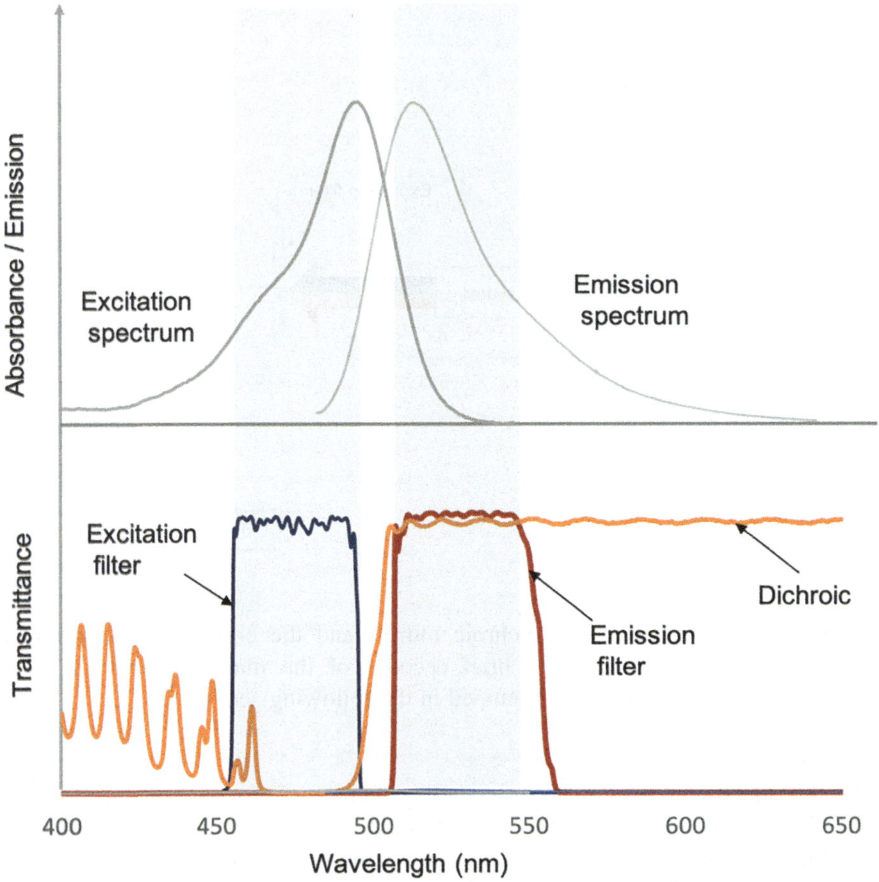

Fig. 9.5 Transmission characteristics of a filter cube corresponding to the fluorescein dye. Top image shows the absorption and emission spectrum of fluorescein. Bottom image shows transmittance of the corresponding excitation filter, dichroic mirror, and the emission filter. Shaded bands show the correspondence of the absorption and emission of the fluorophore to the filter characteristics

boundary. This is achieved by the multiple layers of dialectic film coated on the dichroic mirror.

The transmission properties of a filter cube corresponding to the fluorescein dye are shown in Fig. 9.5 .

Objective

The objective is one of the most important optical components in the microscope. As will be discussed subsequently, the resolution of the microscope is determined by the properties of the objectives numerical aperture (NA). Therefore, a proper selection of objective is highly essential to obtain improved image quality. Objects with high NA have better light-gathering ability, which improves the quality of the image. Both dry and immersion (water/oil) objectives are used in a fluorescence microscope. Immersion oil objectives provide better imaging capabilities as it provides a refractive index-matched medium between the objective and the microscope slides holding the specimen.

Detector

The fluorescence image of the sample can be directly visualized through eye. However, in order to record and store the image for future use, a detector attached with a computer is used. Furthermore, our eye can only visualize the range of wavelengths typically from 400 nm to 700 nm. However, a charge-coupled device (CCD) can in principle detect fluorescent signals from 400 nm up to 1000 nm. In majority of the epi-fluorescence microscopes, a charge-coupled device (CCD) or a complementary metal-oxide semiconductor(cMOS) camera is used for image recording. In advanced microscopic techniques such as confocal microscopy, photomultiplier tubes (PMTs) are used for detection, or Gallium-Arsenide photo detectors (GASP) are used.

9.6 Resolution

The most important parameter in optical microscopy is the resolution of the microscope. It refers to the ability of the instrument to distinguish two very closely spaced spatial points in the specimen. Because of diffraction, light from a point source when passed through an aperture forms spatially extended pattern known as airy pattern. It consists of a central bright disk also known as airy disk and a series of concentric dark and bright rings (Fig. 9.6). If two-point sources are very closely spaced, the airy

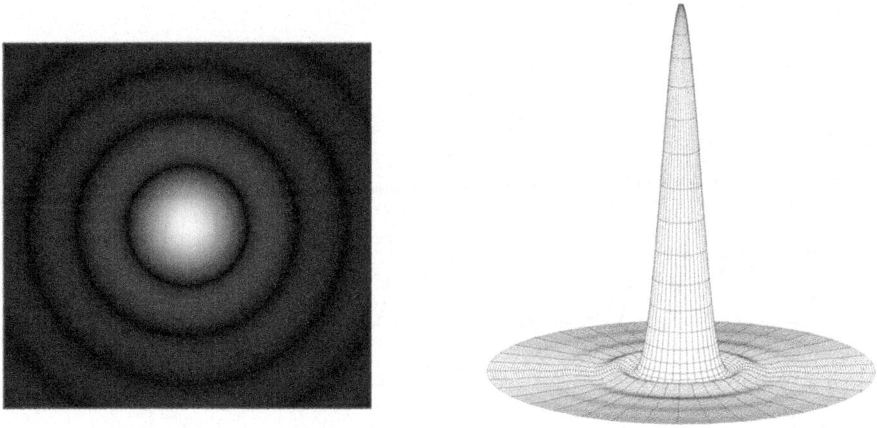

Fig. 9.6 Airy pattern formed by point sources when passed through an aperture. Left image shows the 2D image of a point source, and the right image shows the typical intensity profile of the airy pattern

disks of both the points will overlap which will limit our ability to distinguish the image as that of a two separate point sources.

Generally accepted criteria for resolution is the Raleigh criterion which says that the two points are said to be resolved when the maximum of the airy disk of the one point lies at the first minimum of the airy pattern of the second point [19]. Accordingly, the lateral resolution of a microscope is given by the following expression.

$$d_{\text{lateral}} = \frac{0.61\ \lambda}{NA}$$

where d is minimum spatial length scale that can be resolved, λ is the emission wavelength, and NA refers to the numerical aperture of the corresponding objective. As it is evident, the NA of the objective is a crucial determinant of the resolution of the microscope. It is defined as the

$$\text{NA} = n\ \text{Sin}\ \alpha$$

where n is the refractive index of the immersion medium, and α is the one half of the opening angle of the given objective as shown in Fig. 9.7. Higher NA essentially means a better light collection ability for the objective lens (Fig. 9.8).

The resolution criterion discussed above is valid for imaging a lateral plane in the sample. When it comes to the resolution along the propagation of light (axial resolution), the expression is modified to

$$d_{\text{axial}} = \frac{2\ \lambda}{NA^2}$$

The lateral resolution is proportional to NA of the objective, whereas the axial resolution is proportional to $(NA)^2$. Hence, axial resolution is inferior to the lateral resolution for a particular objective and wavelength.

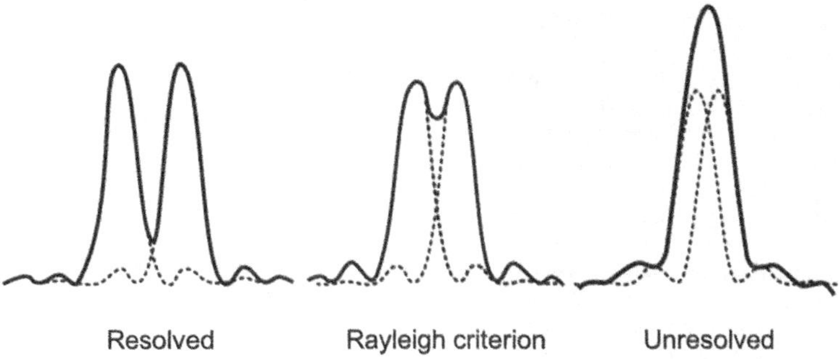

Fig. 9.7 Rayleigh criteria for resolution

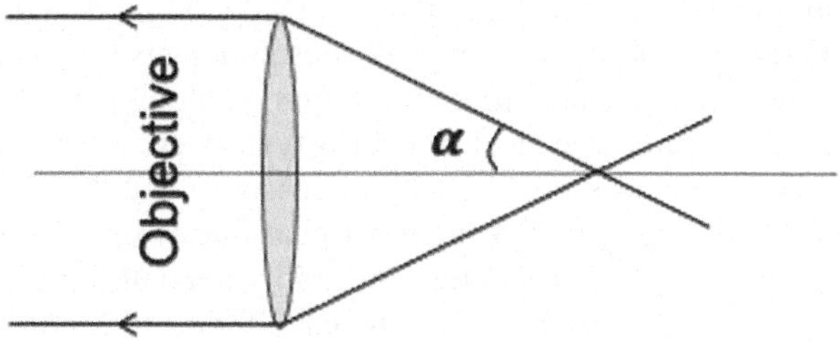

Fig. 9.8 The numerical aperture of objective is Sin of the half of the opening angle of the objective

9.6.1 Nyquist Criterion

While recording an image in a microscope, one important criterion to consider is the sampling. An optical microscope is an analogous system which produces a continuous wave signal (i.e. the emitted fluorescence intensity from the samples through the optical components of the microscope). This signal is recorded as a digital output using a camera (CCD, CMOS, PMT, etc.). Therefore, for the digital image to be meaningful, the signal must be sampled appropriately so as to record the smallest resolvable features present in the sample. According to Shannon's sampling theorem, an analog signal can be reconstructed appropriately if it is sampled with Nyquist criterion that is: Sampling interval must be at least twice the maximum frequency measured [20, 21]. For the case of optical microscopy, Nyquist criterion can be stated as the smallest resolvable feature in the sample (Rayleigh criterion) should be imaged by at least 2.3 pixels in the imaging detector. For example, the Nyquist criterion demands the following minimum condition for the pixel size of the detector for a particular objective magnification.

$$\text{Mag (Objective)} \times \frac{0.61\ \lambda}{NA} = 2.3 \times \text{pixel size of the camera}$$

In wide-field fluorescence imaging, the detector (Camera) has fixed pixel size. Hence, we should be careful in using objective magnification so as to satisfy the Nyquist criterion. In confocal microscopy where we use the PMTs, the pixel size can be adjusted, thereby giving us the advantage of always satisfying the Nyquist criterion.

9.7 Advanced Microscopic Techniques

While a basic epi-fluorescence microscope is immensely useful in visualization and quantification of fluorescent samples, it is very much limited when it comes to the resolving features that are very small, for example, two actin filaments of a cell spaced very close to each other. From the resolution criterion discussed above, it is apparent that the theoretical resolution that can be achieved in a basic fluorescence microscope is not adequate for several imaging purpose. A simple estimation shows that for $\lambda = 500$ nm and typical NA $= 1$, the resolvable length scale turns out to be around 315 nm (refractive index is taken to be 1 that of air). In practice, because of optical aberrations, achieving this resolution also becomes challenging. The problem is more severe when the depth resolution along the z-direction (perpendicular to the sample plane) is considered. In a typical epi-fluorescence microscopic technique (also known as the wide-field microscopy), the fluorescence signal is collected not only from the point of interest (focal plane) but also from the out of focal plane resulting in glare. This limits the ability to control the depth of field. This limitation can be overcome by using spatial filtering techniques, which is achieved using pinholes specifically placed to get rid of the out of focus signals, thereby eliminating the glare. This essentially forms the basis of confocal microscopy. The principles of confocal microscopy are beyond the scope of this current chapter, and the readers are advised to refer to many excellent texts available in that context [22]. Similarly, other advanced fluorescent techniques have been developed which can enhance both the visualization and quantification in fluorescence signals manifolds. Foster resonance energy transfer (FRET), fluorescence lifetime imaging (FLIM), total internal reflection microscopy (TIRF), spinning disk microscopy, and structured illumination microscopy (SIM) are examples of such advanced techniques.

9.8 Application of Fluorescence Microscopy in Biological and Biophysical Research

The fluorescence microscopy has been used extensively to probe both biological and material samples. In this section, we give a brief account of major applications of fluorescence imaging techniques mainly in biological and biophysical research. In particular, fluorescence microscopy is used extensively to study the intracellular distribution, dynamics, and molecular mechanisms of several macromolecules in cells and tissues.

9.8.1 Immunofluorescence and Live Cell Imaging

The techniques of immunofluorescence first developed by Albert Coons in the 1940s permit the visualizations of many cellular components of various cell and tissues. Specifically, it uses the basic immune chemistry of antigen–antibody reaction to tag the fluorescence probe to virtually any cellular proteins [23, 24]. Briefly, the specimen (Cells/tissue) is fixed using a standard protocol and then subsequently labeled with an appropriate antibody tagged with fluorescent probe. In direct immunofluorescence, only one antibody is used (primary). It recognizes the target antigen in the specific region and binds it making. In secondary/indirect staining, two antibodies are used. The unlabeled primary antibody binds to the target molecule, and the secondary antibody tagged with a fluorophore binds to the primary antibody. Cell and tissue morphology along with the organization of specific proteins can be studied using such immunofluorescent techniques.

The major disadvantage of immunofluorescent method is that the cells/tissues need to be fixed. Therefore, the dynamical characteristics of the living system cannot be inferred directly from immunostained images. In order to get that information, live cell imaging is performed.

Fluorescence microscopy is one of the primary imaging tools to probe living cells. Depending upon the target proteins/organs, one or more fluorescent proteins are expressed in the cells. In other method, the fluorescent probe can also be added as tracer to tag a protein/organelle of interest. In either of the method, we have access to multitude of fluorescence probes to selectively image a target molecule of a living cell. One of the primary aspects that is studied using fluorescence microscopy is the dynamic intercellular organization in response to mechanical, chemical stimuli. One such example is the organization the acto-myosin network in rat embryonic fibroblast cells as a function of substrate rigidity [25].

Some fluorescent dyes can permeate a living cell, thereby giving us the ability to tag an intracellular target of interest. Whereas the elective permittivity of dyes like fluorescein has been used to study sensory neurons of Caenorhabditis elegans (Perkin 1986). Polar dyes cannot permeate a cell membrane; hence, they can be effectively used to mark the cellular integrity. Altogether, fluorescent proteins can be used as both biomarkers and biosensors, thereby revealing useful details regarding many molecular and cellular processes [26].

9.8.2 Reconstituted Lipid Membranes

Fluorescence technique has been extensively used in systems of reconstituted lipid membranes especially in the form of giant unilamellar vesicles (GUVs) and supported lipid bilayers (SLBs) [27–30]. Majority of the studies in such systems are focused on investigating the phase separation of lipid bilayer. Because of biased partitioning of fluorescently labeled lipids in different phases, it is easy to visualize the domains. Such studies have led to a better understanding of the lipid raft hypothesis. The fluorescence microscopy data have been used to construct the phase diagrams of lipid mixtures. Another aspect which is often studied is the lateral structure of lipid bilayer. Many investigations have been reported in regards to the temperature-dependent lateral structures in giant vesicles of different phospholipids, binary and ternary mixtures of lipids and sterol [30, 31]. Fluorescence microscopy has also been used to study the lipid–protein interactions. Particularly, a variety of peptides and toxins that are harmful to living cells have been investigated in vesicular membrane [32–36].

9.9 Conclusion and Future Perspective

Since its first realization, the fluorescence microscopy has helped tremendously in obtaining fundamental insights into the structure and functions of biological systems. In particular, the imaging ability at high resolutions has helped us understand the steady-state structure and organization of intracellular components. In addition, the live cell imaging has revealed the dynamical aspects such as mobility, turnover, and various biochemical and biomechanical responses. The development of single-molecule techniques has given us ultrafine structures beyond the optical resolution limit. Such methods are still being refined and may eventually lead us to a simultaneous imaging of ultrafast processes at ultrafine structural details.

Acknowledgments This work was supported by DST-SERB Early career research award (ECR/2016/001986). We also acknowledge the members of "Physical and Chemical Biology Laboratory" at Indian Institute of Technology Palakkad for support.

References

1. Herschel Sir JFW (1845) On a case of superficial colour presented by a homogeneous liquid internally colourless. Phil Trans R Soc London 135:143–145
2. Stokes GG (1852) On the change of refrangibility of light. Phil Trans R Soc London 142:463–562
3. Valeur B, Berberan - Santos M.N. (2011) A brief history of fluorescence and phosphorescence before the emergence of quantum theory. J Chem Educ 88:731–738
4. Berberan-Santos MN (2001) Pioneering contributions of Jean and Francis Perrin to molecular fluorescence, in new trends in fluorescence spectroscopy. In: Valeur B, Brochon JC (eds) Applications to chemical and life sciences. Springer - Verlag, Berlin, pp 7–33
5. Nickel B (1998) From Wiedemann ' s discovery to the Jablonski diagram. EPA Newslett 64:19–72
6. Lackowicz JR (ed) (1999) Principles of fluorescence spectroscopy, 2nd edn. Plenum Press, New York
7. Förster T (1948) Intermolecular energy migration and fluorescence (Transl RS Knox). Ann Phys (Leipzig) 2:55–75
8. Shimomura O, Johnson FH, Saiga Y (1962) Extraction, purification and properties of aequorin, a bioluminescent protein from the luminous hydromedusan, Aequorea. J Cell Comp Physiol 59(3):223–239
9. Ormo M, Cubitt AB, Kallio K, Gross LA, Tsien RY, Remington SJ (1996) Crystal structure of the *Aequorea victoria* green fluorescent protein. Science 273:1392–1395
10. Chalfie M, Tu Y, Euskirchen G, Ward WW, Prasher DC (1994) Green fluorescent protein as a marker for gene expression. Science 263:802–805
11. Sahoo H (2012) Fluorescent labeling techniques in biomolecules: a flashback. RSC Adv 2: 7017–7029
12. Demchenko AP (1981) Ultraviolet spectroscopy of proteins. Springer-Verlag, New York
13. Longworth JW (1971) Luminescence of polypeptides and proteins. In: Steiner RF, Welnryb I (eds) Excited states of proteins and nucleic acids. Plenum, New York, pp 319–484
14. Permyakov EA (1993) Luminescent spectroscopy of proteins. CRC Press, London. Matz MV, Fradkov AF, Labas YA, Savitsky AP, Zaraisky AG, Markelov ML, Lukyanov SA. 1999. Fluorescent proteins from non- bioluminescent Anthozoa species. Nature Biotechnol 17:969–973
15. Fradkov AF, Chen Y, Ding L, Barsova EV, Matz MV, Lukyanov SA (2000) Novel fluorescent protein from discosoma coral and its mutants possesses a unique far-red fluorescence. FEBS Lett 479:127–130
16. Karasawa S, Araki T, Yamamoto-Hino M, Miyawaki A (2003) A green-emitting fluorescent protein from Galaxeidae coral and its monomeric version for use in fluorescent labeling. J Biol Chem 278(36):34167–34171

17. Lukyanov KA, Fradkov AF, Gurskaya NG, Matz MV, Labas YA, Savitsky AP, Markelov ML, Zaraisky AG, Zhao X, Fang Y, Tan W, Lukyanov SA (2000) Natural animal coloration can be determined by a nonfluorescent green fluorescent protein homolog. J Biol Chem 275(34): 25879–25882
18. Lippincott-Schwartz J, Patterson GH (2003) Development and use of fluorescent protein markers in living cells. Science 300:87–91
19. Born M, Wolf E (1999) Principles of optics. Cambridge Univ. Press, Cambridge
20. Nyquist H (2002) Certain topics in telegraph transmission theory proc. IEEE 90:280–305
21. Shannon CE (1998) Communication in the presence of noise proc. IEEE 86:447–457
22. Pawley JB (2006) The handbook of biological confocal microscopy, 3rd edn. Springer, New York
23. Jackson P, Blythe D (2008) Immunohistochemical techniques. In: Bancroft JD, Gamble M (eds) Theory and practice of immunohistological techniques, pp 433–472
24. Im K, Mareninov S, Diaz MFP, Yong WH (2019) An introduction to performing immunofluorescence staining. Methods Mol Biol 1897:299–311
25. Gupta M et al (2015) Adaptive rheology and ordering of cell cytoskeleton govern matrix rigidity sensing. Nat Commun 6:1–9
26. Stepanenko et al (2008) Fluorescent proteins as biomarkers and biosensors: throwing color lights on molecular and cellular processes. Curr Protein Pept Sci 9(4):338–369
27. Korlach J, Schwille P, Webb WW, Feigenson GW (1999) Characterization of lipid bilayer phases by confocal microscopy and fluorescence correlation spectroscopy. Proc Natl Acad Sci U S A 96(15):8461–8466
28. Bagatolli LA, Gratton E (2000) Two photon fluorescence microscopy of coexisting lipid domains in giant unilamellar vesicles of binary phospholipid mixtures. Biophys J 78(1): 290–305
29. Bagatolli LA, Gratton E (1999) Two-photon fluorescence microscopy observation of shape changes at the phase transition in phospholipid giant unilamellar vesicles. Biophys J 77(4): 2090–2101
30. Kahya N, Scherfeld D, Bacia K, Poolman B, Schwille P (2003) Probing lipid mobility of raft-exhibiting model membranes by fluorescence correlation spectroscopy. J Biol Chem 278(30): 28109–28115
31. Veatch SL, Keller SL (2005) Miscibility phase diagrams of giant vesicles containing sphingomyelin. Phys Rev Lett 94(14):148101
32. Henriques ST, Quintas A, Bagatolli LA, Homble F, Castanho MA (2007) Energy-independent translocation of cell-penetrating peptides occurs without formation of pores. A biophysical study with pep-1. Mol Membr Biol 24(4):282–293
33. Ambroggio EE, Kim DH, Separovic F, Barrow CJ, Barnham KJ, Bagatolli LA, Fidelio GD (2005) Surface behavior and lipid interaction of Alzheimer beta-amyloid peptide 1–42: a membrane-disrupting peptide. Biophys J 88(4):2706–2713
34. Ambroggio EE, Separovic F, Bowie JH, Fidelio GD, Bagatolli LA (2005) Direct visualization of membrane leakage induced by the antibiotic peptides: maculatin, citropin, and aurein. Biophys J 89(3):1874–1881
35. Hasper HE, Kramer NE, Smith JL, Hillman JD, Zachariah C, Kuipers OP, de Kruijff B, Breukink E (2006) An alternative bactericidal mechanism of action for lantibiotic pep- tides that target lipid II. Science 313(5793):1636–1637
36. Huang HW, Chen FY, Lee MT (2004) Molecular mechanism of peptide-induced pores in membranes. Phys Rev Lett 92(19):198304

Analysis of Biomolecular Dynamics Under Fourier Transform Infrared Spectroscopy

10

Sanjeev Kumar Paikra and Monalisa Mishra

Abstract

The status of biomolecules reflects the physiological condition of the body. Thus, biomolecules are altered with the physiological status. Biomolecules like proteins, carbohydrates, and nucleic acids are analyzed using several techniques to check the health of an organism. Among several techniques, Fourier-transform infrared spectroscopy (FTIR) is used to determine the status of the biomolecules. FTIR is used to determine the conformational change of the protein. FTIR can determine the biomolecular composition of the human body fluid, cell, and tissue. FTIR spectra provide the composition of carbohydrate, lipid, protein, and nucleic acid in a shorter period. This chapter discusses the analysis of protein and other biomolecular structures in the dried state as well as an aqueous solution using FTIR.

Keywords

FTIR-spectroscopy · Biomolecules · Protein · Nucleic acid

S. K. Paikra
Neural Developmental Biology Laboratory, Department of Life Science, National Institute of Technology (NIT) Rourkela, Rourkela, India

M. Mishra (✉)
Neural Developmental Biology Laboratory, Department of Life Science, National Institute of Technology (NIT) Rourkela, Rourkela, India

Center of Nanomaterials, National Institute of Technology (NIT) Rourkela, Rourkela, India
e-mail: monalisam@nitrkl.ac.in

© The Author(s), under exclusive license to Springer Nature Singapore Pte Ltd. 2022

H. Sahoo (ed.), *Optical Spectroscopic and Microscopic Techniques*,
https://doi.org/10.1007/978-981-16-4550-1_10

10.1 Introduction

Fourier transform infrared spectroscopy (FTIR) is used in structural elucidation of different proteins and polypeptides. The structural properties of biomolecules are analyzed without any labeling. It is an indispensable tool to detect the molecular structure of various cells and tissues. The infrared radiation is used as the light source [1]. The IR absorption by sample depends on the vibrational frequency of the molecules, intermolecular or intramolecular interaction, strength, and polarity of the bond present within it. Various bonds for amide A and B and amide I-VII of proteins can be distinguished in this instrument [2–4], which is essentail to detect the secondary structure of any protein [5, 6]. Secondary structure of a protein provides atomic level information about it. Various techniques like NMR, circular dichroism, mass spectrometer, and X-ray crystallography were used to detect the secondary structure of proteins. However, high-quality crystalline sample is needed, which is not so easy to isolate. NMR can analyze the secondary structure of low-molecular-weight protein [7]. Circular dichroism can detect the secondary structure for optically active clear solution [8, 9]. In this context, FTIR can detect the conformational change of any protein irrespective of aqueous, solid, or crystal state [10].

High-resolution spectra of the secondary structure of a protein can be obtained by band narrowing method, mathematical method, and spectral subtraction [11–13]. The alteration in bond strength is detected easily by vibrational spectroscopy [14]. Molecular vibration also depends on the geometry of the molecules [15]. Hydrogen bonding stabilizes the protein structure which can be detected by vibrational spectroscopy by lowering the frequency of stretching vibration [16]. Taking detecting ability into account, FTIR is used as a tool of potential clinical importance in the diagnosis of several pathological complications. FTIR is mainly used in the field of protein folding and misfolding to detect the diseased state [17]. Although this is not the hallmark for disease diagnosis, it can be used as a complementary tool along with microscopy and spectroscopy.

10.2 Modified Techniques to Detect Biomolecular Dynamics Under FTIR

Sometimes, it is difficult to rely only on vibrational mode of a molecule of interest to reveal its structure. In such cases, the site-specific or particular molecule-specific probe called Infrared probe (IR probe) was designed [18]. IR probe provides valuable information about the protein dynamics and confirmation [19–21], nucleic acid structure [22], and the electric field in the enzymes [23]. The IR probe is sensitive to the electrostatic field [24] and the hydrogen bonding [25] present in the molecule. The extrinsic IR probe also causes minimum damage to the original molecules. Some of the probes are designed for specific sites. Site-specific IR probe of protein is categorized into two different types: (i) side chain based and (ii) backbone based. The side chain of proteins has much diversity, which is the basis to design side chain-based IR probes. It can provide information about the protein folding and enzymatic reaction. $C \equiv N$ stretching vibration is commonly observed in proteins. A number of nitriles comprising unnatural amino acid are available in the market, which can be used as probe [26]. Most commonly used nitrile vibration probe is thiocyanate (SCN), which can be incorporated into protein by chemical modification of cysteine side chain [27, 28]. Carbonyl stretching vibration of carbon monoxide is used to study protein confirmation. Various reasons to use it to study protein conformation include (1) large extinction coefficient and (2) sensitivity toward both electrostatic and hydration environment. Isotope editing method [29] introduces amide I probe within the protein structure in a specific manner by replacing $C^{12}{=}O^{16}$ with either $C^{13}{=}O^{16}$ or $C^{13}{=}O^{18}$ [30–32]. Isotope-amended C=O group gives the vibrational mode for specific chemical environment, structural, and conformational changes. Vibrational coupling between multiple labeled carbonyl groups provides evidence about the secondary/tertiary structure of the protein [33]. The basic spectroscopic characteristics of the commonly used probe are shown in Table 10.1. These probes can offer a structural and spatial resolution of different protein functional events. The use of unnatural amino acid having functional groups such as nitrile, ester, azide, and carbonyl helps to incorporate the probes into proteins [33]. However, all these probes have some pros and cones. Thus, the choice of the probe depends on the chemical nature of the sample, and the site-specific probe can address many questions relevant to biological process, i.e., protein folding, drug–ligand binding, enzymatic reaction, and protein–protein interaction [18]. The triple bond IR probe has also been designed for the spectroscopic study of the structure and dynamics of different proteins and biomolecules [34].

Table 10.1 Site-specific IR probe [18]

Functional group	Name of probe	IR range (cm^{-1})	Molar extinction coefficient in H_2O (M^{-1} cm^{-1})	Ref.
Nitriles	*p*-cyano-phenylalanine	2220–2250	220	[35]
	Cyano-cysteine	2150–2180	120	[36]
	5-cyano-tryptophan	2210–2240	160	[37]
	Cyanate	2220–2300	800	[38]
Azides	Azidohomoalanine	2100–2140	350–400	[39]
	p-azido-phenylalanine	2100–2140	610	[40]
	3-picolyl azide adenine dinucleotide	2080–2160	2000	[41]
Carbonyls	Ketone carbonyl	1660–1700	1800	[42]
	Ester carbonyl	1690–1770	290	[43]
	Carboxylic acid	1700–1775	280	[44]
	Carboxylate	1555–1600	820	[45]

10.3 Principle and Methodology of Fourier Transform Infrared Spectroscopy

When infrared light or radiation hits the molecules, the bond in the molecules absorbs the energy of IR and responds by vibrating. It causes the molecular vibration of different organic or biomolecules. The normal mode of vibration can be seen in infrared-active molecules (absorb the IR light) due to change in dipole-moment that occurs during the molecular vibration [3]. Molecular vibration is divided into two types:

(a) *Stretching vibration* [46]: It involves the movement of the atom along the bond axis leading to the rise or fall in bond length at regular intervals. This type of vibration mainly corresponds to the one-dimensional motion of the molecules. Stretching vibration is generally detected in the mid infrared region (4010–1000 cm^{-1}).
(b) *Bending vibration* [46]: It involves the alteration of bond angle by mutual atoms or groups of atoms. For example, twisting, rocking, and torsional vibration are bending modes of vibration. The bond length will change only if the center of the gravity of a molecule resists displacement. Two-dimensional motions of the molecules could be observed. At very high frequency, stretching vibration can be observed. Stretching vibration is of two types: symmetrical and asymmetrical. Symmetrical mode of vibration in a molecule leads to a change in the size of the molecules due to an increase or decrease in distance between two atoms or molecules. Asymmetric vibration exhibits the interatomic distance between two atoms or molecules, which is present in the alternate or opposite direction. Two atoms at the same time could not move away or toward the central atom during asymmetric vibration but can move during symmetric vibration. Symmetric vibration cannot be detected in IR spectroscopy. Different vibration modes are shown in Fig. 10.1 below.

IR absorption can be easily observed in functional group which has permanent dipole [47]. Carbonyl group of protein or polypeptide largely contributes for infrared absorption spectra. Molecular vibration depends on the bond strength, chemical bond, and mass of the atom [48]. IR absorption associated with different biomolecules is listed in Table 10.2. Vibration frequency is very high for triple or double bonds in comparison to single bond [49]. Vibration frequency is very sensitive to the electronegativity of the atoms, the chemical environment of the functional group, and hydrogen bonding interaction [50]. Absorption in IR range depends on absorption coefficient of sample, concentration, and path length [50].

$$A_\lambda = \varepsilon_\lambda \times b \times c$$

where A_λ is absorbance at the specific wavelength λ, ε_λ is absorption coefficient of sample at that wavelength, c is the concentration of the sample (Mol/L), and b is path length.

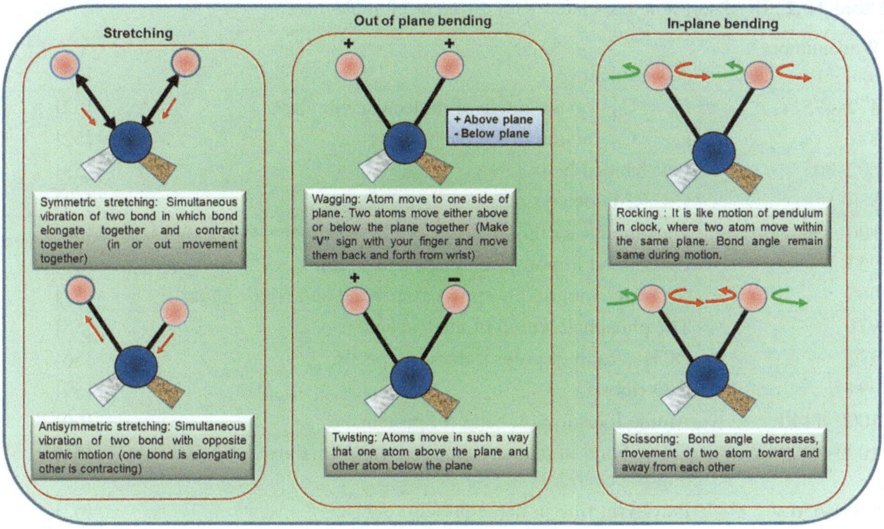

Fig. 10.1 Vibrational mode of molecules

Absorption of IR radiation brings the molecular vibration in covalently bonded molecules. Number of molecular vibration is directly proportional to the size of the molecules, and this is the reason that the large molecules have very complex spectrum [51]. The complex biological system composed of cell, tissue, and biological fluid contains different proteins, nucleic acid, and many micro/macromolecules which provide chemical integrity to the structure. The FTIR spectra of the complex molecules derived from *Drosophila melanogaster* (model organism to study the disease state) are shown in Fig. 10.2.

10.4 Instrumentation in FTIR Spectroscopy

Optical spectroscopy employed different materials that can absorb IR radiation. For example, glass and quartz which can absorb IR radiation [76]. IR spectroscopy apparatus consists of IR radiation sources, monochromators, sample holding cells, and detector systems. FTIR spectroscopy is the instrument that passes IR radiation through organic or biomolecules and produces a spectrum that contains the plot of transmitted or absorbed light on the vertical (y-axis), and wavelength of radiation on the horizontal (x-axis). IR radiation source is composed of inert material that gets heated at high temperatures for the emission under IR radiation. IR radiation should be intense enough to detect the steady and desired wavelength. The most commonly used IR radiation sources are Incandescent lamp, Nernst glower, carbon dioxide laser, Globar sources, and mercury arc.

Table 10.2 IR absorption associated with different biomolecules

Wavenumber (cm^{-1})	Assignment	Ref.
472–475	$C_\alpha = C_{\alpha'}$ torsion mode of molecular vibration	[52]
521	$C_\alpha = C_{\alpha'}$ torsion in the phenyl ring	[52]
802–805	Left handed helix DNA	[53]
889	C-C, C-O deoxyribose	[54]
900–1300	Phosphodiester bond	[55]
963	C=O bond in polysaccharides and pectin	[56]
965	C-O stretching in phosphodiester bond and ribose sugar	[57]
971	The phosphate group of nucleic acid	[58]
972	OCH_3 group of polysaccharides, pectin	[56]
994	C-O ribose	[59]
1000–1140	Amide I absorption band for protein	[60]
1000–1200	C-OH bond in mannose and galactose (mainly in oligosaccharide)	[61]
1000–650	Porphyrin ring in heme protein	[62]
1009	Stretching in C-O deoxyribose	[59]
1024	C-O stretch in glycogen	[63]
1025	CH_2OH group in carbohydrate (glucose, fructose, glycogen)	[64]
1029	$O\text{-}CH_3$ stretching	[52]
1040	Stretching vibration in C-O ribose	[52]
1070	C-O-C stretching vibration in nucleic acid/phospholipid	[65]
1056	C-O stretching deoxyribose	[65]
1078	Vibration mode in phosphate group	[65]
1150	C-O stretching in carbohydrate	[63]
1151	C-O and C-C stretching in glycogen	[57]
1153	Stretching vibration due to hydrogen bonding in the C-OH group	[66]
1159–72	Vibration mode in C-O belongs to carbohydrate or protein	[58]
1180–1300	Amide III band	[67]
1236	Amide III band specific for collagen	[68]
1330	CH_2 wagging	[69, 70]
1400	COO- stretching in amino acid (aspartate and glutamate)	[65]
1403	Symmetric bending mode in CH_3 group of protein	[71]
1418	C-H deformation	[53]
1450	Methylene group in biomolecules	[63]
1469	CH_2 bending mode in acyl chain of the lipid	[66]
1470	CH_2 bending mode in methylene group of lipid	[72]
1480–1543	Amide II	[56]
1480–1600	Amide II band corresponds to C-N stretching bond and bending of C-N-H	[67]
1528	C=N (adenine, cytosine)	[53]
1540	Amide II in protein (β sheet)	[67]
1577	C-C stretch in the phenyl ring	[52]

(continued)

Table 10.2 (continued)

Wavenumber (cm^{-1})	Assignment	Ref.
1592	C=N (adenine)	[52]
1600–1800	C=O stretch in lipid	[65]
1657	α helical structure of amide I	[67]
1660	C=C (cis) in lipid or fatty acid	[56]
1665	Amide I	[56]
1666	C=O stretching in pyrimidine	[65]
1700–1800	Fatty acid esters	[61]
1736	C=O stretching in lipid	[65]
1739	C=O stretching in polysaccharide/hemicellulose	[73]
1745	C=O vibration of ester group in triglycerides	[74]
1750	C=C vibration in lipid or fatty acid	[74]
2800–3500	Stretching vibration in CH_2 and CH_3 of lipid, cholesterol	[64]
2922	Asymmetric stretching in CH_2 group of acyl chain in lipid	[65]
3000–3700	O-H stretching in the water molecule	[65]
3007–3010	=CH group of olefin/unsaturated fatty acid	[74]
3300, 3313, 3328	Amide A band due to N-H stretching in protein/ nucleic acid	[53, 69, 75]

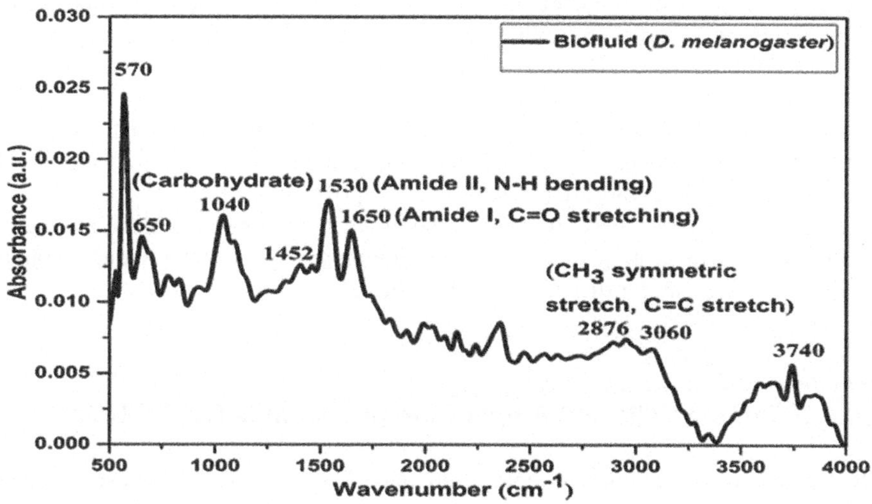

Fig. 10.2 FTIR spectra of biofluid derived from *Drosophila melanogaster*

10.4.1 IR Radiation Sources in Detail

1. *Nernst glower* [77]*:* It is composed of rare earth oxide (zirconia, yttria, thoria) shaped in the form of a hollow cylinder. The material is very fragile in nature, and

it must be preheated before use to make it conductive. It is nonconductive at room temperature. One end of this cylinder is covered with a platinum lid that conducts the electricity. It provides maximum radiation of about 7100 cm^{-1}.
2. *Globar source* [78]*:* It consists of a rod of silicon carbide having length 50 mm and diameter 5 mm which is heated at about 1500 K for electrical conductivity. Maximum emission is observed at 5200 cm^{-1}. It has a positive temperature coefficient which can be controlled by a flexible transforming method. Its intensity is lower than Nernst glower, which is considered as the major disadvantage.
3. *Carbon dioxide laser* [79]*:* It is a good source of infrared for measuring the level of atmospheric pollutants through IR spectroscopy.
4. *Incandescent lamp* [80]: It is used as an IR source in near-infrared instruments. It is not applicable for far IR instruments because the glass (IR source) enclosed has very low spectral emission.
5. *Mercury Arc* [81]: Mercury arc lamp composed of quartz is used for the far IR region.

10.4.2 Monochromator

A monochromator is a very important component of the FTIR instrument, which can separate or disperse the broad range of IR radiation into individual IR frequency with respect to IR absorbed or transmitted by the sample. It can produce monochromatic light. Monochromators are mainly of two types:

1. *Prism monochromator:* Prism can be used as a disseminative element, and it is composed of different metal halide salt, glass, and quartz which transmit the IR radiation. Sodium chloride is widely used as prism salt.
2. *Grating monochromator:* It replaces the prism by grating to achieve the higher dispersion. The grating is composed of aluminum having many parallel straight lines on the plane for dispersion of IR radiation.

10.4.3 Sample Cells and Preparation of a Sample for Analysis

IR spectroscopy can analyze both solid and liquid samples. However, sample preparation plays a vital role to obtain a good-quality spectrum. The sample to be examined under IR must be transparent to the infrared radiation. Thus, solid sample should be dissolved in a nonaqueous solvent with utmost care. The chosen solvent should not be absorbed in the spectral range of IR. The solvent spectra need to be subtracted by using the software available with the instrument. If the sample is amorphous in nature, the solid thin film can be prepared by mixing it with KBr or NaCl. There are two different techniques used for such sample: (i) mull technique and (ii) pressed pellet technique.

1. Mull technique: It is used to detect the sample which is amorphous in nature. In this technique, the finely grounded sample is mixed with Nujol oil, and a thick paste was made. This thick paste is mounted on the IR transmitting window, and the spectrum is taken. Nujol oil is transparent in the IR region.
2. *Pressed pellet technique*: It is the most commonly adopted technique in which a finely grounded solid sample is mixed with potassium bromide powder which is then pressed under high pressure (near about 25,000 p sig) to form a thin film with 1–2 mm of thickness and 1-cm diameter. IR radiation can pass through the pellet and can be stored for a longer time. The major drawback includes the presence of a band at 3450 cm^{-1} due to the OH group present in the KBr pellet.

10.4.4 Detectors Used in FTIR Spectroscopy

Detector responds toward different frequencies [82]. Different types of detectors are bolometers, thermocouple, thermistors, golay cells, photoconductivity cell, semiconductor detector, pyroelectric detector, and Fourier transform detector system. Fourier transforms detection system is the most widely accepted system for the analysis of biological sample [83].

10.4.5 Optical Arrangement of Fourier Transform Infrared Spectroscopy

The schematic diagram of the equipment is shown in Fig. 10.3. FTIR system has four optical arms perpendicular to each other. A beam splitter is placed at the point of intersection of four different arms. Radiation passes through the first arm and then separates by a beam splitter into two perpendicular half-beam of equal intensity that passes down to other arms of the spectrometer. At last end of this arm, two half-

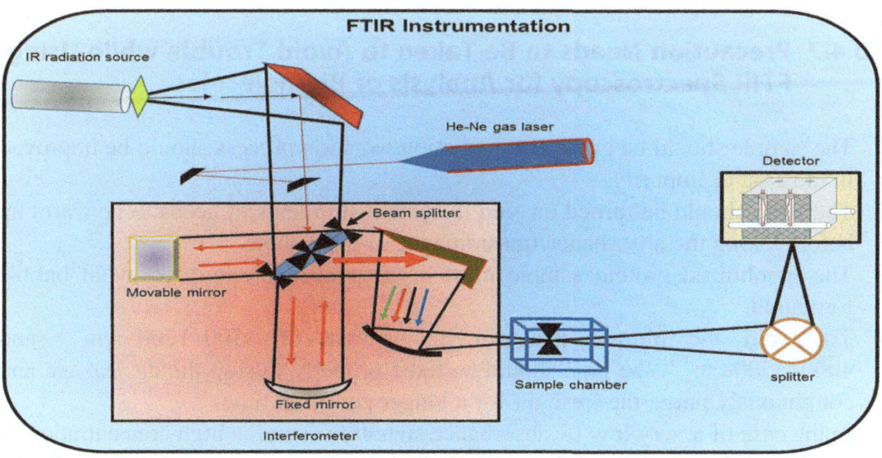

Fig. 10.3 Schematic diagram of instrumentation in FTIR spectroscopy

beams are reflected by the mirror to the beam splitter, where they reconstituted and reflected together onto the detector. Detector signal in the spectrometer is related to the Fourier transformation of the spectrum. The heart of this instrument is an interferometer (Michelson interferometer) which contains a fixed movable mirror. Instrument measures the IR intensity which is relatable to the position of the movable mirror. FTIR performs the two Fourier transformations, one by interferometer and another by computer. Interferometer produces the Fourier transformation of the spectrum by the help of a monochromatic source of light. The advantage of the whole setup is fast data collection and high light intensity at detector which leads to a high signal-to-noise ratio.

The attenuated total reflectance (ATR) technique in comparison to the transmission experiment is helpful to avoid handling problems caused by short path length. In ATR, the sample is kept on a crystal with a refractive index larger than that of the sample. After one or two reflection, light leaves the crystal, and it is focused on the detector. During reflection at the interface between sample and crystal, light focuses on the sample.

10.4.6 Advantage

1. FTIR spectroscopy analysis of protein is not limited by the physical state of protein or size of the protein. Sample can be analyzed in the form of solid pellet or powder, detergents, an aqueous solution of protein, hydrated film, organic solvents, and micelles.
2. Protein and peptide which are insoluble at high concentration and protein which shows the structural changes at the high concentration can also be analyzed in transmission mode.
3. Spectra could be obtained for any protein of any molecular weight in a wide range of chemical environment, and it requires less time and less sample [5].
4. The protein sample doesn't need any proteolytic digestion.

10.4.7 Precaution Needs to Be Taken to Avoid Trouble while Using FTIR Spectroscopy for Analysis of Biomolecules

1. The sample should be pure. The extraction/isolation process should be improved in the case of impurity.
2. The lamp should be turned on, and in some instruments, it needs to be warm up before taking the absorbance/transmittance of the sample.
3. The lyophilized protein sample must be suspended properly to avoid bubble formation.
4. To avoid the blurred peak in the spectra of 1500–1200 cm^{-1} and 4000–3500 cm^{-1}, the pipe should be fixed properly to stop the air leakage and continuously purge the fresh air for a longer period of time.
5. In the case of a very low IR absorbance signal/no signal, a high concentration of protein needs to be taken. The qualitative or quantitative analysis is required to be

done by SDS-PAGE/Bradford assay before the spectral analysis. The protein sample should be more than 95% pure.

6. For the determination of the secondary structure of the protein, the concentration of the sample solution should be more than 3 mg/ml (in case of water as a solvent) [84].
7. Water subtraction is a very important step for getting a correct spectrum. Water gives the IR absorbance band in three different regions: 3400, 2125, and 1645 cm^{-1}. Sometime water absorption band overlaps the protein absorption band. To avoid this trouble, thin film (<10 μm) of protein solution is required to be prepared by which water absorbance can be easily subtracted.
8. Different parameters like cell path length, buffer concentration, temperature, pH, number of scans, and spectral resolution are kept constant for sample and solvent.
9. The second derivative analysis is the most preferably used to separate the overlapping band. It can be performed by using the seven-point Savitzky-Golay second derivative function. This function can be used digitally in origin software.

10.5 Biomedical Importance of FTIR Spectroscopy

Biological fluid is very important component to diagnose the disease state as well for forensic investigation. Here, we had described the importance of ATR-FTIR for the analysis of various components present in biological fluid. We put more emphasis on carbohydrate, lipid, protein, and nucleic acid content of biological fluid (Fig. 10.4).

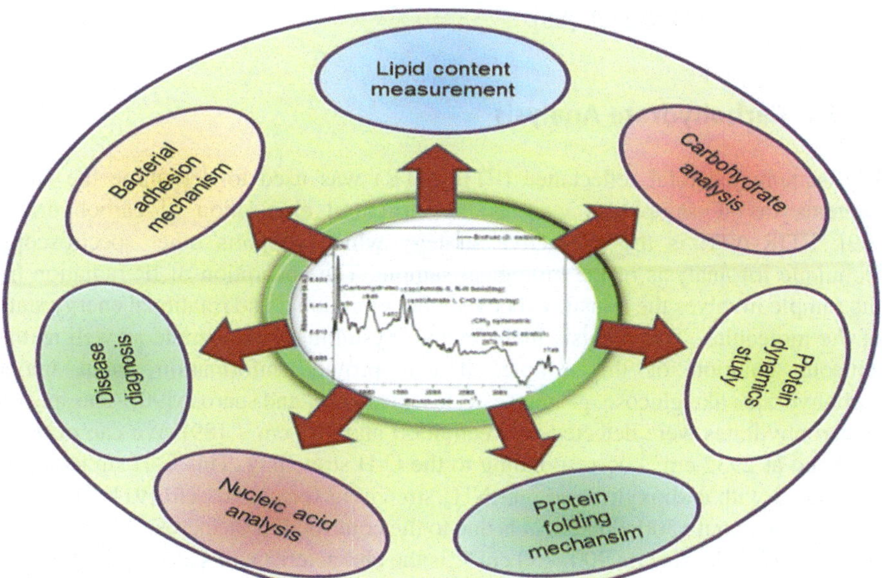

Fig. 10.4 Biomedical application of FTIR spectroscopy

10.5.1 Measurement of Lipid Content

It has been already reported that the concentration of total protein, glucose, protein albumin, cholesterol, and triglycerides in the biological fluid or sample can be calculated by using the IR spectra [85]. It is not a universal approach for quantitative analysis of lipid, but it can be used to detect the level of lipid, when it is present in high concentration. Low extinction coefficient and low concentration of lipids can't be easily detected [86]. FTIR spectroscopic technique focused on the quantification of five different lipids, cholesterol (CH), phosphatidylethanolamine (PE), galactocerebrosides (GC), sphingomyelin (SM), and phosphatidylcholine (PC), which are present in abundance in the brain [87]. Primary brain tumor (glioma type) was identified by the decrease in lipids such as CH, PE, and GC and increase in PC and CE contents which can be diagnosed by FTIR spectroscopy [87]. Diagnosis of a tumor can also be done by employing chromatography, magnetic resonance imaging, Raman spectroscopy technique, and mass spectra which gave strong evidence about the presence of lipid. When normal brain tissue transforms into a tumor, the lipid composition changes. Lipid can be extracted from white matter and gray matter of the brain tissue, to record the IR spectra for quantification by comparing the result obtained from IR spectra with the result obtained from chromatography technique [87]. The methodology involves the comparison of the spectra of lipid extract derived from brain tissue to the spectra of reference lipid. The recorded spectra can be analyzed by partial least square regression which predicts the composition of lipid [87]. PLS is based on the factorization of matrices which can be done by principal components analysis. It is developed for independent variables (concentration) and dependent variables (extinction coefficient). PLS regression method was explained by Otto et al. [88].

10.5.2 Carbohydrate Analysis

FTIR-attenuated total reflectance (FTIR-ATR) was used to determine the sugar contents in the sample as well as the structural elucidation of carbohydrates [89]. FTIR-ATR is the alternative strategy which supports other spectroscopy technique for analysis of the biological sample. The absorption of IR radiation by the sample involves the transition between the vibrational and rotational energy state of the molecules. The analysis includes the crystalline solid sample as well as the aqueous solution of the sample. It can provide information about major carbohydrates like glucose, galactose, xylose, fructose, and sucrose [90]. The spectra of carbohydrates were detected between 4000 and 400 cm^{-1} [89]. We can observe the band at 2932 cm^{-1} corresponding to the C-H stretching. This C-H stretching is associated with carboxylic acid and NH_3 stretching for amino acid [91]. The band observed at about 1500–900 cm^{-1} is due to the contribution of sucrose, fructose, and glucose [92]. The spectra 900–750 cm^{-1} is the characteristic of saccharide [92]. The peak at 918 cm^{-1}, 1043 cm^{-1}, 1254 cm^{-1}, and 1110 cm^{-1} corresponds to C-H bending, C-O stretch in C-OH group, and C-C stretch and stretch of C-O band of the C-O-C linkage, respectively, in the carbohydrate structure [93]. The peak at

1321 cm^{-1} is due to the O-H bending of the C-OH group. The peak at 1411 cm^{-1} corresponds to the C-H bending of carbohydrates [94].

The IR spectral range of carbohydrate includes 950–750 cm^{-1}, where we can distinguish between α and β confirmation of monosaccharide or polysaccharide [95], and it can be specifically observed at 850 and 890 cm^{-1}, respectively [95]. Spectral features of the glycosidic bond of carbohydrates are studied in detail by Sekkal et al. [96]. Acquired data from the FTIR instrument can be further processed by the chemometric method of analysis which is based on principal components analysis, discriminant analysis, principal least square, artificial neural network, and principal variables [97–100].

10.5.3 Protein Dynamics Study

FTIR is the major technique for structural characterization of a different protein, peptide, and its chemical/biological interaction with membrane protein [101]. It is used in a study of the secondary structure of the protein. Misfolding of any protein having a deleterious effect on the activity of enzymes, aggregation of protein, and formation of fibril is also called amyloids [102]. This technique helps us to understand this folding mechanism, to avoid or mitigate this physiological complication. Folding of the amino acid chain in the protein determines its structure and functionality. Protein stability in aqueous ionic liquid solution depends on the protein ionic liquid combination. This ionic liquid was used to minimize protein aggregation and improve the protein crystallization as well as stability [103]. Here FTIR is the suitable screening method for the characterization of protein stability. It quantifies the wide array of protein ion combinations. Structural dynamics are essential to biomolecules and proteins to carry out proper biological functions. Structural dynamics influence the biological activity of proteins. Proteins contain the amino acid side chains, and IR absorption by these side chains can provide the molecular information about the proteins. IR absorption range of these amino acids is listed in Table 10.3.

The assignment of amide band position corresponding to the secondary structure of proteins is listed in Table 10.4.

10.5.4 Monitoring the Mechanism of Action of Protein by Time-Resolved FTIR Spectroscopy

Time-resolved FTIR spectroscopy (TRS) is used to detect the kinetics of single chemical reactions by the step scan approach. It can reveal the molecular details of the mode of action of a protein in the body or biological system [122]. This technique can be used to analyze different proteins and membrane protein having size up to 100,000 daltons and to provide the information regarding the charge scattering, hydrogen bonding, protonation state, and time-dependent protein reaction [123]. A small protein of the size 20 kD has 10^4 vibrational modes, and that is why absorption spectra alone can't provide the information regarding the individual band [122]. FTIR can only provide the information regarding the amide I (C=O stretch)

Table 10.3 The functional group associated with amino acid side chain and its IR range

Wave number (cm^{-1})	Assignment	Ref.
2551	Cysteine (-SH) stretching mode	[104]
1716	Aspartic acid (C=O)	[105]
1712	Glutamic acid (C=O)	[105]
1677	Asparagine (C=O) it is observed up to 1704 cm^{-1} in proteins	[106]
1652–1690	Arginine ($CN_3H_5^+$). In the case of an aqueous solution of protein, it is observed at 1695 cm^{-1}	[107]
1668–1688	Glutamine (C=O)	[107]
1615–1660	Arginine ($CN_3H_5^+$) peak position depends on the salt bridge between arginine and other amino acid residue	[108]
1631	Histidine (C=C) is for 4-methylimidazole at 1633 cm^{-1}	[109]
1625–1630	Lysine (NH_3^+)	[109]
1615–1620	Tyrosine (C-C and C-H)	[109]
1622	Tryptophan (C=C)	[109]
1575–1580	Aspartic acid (COO^-)	[110]
1526	Lysine (NH_3^+)	[111]
1400–1465	Proline (CN). It is very sensitive to protein backbone confirmation	[112]
1180–1420	Serine (COH, COOH, CO). Band position is very sensitive to hydrogen bonding	[16]
1160–1250	Glutamic and aspartic acid (C-O). Aqueous solutions have peaked at 1250 cm^{-1}	[113, 114]
1092	Tryptophan (CH, NC) indole ring	[16]

Table 10.4 Amide band position in secondary structure of protein [115]

Conformation	Band position for amide group (Wavenumber cm^{-1})	Ref
α-Helix β-Sheet Random coil	Amide I band 1649–1657 1621–1648 1654	[116, 117]
α-Helix β-Sheet	Amide II 1545 1530	[118, 119]
α-Helix β-Sheet Unordered turn	Amide III 1293–1331 1223–1251 1244–1270	[120, 121]

and amide II (NH band coupled with C-N stretch), which is found in almost every amino acid, and that is why the TRS technique can be adopted as the emerging tool to get the detail molecular information [122]. Time-resolved FTIR instrument comprises of a sample chamber, light sources, interferometer, and detector system. Multiple spectral bands within the spectra create confusion about the confirmation of the functional group. To get a conclusion from the spectra, each band is assigned to an individual functional group. For example, the carbonyl group vibration of glutamic or aspartic acid shows absorption between 1700 and 1770 cm^{-1}.

Table 10.5 Spectral assignment of IR band in nucleic acid [124]

Wavenumber (cm^{-1})	Molecular assignment	Ref.
1715	G-G-C/triple-strand DNA, where guanine is important for hoogsteen third-strand binding (C=O stretching)	[132]
1712	T-A-T/tsd, where thymine is important for hoogsteen third-strand binding (C=O stretching)	[132]
1698–1691	U, T/ssd (C=O stretch in uracil or thymine)	[132, 133]
1671–1655	T/ssd (C=O stretch in thymine and the peak intensity decrease in duplex DNA)	[131]
1655–1647	C/ssd and dsd (C=O stretch of cytosine and peak intensity decrease in duplex DNA)	[132]
1632–1622	A/ssd and dsd (C=N and C=C ring vibration of adenine and peak intensity decrease in double-stranded DNA)	[124]
1568–1564	G/ssd and dsd (C=N ring vibration of guanine)	[134]

Abbreviation used: *G* guanine, *A* adenine, *C* cytosine, *U* uracil, *tsd* triple-strand DNA, *dsd* double-strand DNA, *ssd* single-strand DNA

10.5.5 Analysis of Nucleic Acid in Aqueous Solution

The spectral range 1800–800 cm^{-1} is assigned to DNA/RNA/nucleic acid content in aqueous solution [124]. The spectral assignment of the IR band in a nucleic acid is tabulated (Table 10.5). IR spectroscopy studies have been already performed on single-stranded [125], double-stranded [126–128], and triple-stranded [129–131] structure of nucleic acid. IR band assignment for nucleic acid components has been obtained by spectral comparison with the reference (model) molecules. An aqueous solution of nucleic acid shows the absorption band at 1800–800 cm^{-1}, but water gives the absorption band at 1645 cm^{-1}, and deuterated water gives the absorption band at 1450 cm^{-1} and 1210 cm^{-1}, and that is why spectra recorded with both water and D_2O and the result can be combined together to get the insight details. DNA double helical structure was studied by identifying the corresponding peak position (970–950 cm^{-1}) and band ratio [59].

10.5.6 To Understand the Bacterial Adhesion Mechanism at Metal Oxide Surface

P-containing group and the carboxyl group of bacterial cell membranes have an affinity to attach with the α-Fe_2O_3 and α-FeOOH at pH 7, which results in Fe-O-P molecular vibration, identified by ATR-FTIR spectroscopy [135]. Adhesion at the metallic surface leads to membrane deformation causing bacterial death due to loss in cell integrity [136]. Understanding the mechanism can be helpful in the fabrication of the novel material with the exquisite property. It has given us an idea about signifying the critical role of P in the induction of bacterial attachment. Interaction of phosphate group with iron oxide surface has been already reported in different spectroscopic studies [137–140]. Membrane phospholipid interaction with iron oxide is facilitated by the inner sphere coordination of the phosphate group [138]. The methodology involves the coating of bacteria on metal oxide deposited on the internal reflection element (IRE, i.e., diamond or germanium or zinc selenide), and spectra were taken. For the validation of the method, standard biomolecules (DNA, glutamic acid, casamino acid, tryptone) were deposited on metal oxide impregnated on the IRE element, and spectra were compared with the spectra obtained from bacteria deposited on metal oxide. Minimum bacterial load of 10^5 cells/ml and minimum concentration of biomolecules solution, i.e., 2.5–8 mg/ml, are required for the deposition on IRE element for getting a better spectra. Reference biomolecules solution and bacterial cell suspension were prepared at pH 7. Methodology can be adopted from work done by Parikh et al. [135]. FTIR spectra of different strains of bacteria deposited on material impregnated on IRE have shown the distinct band which is corresponding to different proteins, carbohydrates, and lipid, especially phosphate group. The interaction of different strains of bacteria with metallic surface shows small variation in band position [135].

10.5.7 Disease Diagnosis

Several biological studies have been already carried to detect the level of different biomolecules in disease state [141]. This technique can be useful in discrimination of normal cells and disease-affected cells. FTIR is helpful in analysis of bacteria [142], cell and cell membrane [143, 144], vesicles [145], etc. Water contents in the sample don't cause any hindrance in deriving the absorption spectra because it can be easily subtracted from the final spectrum. The major limitation of FTIR in the analysis of biological samples is a very slow acquisition of data and the weak sensitivity of detectors [141].

Vibrational spectroscopic technique can be employed for screening and diagnosis of prostate cancer [146, 147], diagnosis of chronic obstructive pulmonary disease (COPD) by analysis of sputum [148], early diagnosis of Parkinson's disease from the blood plasma [149], diagnosis of leukemia from blood plasma [150], dementia [151], diagnosis of HIV/AIDS from the human serum analysis [152], breast cancer tumor [153], Alzheimer's disease [154, 155], malignant biliary strictures [156], cysteine stones [157], analysis of tears for diagnosis of eye disease [158], colorectal cancer [159], periodontitis [160], brain tumor severity from the serum sample [161], diagnosis of malaria parasitaemia by analyzing whole blood [162], identification of triglyceride dependent obesity [163], protein aggregation in living cells [164], diagnosis of diabetes by analysis of saliva of pregnant women in first, second, and third trimesters of pregnancy period [165], diagnosis of herpes virus infection in live cells [166], kidney stones [167, 168], and many more reported in literature. The diseased state is identified by the alteration in the level of different biomolecules associated with the body's physiological system. The spectral changes in diseased cells or tissue or biological fluid in comparison to normal cells were tabulated (Table 10.6). Biological samples are analyzed under the mid-IR range

Table 10.6 Spectral changes in the biomolecules in a various diseased state

Diseased state	Spectral changes in disease state	Ref.
Alzheimer's disease (AD)	The ratio of α helix β sheet structure of protein and lipid protein balance is a potential marker of AD. FTIR spectra of cerebrospinal fluid obtained from AD patients have shown the shift in peak position which is corresponding to the change in lipid, protein, and DNA contents in comparison to control subject	[169]
Cervical cancer	Spectra of epithelial cells represent the glycogen band at 1021 cm^{-1} and 1150 cm^{-1}. A reduction in band intensity in the case of the diseased state. Band intensity change in spectral region 1300–900 cm^{-1} is assigned for the phosphodiester bond observed due to malignancy	[63]
Lung cancer cells in pleural fluid	The noteworthy change in the spectral region (1030 and 1080 cm^{-1}) was observed in cancer cells and tuberculous cells in comparison to normal cells	[66, 170]
Fibroadenoma and carcinoma	• Band in the spectral region (3000 to 3600 cm^{-1}) shift to lower frequency • In a malignant tumor, α helix amide I frequency diminished and frequency of β sheet amide I found increased • The band positions at 1204 and 1278 cm^{-1} attributed to the collagen have shown distinctly strong frequency as compared to normal cells	[67]
Malignant non-Hodgkin's lymphoma	• An increase in lymphoma grade was observed due to a change in lipid and protein contents in the region of 2800–3000 cm^{-1} by elevated the CH_3:CH_2 ratio • Acclivity in malignancy grade was also identified by an increase in ribose contents with rising RNA:DNA ratio (996:966 cm^{-1})	[171]
Glioblastoma tumor	• Raman and FTIR spectra of brain tumor tissue show the difference in chemical composition, in comparison to healthy brain tissue. The difference in the lipid content is the major spectroscopic marker for the identification of glioblastoma • In the control sample, maximum absorbance was observed which is corresponding to asymmetric PO_3^{2-}, the group from phospholipid in comparison to glioblastoma tissue	[172]
Chronic kidney disease (CKD)	• The band position at 3364 cm^{-1} attributed to amide A has shown a decrease in band area as compared to control human saliva • Band position at 2063 cm^{-1} indicates the C-N stretching, and it has shown a strong decrease in peak area in case of the disease state • The band position at 1635 cm^{-1} assigned for amide I which doesn't show any alteration in CKD • The band position (990 cm^{-1}) for polysaccharide functional group doesn't show any change in CKD	[173]
Hypercholesterolemia	The serum sample derived from the subject has shown the high intensity of olefinic double bond (=CH) in FTIR spectra within the range of 3012 cm^{-1}	[174]

(400–4000 cm^{-1}), which corresponds to molecular vibration mode associated with molecular components. It reveals the information about the structural and functional complexity of components.

10.6 Conclusion

FTIR-spectroscopy in protein dynamics study is the early growing stage. The current chapter has attempted to explore the molecular information of different biomolecules found in the animal body and its alteration in the diseased state. FTIR and other spectroscopic techniques are complementary to each other for the diagnosis of disease. This method can be used with the principle component analysis, linear discriminant analysis, and PLS regression method (partial least square regression) for the analysis of biological samples. It can be performed using origin lab software. We also emphasize the basic principle behind the operation of FTIR for different types of samples. FTIR-spectroscopy has application in the measurement of lipid profile, carbohydrate contents, protein dynamics study as well as nucleic acid analysis. Site-specific side chain vibrational probe is a useful tool for the enhancement of the resolution of the IR spectra of any molecule. These site-specific probes are used to study complex biological systems that require many probes in a single experiment. This technique may be helpful in elucidating the protein aggregation/folding/misfolding mechanism. It is expected that the advancement in vibrational spectroscopy and computational method can be helpful in addressing many issues related to the disease diagnosis and mechanism of disease progression.

Acknowledgments SKP is thankful to MHRD for financial support. MM lab is supported by Grant No. BT/PR21857/NNT/28/1238/2017, EMR/2017/003054, Odisha DBT 3325/ST(BIO)-02/2017.

References

1. Doyle WM (1992) Principles and applications of Fourier transform infrared (FTIR) process analysis. Process Control Qual 2(1):11–41
2. Wea G (2009) FTIR analysis of protein structure. Course manual Chem 455:1–8
3. Krimm S, Bandekar J (1986) Vibrational spectroscopy and conformation of peptides, polypeptides, and proteins. In: Advances in protein chemistry, vol 38. Elsevier, Amsterdam, pp 181–364
4. Surewicz WK, Mantsch HH (1988) New insight into protein secondary structure from resolution-enhanced infrared spectra. Biochim Biophys Acta Prot Struct Mol Enzymol 952: 115–130
5. Kong J, Yu S (2007) Fourier transform infrared spectroscopic analysis of protein secondary structures. Acta Biochim Biophys Sin 39(8):549–559
6. Jiang Y, Li C, Nguyen X, Muzammil S, Towers E, Gabrielson J, Narhi L (2011) Qualification of FTIR spectroscopic method for protein secondary structural analysis. J Pharm Sci 100(11): 4631–4641
7. Lynch I, Dawson KA, Linse S (2006) Detecting cryptic epitopes created by nanoparticles. Sci STKE 327:pe14

8. Matsuo K, Yonehara R, Gekko K (2005) Improved estimation of the secondary structures of proteins by vacuum-ultraviolet circular dichroism spectroscopy. J Biochem 138(1):79–88
9. Greenfield NJ (2006) Using circular dichroism spectra to estimate protein secondary structure. Nat Protoc 1(6):2876
10. Arrondo JLR, Goñi FM (1999) Structure and dynamics of membrane proteins as studied by infrared spectroscopy. Prog Biophys Mol Biol 72(4):367–405
11. Griebenow K, Klibanov AM (1996) On protein denaturation in aqueous– organic mixtures but not in pure organic solvents. J Am Chem Soc 118(47):11695–11700
12. Goormaghtigh E, Raussens V, Ruysschaert J-M (1999) Attenuated total reflection infrared spectroscopy of proteins and lipids in biological membranes. Biochimica et Biophysica Acta (BBA)/Reviews on Biomembranes 2(1422):105–185
13. Liu K-Z, Shaw RA, Man A, Dembinski TC, Mantsch HH (2002) Reagent-free, simultaneous determination of serum cholesterol in HDL and LDL by infrared spectroscopy. Clin Chem 48(3):499–506
14. Deng H, Callender R (1999) Raman spectroscopic studies of the structures, energetics and bond distortions of substrates bound to enzymes. In: Methods in enzymology, vol 308. Elsevier, Amsterdam, pp 176–201
15. Gillespie RJ, Hargittai I (2013) The VSEPR model of molecular geometry. Courier Corporation, North Chelmsford
16. Lin-Vien D, Colthup NB, Fateley WG, Grasselli JG (1991) The handbook of infrared and Raman characteristic frequencies of organic molecules. Elsevier
17. Zandomeneghi G, Krebs MR, McCammon MG, Fändrich M (2004) FTIR reveals structural differences between native β-sheet proteins and amyloid fibrils. Protein Sci 13(12):3314–3321
18. Ma J, Pazos IM, Zhang W, Culik RM, Gai F (2015) Site-specific infrared probes of proteins. Annu Rev Phys Chem 66:357–377
19. Ye S, Zaitseva E, Caltabiano G, Schertler GF, Sakmar TP, Deupi X, Vogel R (2010) Tracking G-protein-coupled receptor activation using genetically encoded infrared probes. Nature 464(7293):1386
20. Taskent-Sezgin H, Chung J, Banerjee PS, Nagarajan S, Dyer RB, Carrico I, Raleigh DP (2010) Azidohomoalanine: a conformationally sensitive IR probe of protein folding, protein structure, and electrostatics. Angew Chem Int Ed 49(41):7473–7475
21. Remorino A, Hochstrasser RM (2012) Three-dimensional structures by two-dimensional vibrational spectroscopy. Acc Chem Res 45(11):1896–1905
22. Krummel AT, Zanni MT (2008) Evidence for coupling between nitrile groups using DNA templates: a promising new method for monitoring structures with infrared spectroscopy. J Phys Chem B 112(5):1336–1338
23. Suydam IT, Snow CD, Pande VS, Boxer SG (2006) Electric fields at the active site of an enzyme: direct comparison of experiment with theory. Science 313(5784):200–204
24. Jeon J, Yang S, Choi J-H, Cho M (2009) Computational vibrational spectroscopy of peptides and proteins in one and two dimensions. Acc Chem Res 42(9):1280–1289
25. Zheng J, Kwak K, Fayer M (2007) Ultrafast 2D IR vibrational echo spectroscopy. Acc Chem Res 40(1):75–83
26. Oh K-I, Lee J-H, Joo C, Han H, Cho M (2008) β-Azidoalanine as an IR probe: application to amyloid Aβ (16-22) aggregation. J Phys Chem B 112(33):10352–10357
27. Bischak CG, Longhi S, Snead DM, Costanzo S, Terrer E, Londergan CH (2010) Probing structural transitions in the intrinsically disordered C-terminal domain of the measles virus nucleoprotein by vibrational spectroscopy of cyanylated cysteines. Biophys J 99(5): 1676–1683
28. Stafford AJ, Walker DM, Webb LJ (2012) Electrostatic effects of mutations of Ras glutamine 61 measured using vibrational spectroscopy of a thiocyanate probe. Biochemistry 51(13): 2757–2767

29. Tadesse L, Nazarbaghi R, Walters L (1991) Isotopically enhanced infrared spectroscopy: a novel method for examining secondary structure at specific sites in conformationally heterogeneous peptides. J Am Chem Soc 113(18):7036–7037
30. Decatur SM, Antonic J (1999) Isotope-edited infrared spectroscopy of helical peptides. J Am Chem Soc 121(50):11914–11915
31. Decatur SM (2006) Elucidation of residue-level structure and dynamics of polypeptides via isotope-edited infrared spectroscopy. Acc Chem Res 39(3):169–175
32. Lin Y-S, Shorb J, Mukherjee P, Zanni M, Skinner J (2008) Empirical amide I vibrational frequency map: application to 2D-IR line shapes for isotope-edited membrane peptide bundles. J Phys Chem B 113(3):592–602
33. Daggett KA, Sakmar TP (2011) Site-specific in vitro and in vivo incorporation of molecular probes to study G-protein-coupled receptors. Curr Opin Chem Biol 15(3):392–398
34. Kim H, Cho M (2013) Infrared probes for studying the structure and dynamics of biomolecules. Chem Rev 113(8):5817–5847
35. Getahun Z, Huang C-Y, Wang T, De León B, DeGrado WF, Gai F (2003) Using nitrile-derivatized amino acids as infrared probes of local environment. J Am Chem Soc 125(2): 405–411
36. Fafarman AT, Webb LJ, Chuang JI, Boxer SG (2006) Site-specific conversion of cysteine thiols into thiocyanate creates an IR probe for electric fields in proteins. J Am Chem Soc 128(41):13356–13357
37. Waegele MM, Tucker MJ, Gai F (2009) 5-Cyanotryptophan as an infrared probe of local hydration status of proteins. Chem Phys Lett 478(4–6):249–253
38. Tucker MJ, Kim YS, Hochstrasser RM (2009) 2D IR photon echo study of the anharmonic coupling in the OCN region of phenyl cyanate. Chem Phys Lett 470(1–3):80–84
39. Bloem R, Koziol K, Waldauer SA, Buchli B, Walser R, Samatanga B, Jelesarov I, Hamm P (2012) Ligand binding studied by 2D IR spectroscopy using the azidohomoalanine label. J Phys Chem B 116(46):13705–13712
40. Bazewicz CG, Liskov MT, Hines KJ, Brewer SH (2013) Sensitive, site-specific, and stable vibrational probe of local protein environments: 4-azidomethyl-L-phenylalanine. J Phys Chem B 117(30):8987–8993
41. Dutta S, Li Y-L, Rock W, Houtman JC, Kohen A, Cheatum CM (2011) 3-Picolyl azide adenine dinucleotide as a probe of femtosecond to picosecond enzyme dynamics. J Phys Chem B 116(1):542–548
42. Fried SD, Bagchi S, Boxer SG (2013) Measuring electrostatic fields in both hydrogen-bonding and non-hydrogen-bonding environments using carbonyl vibrational probes. J Am Chem Soc 135(30):11181–11192
43. Pazos IM, Ghosh A, Tucker MJ, Gai F (2014) Ester carbonyl vibration as a sensitive probe of protein local electric field. Angew Chem Int Ed 53(24):6080–6084
44. Barth A (2000) The infrared absorption of amino acid side chains. Prog Biophys Mol Biol 74(3–5):141–173
45. Bath A, Zscherp C (2002) What vibrations tell about proteins. Q Rev Biophys 35:369–430
46. Wilson EB, Decius JC, Cross PC (1980) Molecular vibrations: the theory of infrared and Raman vibrational spectra. Courier Corporation, North Chelmsford
47. Hochstrasser RM (2001) Two-dimensional IR-spectroscopy: polarization anisotropy effects. Chem Phys 266(2–3):273–284
48. Brenig W, Küchenhoff S, Kasai H (1990) Molecular vibration dynamics in molecule-surface interactions. Applied Physics A 51(2):115–120
49. Le Ru EC (2019) Snapshots of vibrating molecules. Nature Publishing Group
50. Banwell CN, McCash EM (1994) Fundamentals of molecular spectroscopy, vol 851. McGraw-Hill, New York
51. Liu KZ, Xu M, Scott DA (2007) Biomolecular characterisation of leucocytes by infrared spectroscopy. Br J Haematol 136(5):713–722

52. Schulz H, Baranska M (2007) Identification and quantification of valuable plant substances by IR and Raman spectroscopy. Vib Spectrosc 43(1):13–25
53. Dovbeshko G, Chegel V, Gridina NY, Repnytska O, Shirshov Y, Tryndiak V, Todor I, Solyanik G (2002) Surface enhanced IR absorption of nucleic acids from tumor cells: FTIR reflectance study. Biopolymers: Original Research on Biomolecules 67(6):470–486
54. Li X, Lin J, Ding J, Wang S, Liu Q, Qing S (2004) Raman spectroscopy and fluorescence for the detection of liver cancer and abnormal liver tissue. In: The 26th annual international conference of the IEEE engineering in medicine and biology society. IEEE, pp 212–215
55. Wood BR, Quinn MA, Tait B, Ashdown M, Hislop T, Romeo M, McNaughton D (1998) FTIR microspectroscopic study of cell types and potential confounding variables in screening for cervical malignancies. Biospectroscopy 4(2):75–91
56. Shetty G, Kendall C, Shepherd N, Stone N, Barr H (2006) Raman spectroscopy: elucidation of biochemical changes in carcinogenesis of oesophagus. Br J Cancer 94(10):1460
57. Chiriboga L, Xie P, Yee H, Vigorita V, Zarou D, Zakim D, Diem M (1998) Infrared spectroscopy of human tissue. I. Differentiation and maturation of epithelial cells in the human cervix. Biospectroscopy 4(1):47–53
58. Fung MFK, Senterman MK, Mikhael NZ, Lacelle S, Wong PT (1996) Pressure-tuning fourier transform infrared spectroscopic study of carcinogenesis in human endometrium. Biospectroscopy 2(3):155–165
59. Dovbeshko GI, Gridina NY, Kruglova EB, Pashchuk OP (2000) FTIR spectroscopy studies of nucleic acid damage. Talanta 53(1):233–246
60. Gazi E, Dwyer J, Gardner P, Ghanbari-Siahkali A, Wade A, Miyan J, Lockyer NP, Vickerman JC, Clarke NW, Shanks JH (2003) Applications of Fourier transform infrared microspectroscopy in studies of benign prostate and prostate cancer. A pilot study. The Journal of Pathology: A Journal of the Pathological Society of Great Britain and Ireland 201(1): 99–108
61. Yoshida S, Miyazaki M, Sakai K, Takeshita M, Yuasa S, Sato A, Kobayashi T, Watanabe S, Okuyama H (1997) Fourier transform infrared spectroscopic analysis of rat brain microsomal membranes modified by dietary fatty acids: possible correlation with altered learning behavior. Biospectroscopy 3(4):281–290
62. Alben J, Choi S, Adler A, Caughey W (1973) Infrared spectroscopy of porphyrins. The porphyrins 3(Part A):323
63. Wood BR, Quinn MA, Burden FR, McNaughton D (1996) An investigation into FTIR spectroscopy as a biodiagnostic tool for cervical cancer. Biospectroscopy 2(3):143–153
64. Huleihel M, Salman A, Erukhimovitch V, Ramesh J, Hammody Z, Mordechai S (2002) Novel spectral method for the study of viral carcinogenesis in vitro. J Biochem Biophys Methods 50(2–3):111–121
65. Fabian H, Jackson M, Murphy L, Watson PH, Fichtner I, Mantsch HH (1995) A comparative infrared spectroscopic study of human breast tumors and breast tumor cell xenografts. Biospectroscopy 1(1):37–45
66. Wang H-P, Wang H-C, Huang Y-J (1997) Microscopic FTIR studies of lung cancer cells in pleural fluid. Sci Total Environ 204(3):283–287
67. Eckel R, Huo H, Guan H-W, Hu X, Che X, Huang W-D (2001) Characteristic infrared spectroscopic patterns in the protein bands of human breast cancer tissue. Vib Spectrosc 27(2):165–173
68. Andrus PG, Strickland RD (1998) Cancer grading by Fourier transform infrared spectroscopy. Biospectroscopy 4(1):37–46
69. Yang Y, Sulé-Suso J, Sockalingum GD, Kegelaer G, Manfait M, El Haj AJ (2005) Study of tumor cell invasion by Fourier transform infrared microspectroscopy. Biopolymers: Original Research on Biomolecules 78(6):311–317
70. McIntosh LM, Jackson M, Mantsch HH, Stranc MF, Pilavdzic D, Crowson AN (1999) Infrared spectra of basal cell carcinomas are distinct from non-tumor-bearing skin components. J Investig Dermatol 112(6):951–956

71. Fujioka N, Morimoto Y, Arai T, Kikuchi M (2004) Discrimination between normal and malignant human gastric tissues by Fourier transform infrared spectroscopy. Cancer Detect Prev 28(1):32–36
72. Laska J, Widlarz J (2005) Spectroscopic and structural characterization of low molecular weight fractions of polyaniline. Polymer 46(5):1485–1495
73. Ruiz-Chica A, Medina M, Sanchez-Jimenez F, Ramirez F (2004) Characterization by Raman spectroscopy of conformational changes on guanine–cytosine and adenine–thymine oligonucleotides induced by aminooxy analogues of spermidine. J Raman Spectrosc 35(2): 93–100
74. Wu JG, Xu YZ, Sun CW, Soloway RD, Xu DF, Wu QG, Sun KH, Weng SF, Xu GX (2001) Distinguishing malignant from normal oral tissues using FTIR fiber-optic techniques. Biopolym Orig Res Biomol 62(4):185–192
75. Naumann D (1998) Infrared and NIR Raman spectroscopy in medical microbiology. In: Infrared spectroscopy: new tool in medicine. International Society for Optics and Photonics, Bellingham, pp 245–257
76. Hermann P, Hoehl A, Ulrich G, Fleischmann C, Hermelink A, Kästner B, Patoka P, Hornemann A, Beckhoff B, Rühl E (2014) Characterization of semiconductor materials using synchrotron radiation-based near-field infrared microscopy and nano-FTIR spectroscopy. Opt Express 22(15):17948–17958
77. Sri KB, Bhairavi B, Swethasri R (2019) Gaschromatography-FTIR spectroscopy-a review. Journal of the Gujarat Research Society 21(3):226–237
78. Buckley S (2019) Broadband High-Brightness Sources for Spectroscopy. Spectroscopy (Santa Monica) 34(1):16–23
79. Mendelson Y, Lin B-c, Peura RA, Clermont AC (1988) Carbon dioxide laser based multiple ATR technique for measuring glucose in aqueous solutions. Appl Opt 27(24):5077–5081
80. Weida MJ, Yee B (2011) Quantum cascade laser-based replacement for FTIR microscopy. In: Imaging, manipulation, and analysis of biomolecules, cells, and tissues IX. International Society for Optics and Photonics, p 79021C
81. Scherzer T, Tauber A, Mehnert R (2002) UV curing of pressure sensitive adhesives studied by real-time FTIR-ATR spectroscopy. Vib Spectrosc 29(1–2):125–131
82. Nasse MJ, Walsh MJ, Mattson EC, Reininger R, Kajdacsy-Balla A, Macias V, Bhargava R, Hirschmugl CJ (2011) High-resolution Fourier-transform infrared chemical imaging with multiple synchrotron beams. Nat Methods 8(5):413
83. Lewis EN, Treado PJ, Reeder RC, Story GM, Dowrey AE, Marcott C, Levin IW (1995) Fourier transform spectroscopic imaging using an infrared focal-plane array detector. Anal Chem 67(19):3377–3381
84. Yang H, Yang S, Kong J, Dong A, Yu S (2015) Obtaining information about protein secondary structures in aqueous solution using Fourier transform IR spectroscopy. Nat Protoc 10(3):382
85. Hoşafçı G, Klein O, Oremek G, Mäntele W (2007) Clinical chemistry without reagents? An infrared spectroscopic technique for determination of clinically relevant constituents of body fluids. Anal Bioanal Chem 387(5):1815
86. Bugay DE, Bushweller CH, Danehey CT Jr, Hoogasian S, Blersch JA, Leenstra WR (1989) Complementary IR/NMR approach for the determination of IR extinction coefficients and thermodynamic parameters for conformers in rapid equilibrium. The Halocyclohexanes. J Phys Chem 93(10):3908–3911
87. Dreissig I, Machill S, Salzer R, Krafft C (2009) Quantification of brain lipids by FTIR spectroscopy and partial least squares regression. Spectrochim Acta A Mol Biomol Spectrosc 71(5):2069–2075
88. Otto M (2016) Chemometrics: statistics and computer application in analytical chemistry. John Wiley & Sons

89. Canteri MH, Renard CM, Le Bourvellec C, Bureau S (2019) ATR-FTIR spectroscopy to determine cell wall composition: application on a large diversity of fruits and vegetables. Carbohydr Polym 212:186–196
90. Leopold LF, Leopold N, Diehl H-A, Socaciu C (2011) Quantification of carbohydrates in fruit juices using FTIR spectroscopy and multivariate analysis. J Spectrosc 26(2):93–104
91. Tewari J, Irudayaraj J (2004) Quantification of saccharides in multiple floral honeys using fourier transform infrared microattenuated total reflectance spectroscopy. J Agric Food Chem 52(11):3237–3243
92. Anjos O, Campos MG, Ruiz PC, Antunes P (2015) Application of FTIR-ATR spectroscopy to the quantification of sugar in honey. Food Chem 169:218–223
93. Gallardo-Velázquez T, Osorio-Revilla G, Zuñiga-de Loa M, Rivera-Espinoza Y (2009) Application of FTIR-HATR spectroscopy and multivariate analysis to the quantification of adulterants in Mexican honeys. Food Res Int 42(3):313–318
94. Bureau S, Ruiz D, Reich M, Gouble B, Bertrand D, Audergon J-M, Renard CM (2009) Application of ATR-FTIR for a rapid and simultaneous determination of sugars and organic acids in apricot fruit. Food Chem 115(3):1133–1140
95. Mathlouthi M, Koenig JL (1987) Vibrational spectra of carbohydrates. In: Advances in carbohydrate chemistry and biochemistry, vol 44. Elsevier, pp 7–89
96. Sekkal M, Dincq V, Legrand P, Huvenne J (1995) Investigation of the glycosidic linkages in several oligosaccharides using FT-IR and FT Raman spectroscopies. J Mol Struct 349:349–352
97. Kačuráková M, Wilson R (2001) Developments in mid-infrared FT-IR spectroscopy of selected carbohydrates. Carbohydr Polym 44(4):291–303
98. Defernez M, Wilson RH (1995) Mid-infrared spectroscopy and chemometrics for determining the type of fruit used in jam. J Sci Food Agric 67(4):461–467
99. Wilson R, Slack P, Appleton G, Sun L, Belton P (1993) Determination of the fruit content of jam using Fourier transform infrared spectroscopy. Food Chem 47(3):303–308
100. Defernez M, Kemsley EK, Wilson RH (1995) Use of infrared spectroscopy and chemometrics for the authentication of fruit purees. J Agric Food Chem 43(1):109–113
101. Tatulian SA (2019) FTIR analysis of proteins and protein–membrane interactions. In: Lipid-protein interactions. Springer, Cham, pp 281–325
102. Ami D, Mereghetti P, Foli A, Tasaki M, Milani P, Nuvolone M, Palladini G, Merlini G, Lavatelli F, Natalello A (2019) ATR-FTIR spectroscopy supported by multivariate analysis for the characterization of adipose tissue aspirates from patients affected by systemic amyloidosis. Anal Chem 91(4):2894–2900
103. Zhou R-B, Lu X-L, Dong C, Ahmad F, Zhang C-Y, Yin D-C (2018) Application of protein crystallization methodologies to enhance the solubility, stability and monodispersity of proteins. CrystEngComm 20(14):1923–1927
104. Tai H, Nishikawa K, Higuchi Y, Zw M, Hirota S (2019) Cysteine SH and glutamate COOH contributions to [NiFe] hydrogenase proton transfer revealed by highly sensitive FTIR spectroscopy. Angew Chem 131(38):13419–13424
105. Dioumaev AK, Braiman MS (1995) Modeling vibrational spectra of amino acid side chains in proteins: the carbonyl stretch frequency of buried carboxylic residues. J Am Chem Soc 117(42):10572–10574
106. Cao Y, Varo G, Klinger AL, Czajkowsky DM, Braiman MS, Needleman R, Lanyi JK (1993) Proton transfer from Asp-96 to the bacteriorhodopsin Schiff base is caused by a decrease of the pKa of Asp-96 which follows a protein backbone conformational change. Biochemistry 32(8): 1981–1990
107. Braiman MS, Briercheck DM, Kriger KM (1999) Modeling vibrational spectra of amino acid side chains in proteins: effects of protonation state, counterion, and solvent on arginine C– N stretch frequencies. J Phys Chem B 103(22):4744–4750

108. Chirgadze YN, Fedorov O, Trushina N (1975) Estimation of amino acid residue side-chain absorption in the infrared spectra of protein solutions in heavy water. Biopolymers: Original Research on Biomolecules 14(4):679–694
109. Hasegawa K, Ono T-a, Noguchi T (2000) Vibrational spectra and ab initio DFT calculations of 4-methylimidazole and its different protonation forms: infrared and Raman markers of the protonation state of a histidine side chain. J Phys Chem B 104(17):4253–4265
110. Nara M, Tasumi M, Tanokura M, Hiraoki T, Yazawa M, Tsutsumi A (1994) Infrared studies of interaction between metal ions and Ca2+−binding proteins marker bands for identifying the types of coordination of the side-chain COO− groups to metal ions in pike parvalbumin (pI= 4.10). FEBS Lett 349(1):84–88
111. Pinchas S, Laulicht I (1971) Infrared spectra of labelled compounds. Wiley-Liss, New York
112. Caswell DS, Spiro TG (1987) Proline signals in ultraviolet resonance Raman spectra of proteins: cis-trans isomerism in polyproline and ribonuclease a. J Am Chem Soc 109(9): 2796–2800
113. Venyaminov SY, Kalnin N (1990) Quantitative IR spectrophotometry of peptide compounds in water (H2O) solutions I. spectral parameters of amino acid residue absorption bands. Biopolymers: Original Research on Biomolecules 30(13–14):1243–1257
114. Sengupta PK, Krimm S (1985) Vibrational analysis of peptides, polypeptides, and proteins. XXXII. α-poly (l-glutamic acid). Biopolymers: Original Research on Biomolecules 24(8): 1479–1491
115. Goormaghtigh E, Ruysschaert J-M, Raussens V (2006) Evaluation of the information content in infrared spectra for protein secondary structure determination. Biophys J 90(8):2946–2957
116. Usoltsev D, Sitnikova V, Kajava A, Uspenskaya M (2019) Systematic FTIR spectroscopy study of the secondary structure changes in human serum albumin under various denaturation conditions. Biomol Ther 9(8):359
117. Arunkumar R, Drummond CJ, Greaves TL (2019) FTIR spectroscopic study of the secondary structure of globular proteins in aqueous Protic ionic liquids. Front Chem 7:74
118. Szyk-Warszyńska L, Raszka K, Warszyński P (2019) Interactions of casein and polypeptides in multilayer films studied by FTIR and molecular dynamics. Polymers 11(5):920
119. Grdadolnik J (2003) Saturation effects in FTIR spectroscopy: intensity of amide I and amide II bands in protein spectra. Acta Chim Slov 50(4):777–788
120. Schweitzer-Stenner R, Eker F, Huang Q, Griebenow K, Mroz PA, Kozlowski PM (2002) Structure analysis of dipeptides in water by exploring and utilizing the structural sensitivity of amide III by polarized visible Raman, FTIR− spectroscopy and DFT based Normal coordinate analysis. J Phys Chem B 106(16):4294–4304
121. Anderle G, Mendelsohn R (1987) Thermal denaturation of globular proteins. Fourier transform-infrared studies of the amide III spectral region. Biophys J 52(1):69–74
122. Gerwert K (1993) Molecular reaction mechanisms of proteins as monitored by time-resolved FTIR spectroscopy. Curr Opin Struct Biol 3(5):769–773
123. Remy A, Gerwert K (2003) Coupling of light-induced electron transfer to proton uptake in photosynthesis. Nat Struct Mol Biol 10(8):637
124. Banyay M, Sarkar M, Gräslund A (2003) A library of IR bands of nucleic acids in solution. Biophys Chem 104(2):477–488
125. Lindqvist M, Sarkar M, Winqvist A, Rozners E, Strömberg R, Gräslund A (2000) Optical spectroscopic study of the effects of a single deoxyribose substitution in a ribose backbone: implications in RNA− RNA interaction. Biochemistry 39(7):1693–1701
126. Lindqvist M, Gräslund A (2001) An FTIR and CD study of the structural effects of G-tract length and sequence context on DNA conformation in solution. J Mol Biol 314(3):423–432
127. Geinguenaud F, Liquier J, Brevnov MG, Petrauskene OV, Alexeev YI, Gromova ES, Taillandier E (2000) Parallel self-associated structures formed by T. C-rich sequences at acidic pH Biochemistry 39(41):12650–12658

128. Mohammadi S, Klement R, Shchyolkina AK, Liquier J, Jovin TM, Taillandier E (1998) FTIR and UV spectroscopy of parallel-stranded DNAs with mixed a• T/G• C sequences and their a• T/I• C analogues. Biochemistry 37(47):16529–16537
129. White A, Powell J (1995) Observation of the hydration-dependent conformation of the (dG) 20. Cntdot.(dG) 20 (dC) 20 oligonucleotide triplex using FTIR spectroscopy. Biochemistry 34(4):1137–1142
130. Ouali M, Letellier R, Sun JS, Akhebat A, Adnet F, Liquier J, Taillandier E (1993) Determination of G* G. cntdot. C triple-helix structure by molecular modeling and vibrational spectroscopy. J Am Chem Soc 115(10):4264–4270
131. Dagneaux C, Liquier J, Taillandier E (1995) FTIR study of a nonclassical dT10* dA10-dT10 intramolecular triple helix. Biochemistry 34(45):14815–14818
132. Liquier J (1996) Infrared spectroscopy of nucleic acids. In: Infrared spectroscopy of biomolecules. Wiley-Liss, New York
133. Tsuboi M (1970) Application of infrared spectroscopy to structure studies of nucleic acids. Appl Spectrosc Rev 3(1):45–90
134. Tsuboi M, Takahashi S, Harada I (1973) Infrared and Raman spectra of nucleic acids—vibrations in the base-residues. In: Structural studies on nucleic acids and other biopolymers. Elsevier, pp 91–145
135. Parikh SJ, Mukome FN, Zhang X (2014) ATR–FTIR spectroscopic evidence for biomolecular phosphorus and carboxyl groups facilitating bacterial adhesion to iron oxides. Colloids Surf B: Biointerfaces 119:38–46
136. Sahoo JK, Paikra SK, Mishra M, Sahoo H (2019) Amine functionalized magnetic iron oxide nanoparticles: synthesis, antibacterial activity and rapid removal of Congo red dye. J Mol Liq 282:428–440
137. Elzinga EJ, Huang J-H, Chorover J, Kretzschmar R (2012) ATR-FTIR spectroscopy study of the influence of pH and contact time on the adhesion of Shewanella putrefaciens bacterial cells to the surface of hematite. Environ Sci Technol 46(23):12848–12855
138. Cagnasso M, Boero V, Franchini MA, Chorover J (2010) ATR-FTIR studies of phospholipid vesicle interactions with α-FeOOH and α-Fe2O3 surfaces. Colloids Surf B: Biointerfaces 76(2):456–467
139. Olsson R, Giesler R, Loring JS, Persson P (2010) Adsorption, desorption, and surface-promoted hydrolysis of glucose-1-phosphate in aqueous goethite (α-FeOOH) suspensions. Langmuir 26(24):18760–18770
140. Fang L, Cao Y, Huang Q, Walker SL, Cai P (2012) Reactions between bacterial exopolymers and goethite: a combined macroscopic and spectroscopic investigation. Water Res 46(17): 5613–5620
141. Petibois C, Desbat B (2010) Clinical application of FTIR imaging: new reasons for hope. Trends Biotechnol 28(10):495–500
142. Erukhimovitch V, Pavlov V, Talyshinsky M, Souprun Y, Huleihel M (2005) FTIR microscopy as a method for identification of bacterial and fungal infections. J Pharm Biomed Anal 37(5): 1105–1108
143. Drogat B, Bouchecareilh M, North S, Petibois C, Déléris G, Chevet E, Bikfalvi A, Moenner M (2007) Acute L-glutamine deprivation compromises VEGF-a upregulation in A549/8 human carcinoma cells. J Cell Physiol 212(2):463–472
144. Petibois C, Déléris G (2004) Oxidative stress effects on erythrocytes determined by FT-IR spectrometry. Analyst 129(10):912–916
145. Serulle Y, Sugimori M, Llinás RR (2007) Imaging synaptosomal calcium concentration microdomains and vesicle fusion by using total internal reflection fluorescent microscopy. Proc Natl Acad Sci 104(5):1697–1702
146. Mackanos MA, Contag CH (2009) FTIR microspectroscopy for improved prostate cancer diagnosis. Trends Biotechnol 27(12):661–663

147. Baker MJ, Gazi E, Brown MD, Shanks JH, Gardner P, Clarke NW (2008) FTIR-based spectroscopic analysis in the identification of clinically aggressive prostate cancer. Br J Cancer 99(11):1859–1866
148. Whiteman S, Yang Y, Jones J, Spiteri M (2008) FTIR spectroscopic analysis of sputum: preliminary findings on a potential novel diagnostic marker for COPD. Ther Adv Respir Dis 2(1):23–31
149. Ahmed SS, Santosh W, Kumar S, Christlet THT (2010) Neural network algorithm for the early detection of Parkinson's disease from blood plasma by FTIR micro-spectroscopy. Vib Spectrosc 53(2):181–188
150. Erukhimovitch V, Talyshinsky M, Souprun Y, Huleihel M (2006) FTIR spectroscopy examination of leukemia patients plasma. Vib Spectrosc 40(1):40–46
151. Lopes J, Correia M, Martins I, Henriques AG, Delgadillo I, da Cruz e Silva O, Nunes A (2016) FTIR and Raman spectroscopy applied to dementia diagnosis through analysis of biological fluids. J Alzheimers Dis 52(3):801–812
152. Sitole L, Steffens F, Krüger TP, Meyer D (2014) Mid-ATR-FTIR spectroscopic profiling of HIV/AIDS sera for novel systems diagnostics in global health. Omics: a journal of integrative biology 18(8):513–523
153. Dimitrova M, Ivanova D, Karamancheva I, Milev A, Dobrev I (2009) Application of FTIR-spectroscopy for diagnosis of breast cancer tumors. Journal of the University of Chemical Technology and Metallurgy 44(3):297–300
154. Kleiren E, Ruysschaert J-M, Goormaghtigh E, Raussens V (2010) Development of a quantitative and conformation-sensitive ATR-FTIR biosensor for Alzheimer's disease: the effect of deuteration on the detection of the Aβ peptide. J Spectrosc 24(1–2):61–66
155. Liao CR, Rak M, Lund J, Unger M, Platt E, Albensi BC, Hirschmugl CJ, Gough KM (2013) Synchrotron FTIR reveals lipid around and within amyloid plaques in transgenic mice and Alzheimer's disease brain. Analyst 138(14):3991–3997
156. Untereiner V, Dhruvananda Sockalingum G, Garnotel R, Gobinet C, Ramaholimihaso F, Ehrhard F, Diebold MD, Thiéfin G (2014) Bile analysis using high-throughput FTIR spectroscopy for the diagnosis of malignant biliary strictures: a pilot study in 57 patients. J Biophotonics 7(3–4):241–253
157. Marickar YF, Lekshmi P, Varma L, Koshy P (2009) Problem in analyzing cystine stones using FTIR spectroscopy. Urol Res 37(5):263–269
158. Travo A, Paya C, Déléris G, Colin J, Mortemousque B, Forfar I (2014) Potential of FTIR spectroscopy for analysis of tears for diagnosis purposes. Anal Bioanal Chem 406(9–10): 2367–2376
159. Dong L, Sun X, Chao Z, Zhang S, Zheng J, Gurung R, Du J, Shi J, Xu Y, Zhang Y (2014) Evaluation of FTIR spectroscopy as diagnostic tool for colorectal cancer using spectral analysis. Spectrochim Acta A Mol Biomol Spectrosc 122:288–294
160. Simsek Ozek N, Zeller I, Renaud D, Gümüş P, Nizam N, Severcan F, Buduneli N, Scott D (2016) Differentiation of chronic and aggressive periodontitis by FTIR spectroscopy. J Dent Res 95(13):1472–1478
161. Hands JR, Dorling KM, Abel P, Ashton KM, Brodbelt A, Davis C, Dawson T, Jenkinson MD, Lea RW, Walker C (2014) Attenuated total reflection Fourier transform infrared (ATR-FTIR) spectral discrimination of brain tumour severity from serum samples. J Biophotonics 7(3–4): 189–199
162. Roy S, Perez-Guaita D, Andrew DW, Richards JS, McNaughton D, Heraud P, Wood BR (2017) Simultaneous ATR-FTIR based determination of malaria parasitemia, glucose and urea in whole blood dried onto a glass slide. Anal Chem 89(10):5238–5245
163. Kucuk Baloglu F, Baloglu O, Heise S, Brockmann G, Severcan F (2017) Triglyceride dependent differentiation of obesity in adipose tissues by FTIR spectroscopy coupled with chemometrics. J Biophotonics 10(10):1345–1355

164. Miller LM, Bourassa MW, Smith RJ (2013) FTIR spectroscopic imaging of protein aggregation in living cells. Biochimica et Biophysica Acta (BBA)-Biomembranes 1828(10): 2339–2346
165. Sultana RR, Zafarullah S, Kirubamani NH (2011) Utility of FTIR spectroscopic analysis of saliva of diabetic pregnant women in each trimester. Indian J Sci Technol 4(8):967–970
166. Salman A, Erukhimovitch V, Talyshinsky M, Huleihil M, Huleihel M (2002) FTIR spectroscopic method for detection of cells infected with herpes viruses. Biopolymers: Original Research on Biomolecules 67(6):406–412
167. Narasimhan K, Kaur B, Suri D, Mahajan J (2009) Diagnosis of renal stones with underlying metabolic abnormalities using FTIR spectroscopy. Indian J Pediatr 76(8):856–856
168. Khanmohammadi M, Ghasemi K, Garmarudi AB, Ramin M (2015) Diagnostic prediction of renal failure from blood serum analysis by FTIR spectrometry and chemometrics. Spectrochim Acta A Mol Biomol Spectrosc 136:1782–1785
169. Depciuch J, Zawlik I, Skrzypa M, Pająk J, Potocka N, Łach K, Bartosik-Psujek H, Koziorowska A, Kaznowska E, Cebulski J (2019) FTIR spectroscopy of cerebrospinal fluid reveals variations in the lipid: protein ratio at different stages of Alzheimer's disease. J Alzheimers Dis 68:1–13
170. Wang X, Shen X, Sheng D, Chen X, Liu X (2014) FTIR spectroscopic comparison of serum from lung cancer patients and healthy persons. Spectrochim Acta A Mol Biomol Spectrosc 122:193–197
171. Andrus PG (2006) Cancer monitoring by FTIR spectroscopy. Technol Cancer Res Treat 5(2): 157–167
172. Depciuch J, Tołpa B, Witek P, Szmuc K, Kaznowska E, Osuchowski M, Król P, Cebulski J (2020) Raman and FTIR spectroscopy in determining the chemical changes in healthy brain tissues and glioblastoma tumor tissues. Spectrochim Acta A Mol Biomol Spectrosc 225: 117526
173. Rodrigues RP, Aguiar EM, Cardoso-Sousa L, Caixeta DC, Guedes CC, Siqueira WL, Maia YCP, Cardoso SV, Sabino-Silva R (2019) Differential molecular signature of human saliva using ATR-FTIR spectroscopy for chronic kidney disease diagnosis. Braz Dent J 30(5): 437–445
174. Safiullah A, Mohan SK, Aea P (2018) FTIR-ATR spectroscopy as a diagnostic tool in the diagnosis of hypercholes-terolemia induced by cholesterol-rich diet in rats. International Journal of Research in Pharmaceutical Sciences 9(3):1056–1062

Raman Spectroscopy in Biology: Perspectives and Emerging Frontiers

11

Pradeep K. Sengupta

Abstract

Raman spectroscopy is based on the **Raman effect**, which refers to inelastic scattering of monochromatic light (in the visible, near-infrared, or ultraviolet range) incident on molecules, which results in a small fraction of the scattered light undergoing changes in frequency relative to the frequency of the incident photons. The frequency change corresponds to the frequency of transition between internal energy levels (usually molecular vibrational levels) of the scattering molecule.

11.1 Introduction

11.1.1 Raman Effect and Its Discovery

Raman spectroscopy is based on the **Raman effect**, which refers to inelastic scattering of monochromatic light (in the visible, near-infrared, or ultraviolet range) incident on molecules, which results in a small fraction of the scattered light undergoing changes in frequency relative to the frequency of the incident photons [1, 2]. The frequency change corresponds to the frequency of transition between internal energy levels (usually molecular vibrational levels) of the scattering molecule.

The seminal discovery of Raman effect was made on February 28, 1928, by Dr. C.V. Raman (then Palit Professor of Physics at the University of Calcutta) with his research student-collaborator K.S. Krishnan. The experiments were carried out in

P. K. Sengupta (✉)
Department of Biophysics, Molecular Biology and Bioinformatics, University of Calcutta, University College of Science, Kolkata, India

© The Author(s), under exclusive license to Springer Nature Singapore Pte Ltd. 2022
H. Sahoo (ed.), *Optical Spectroscopic and Microscopic Techniques*,
https://doi.org/10.1007/978-981-16-4550-1_11

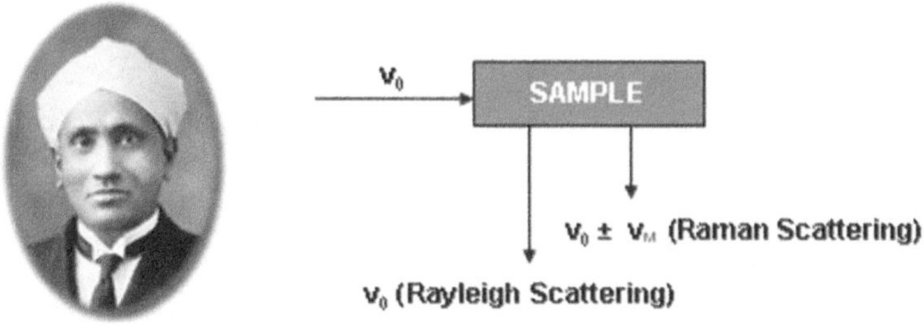

C. V. Raman(1888-1970)

Fig. 11.1 Raman and Rayleigh scattering processes. (A photograph of Prof. C.V. Raman is included)

Raman's laboratory (which was equipped with simple, rudimentary facilities) at the Indian Association for the Cultivation of Science (IACS), which was then located at 210 Bowbazar Street, Kolkata (then Calcutta), India. For this great discovery, Raman won the Nobel Prize in Physics in 1930, becoming the first Asian recipient of the Nobel Prize in a field of science. Figure 11.1 presents a simple pictorial depiction of the Raman scattering process.

The emergence of Raman spectroscopy as an analytical tool signified an important new addition to the repertoire of vibrational spectroscopy, providing an important and potentially more versatile alternative to infrared (IR) spectroscopy (based on absorption of IR radiation, inducing transitions between vibrational energy levels). Raman envisioned the immense future potential of the new spectroscopic technique he had just discovered. Indeed, in his Nobel lecture delivered at Stockholm on December 11, 1930, he made the prophetic remark '. . . *the new field of spectroscopy has practically unrestricted scope in the study of problems relating to the structure of matter*'. In a little over nine decades that have elapsed since Raman's original discovery, with many important technological breakthroughs having taken place, Professor Raman's dreams have been amply fulfilled.

11.1.2 Historical Perspective

Professor Raman's fascination for light scattering started in 1921 during a voyage through the Mediterranean Sea on his return trip to India from Europe, where he went to participate in the Congress of Universities of the British Empire held at Oxford, England. Raman, who possessed a lifelong interest in natural phenomena, especially natural colors, was curious about the origin of the beautiful blue opalescence of the Mediterranean. The prevailing idea then was based on Lord Rayleigh's belief that the blue color of the sea was due to reflection of the blue sky, which had been previously explained by Lord Rayleigh as being due to elastic (Rayleigh) scattering by atmospheric gas molecules. Raman was skeptic about Lord Rayleigh's explanation regarding the role of the sea water 'simply as a reflecting mirror for the blue sky', and since it was Raman's practice to carry some small optical equipment with him, he could conduct some simple experiments on board the ship. Using a Nicol prism polarizer, he was able to quench the surface reflection of the sky, while the blue color of the sea became more vivid. Hence, he concluded that contrary to Lord Rayleigh's belief, the color of the sea was fundamentally due to the same reason as that of the sky, namely, in case of the sea, due to elastic scattering of light by water molecules. Upon his return to Kolkata, Raman and his students immediately switched the emphasis of their research activities to the scattering of light by different liquids.

In 1923, the Austrian physicist Adolf Smekal theoretically predicted the change of frequency of light due to scattering by molecules. However, experimental observation of such a phenomenon presented a formidable challenge because of the extreme feebleness of inelastically scattered light. In fact, only 1 in about 10^6 photons participates in inelastic scattering, while the bulk of the photons (10^{-4}–10^{-3} parts of the incident light) undergo elastic scattering (Rayleigh scattering). While Raman's work on the scattering of light by liquids was in progress, A.H. Compton (USA) discovered that X-rays lose energy and thus shift to longer wavelengths [3], due to inelastic collisions with electrons (Compton Effect), and for this discovery, Compton received the Nobel Prize in Physics in 1927. Compton's success acted as a strong incentive for Raman to intensify his quest for the 'optical analogue of the Compton effect', due to scattering of visible light by molecules.

The set-up that Raman used for his initial findings [1, 2, 4] was striking in simplicity and elegance, and stands out as a remarkable testimony to his ingenuity and experimental prowess. Raman chose bright sunny days and used sunlight (brought to a strong focus by a telescope objective combined with an additional lens) as the source of light, complementary optical filters (which comprised a combination of blue-violet and yellow-green filters) as the analyzer to detect the wavelength change of the scattered light, some 60 liquid samples (freed from dust and fluorescent impurity by careful filtration and repeated distillation), which he took one by one in a glass container, while his own dark-adapted eyes served as the detector. As expected, the complementary filter combination completely extinguished the track of light through the sample, but to Raman's great delight, for all of the samples, the light track reappeared when the yellow-green filter was transferred to a place between the sample and the observer's eyes. This unequivocally proved modification of the frequency of light due to scattering by the liquids. Thus, after nearly 7 years of dedicated work on light scattering, Raman succeeded in making the first experimental observation of the inelastic scattering of photons in the optical region, and described his novel observations in a one-page paper (entitled '*A New Type of Secondary Radiation*') which he immediately cabled to Nature [1]. In order to obtain quantitative measurements of the spectral shifts, for his subsequent studies, Raman used an improved set-up which now consisted of filtered light from a mercury arc lamp, as the source of monochromatic radiation and a baby Hilger spectrograph involving photographic recording of the spectra, permitting measurements of the Raman spectral shifts [2]. Raman and his colleagues showed the applicability of the Raman scattering method to materials in various states, solids, liquids, and gases, proving the universality of the technique, with the ability to provide unique fingerprints of a broad range of molecules of interest.

11.1.3 Invention of Lasers and Other Technological Breakthroughs: Impact on the Growth and Development of Raman Spectroscopy

Owing to the initial lack of adequate technology, especially the nonavailability of intense light sources, for well over the first three decades since the original discovery of the Raman Effect (in 1928), the application of Raman spectroscopy was limited to relatively simple molecules. Strong Rayleigh background scattering from complex molecules prevented detection of the feeble Raman signals for most of their vibrational modes. The invention of laser sources, beginning with the ruby laser in 1960, followed by more efficient lasers, namely gas lasers and compact miniaturized diode lasers (which are excellent sources of highly intense, nearly monochromatic light covering ranges from the UV, spanning the visible range, and extending to near infrared), changed the scenario and revolutionized Raman spectroscopy. This aspect, along with other factors, namely the development of high-quality monochromators and sensitive light detectors comprising photomultiplier tubes (PMTs) and more recently, highly sensitive charged coupled detectors (CCDs), and inclusion of microscopy and imaging techniques have enormously advanced the scope and range of applications of Raman spectroscopy [5]. At least 25 variants of Raman spectroscopy have been developed with various goals [6], e.g., (a) to enhance the signal sensitivity (Resonance Raman and Surface-Enhanced Raman (SERS) techniques), (b) to selectively probe specific sites (e.g. functional groups of macromolecules (Resonance Raman)), and (c) to improve spatial resolution (Raman microscopy). Comprehensive, recent reviews have been published highlighting the theory and methodology of advances in various Raman spectroscopy techniques, with relevant applications [6–10]. Because of its noninvasive nature, the use of Raman spectroscopy has proliferated to diverse fields encompassing broad areas of chemistry, physics, and life sciences, nanotechnology, materials science, biomedical applications including diagnosis of malignancies in cancer cells and tissues and other diseases, pharmaceutical science, agricultural and food technology, and forensic science.

11.1.4 Scope of This Chapter

Since Raman spectroscopy is based on scattering of light, it is noninvasive in nature. Therefore, the field of biology, including the more recently developing biomedical field, has probably most significantly benefited from developments of Raman spectroscopy. The ease of studying solutions of biomolecules in water (for which Raman scattering is relatively weak, thus posing negligible interference in signal gathering), and opportunities for a wide range of applications, encompassing in vitro, ex vivo, and in vivo studies, have opened up new frontiers. The object of this chapter is to provide an overview on Raman spectroscopy, including theory and basic instrumentation, along with representative applications in biology, emphasizing biophysics, structural biology, and related aspects.

11.2 Principles: Basic Mechanisms of Photon–Molecule Interactions in Rayleigh Scattering Different Types of Raman Scattering and in IR Absorption

Raman and infrared spectroscopy are two alternative approaches of vibrational spectroscopy, which are useful for gaining information on molecular vibrations. A nonlinear molecule containing N atoms possesses $3N$-6 fundamental vibrations. This number is $3N$-5 for a linear molecule. The energies of these vibrations correspond to the infrared region of the electromagnetic spectrum.

Raman spectroscopy involves inelastic scattering of light, usually in the visible, ultraviolet, or near-infrared region. On the other hand, infrared (IR) spectroscopy involves direct absorption of electromagnetic radiation in the infrared region. In order for a molecular vibration to be Raman active, the vibration must be accompanied by a change in polarizability, α (which can be looked upon as the deformability of the electron cloud of the molecule by the electric field) so that the first derivative of α with respect to the normal coordinate Q (representing stretching or deformation vibration) must be nonzero. Thus, what happens in a Raman process is a change in the induced dipole moment, **P**, which is related to the electric field vector **E** as follows:

$$\mathbf{P} = \alpha \cdot \mathbf{E}$$

In order to absorb infrared radiation, a molecular vibration must cause a change in the dipole moment (μ), and accordingly, the first derivative of $\boldsymbol{\mu}$ with respect to the normal coordinate, Q, must be nonzero. Since the presence of a dipole moment and its change are essential for a molecular vibration to be IR active, the vibrational spectra of homonuclear diatomic (like O_2, N_2, H_2) could not be studied via IR absorption, owing to the absence of dipole moments. The discovery of the Raman technique in 1928 opened the door to exploration of the vibrational spectra of such molecules.

The essential differences between IR and Raman processes, Rayleigh scattering, and different types of the basic Raman scattering process (normal, pre-resonance, and Resonance Raman) are shown in the energy level diagram (Fig. 11.2) depicting the photon–molecule interactions involved.

At ambient temperature, the bulk of the molecules are in the ground vibrational level of the ground electronic state. For Rayleigh scattering, an incident photon with the monochromatic frequency, ν_o, momentarily raises the molecule to a virtual level represented by a dashed line (the level is 'virtual' in the sense that it is not a molecular eigenstate and has an energy corresponding to the energy of the exciting

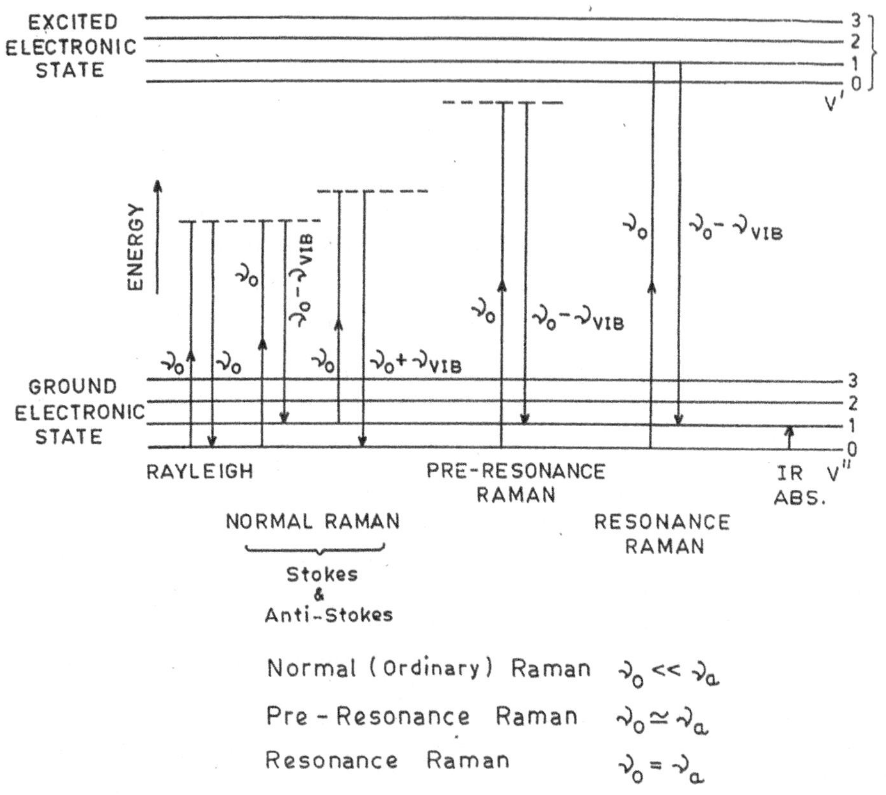

Fig. 11.2 Energy level diagram showing the transitions involved in Raman scattering, and other photon–molecule interactions of related interest differences between Norman, pre-resonance, and Resonance Raman scattering are indicated [4, 11]

photon). A second scattered photon with the same frequency as that of the incident photon appears, and therefore, this is an elastic scattering process (Rayleigh scattering). The scattered photons with modified frequencies, $\nu_o - \nu_{vib}$ and $\nu_o + \nu_{vib}$ are called Stokes and anti-Stokes Raman lines, respectively. For IR absorption and a particular vibration with frequency ν_{vib}, a photon with energy $h\nu_{vib}$ is absorbed, raising the molecule to a vibrationally excited level in the electronic ground state. Both elastic and inelastic scattering processes are two photon processes. Owing to the vastly greater population of molecules in the ground vibrational level (compared to those in vibrationally excited state), Stokes lines are significantly stronger than the anti-Stokes line, and therefore generally used in conventional Raman spectra. In Resonance Raman spectroscopy, the exciting frequency falls in the region of the electronic absorption spectrum (matching the absorption frequency ν_a of the molecule). In this situation, the polarizability, α, becomes frequency dependent, resulting in tremendous (>10^3 fold) increase in intensity of the Raman scattered light, resulting in its enhanced analytical sensitivity as compared to the weak normal Raman spectra. Besides higher sensitivity, another advantage of Resonance Raman is selectivity, permitting selective excitation of functional groups in macromolecules (e.g. the retinal part of the visual pigment rhodopsin, or specific components of molecular components like drug–protein or drug–DNA complexes), enabling selectively probing the structural changes in the component of particular interest [11–13].

11.3 Instrumentation

11.3.1 Basic Set-Up Used in a Raman Spectrometer

Two different approaches are used to collect basic Raman spectra, and consequently, two major types of Raman spectrometers are commercially available, and the choice is largely dependent on user requirement and preferences. These two types of instruments are: (a) Dispersive Raman spectrometer, and (b) Fourier Transform Raman spectrometer.

11.3.1.1 Dispersive Raman Spectrometer

A block diagram showing the basic components of a conventional, dispersive Raman spectrometer is shown in Fig. 11.3.

The components are: source of monochromatic light (Laser), sample compartment, monochromator (to analyze the Raman scattered light), and detector (Photomultiplier (PMT) or charge coupled detector (CCD)), computer, and recorder. The collections as well as processing of the spectra are controlled by a computer (PC). A variety of laser sources can be used as the excitation source. These include: argon ion (514.5 and 488.0 nm), Krypton ion (530.9 and 547.1 nm), He: NE (632.8 nm), Nd: YAG (1064 nm fundamental, with second, third, and fourth harmonics being available), and diode laser (630 and 780 nm).While in previous decades gas lasers were predominantly used for visible range excitation, they are being gradually replaced by solid state diode-based lasers, in view of the superior spectral quality and stability of these new generation lasers. Generally, the following three different types of CW (continuous wave) solid-state, diode laser light sources are commercially available: (1) Diode-pumped single longitudinal mode (SLM) lasers, which are readily available in compact form with a choice of a wide range of wavelengths from UV to near-IR range, (2) single-mode diode lasers, and

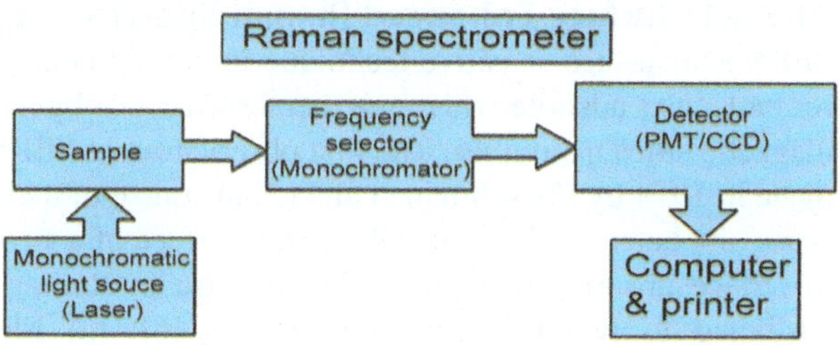

Fig. 11.3 Block diagram showing the components of a conventional, dispersive type Raman spectrometer. (Adapted in slightly modified form, from Refs. [7, 14])

(3) Volume Bragg grating (VBG) frequency stabilized diode lasers. Photomultiplier tubes, used in earlier decades, have been largely replaced by CCD detectors in modern Raman spectrometers.

11.3.1.2 Fourier Transform Raman Spectrometer

As an alternative to conventional Raman spectrometers, which have an older history, Fourier transform Raman Spectrometers have been an important part of the repertoire for workers in the field of Biological Raman spectroscopy. Motivation for the development of Fourier Transform Raman (FT-Raman) systems, which were first developed in the 1980s, largely came from the need to avoid ambient fluorescence from the sample, which can be avoided by near-IR excitation, using the 1064 nm line of an Nd: YAG laser. Since not too many species have electronic absorption band in the near-IR, and therefore, excitation in this region precludes the possibility of an interfering fluorescence background, which can swamp the intrinsically weak Raman signals. However, owing to the inverse fourth power dependence of the scattering intensity on the incident wavelength (which is true for both Raman and Rayleigh scattering), the Raman signals obtained with near-IR excitation are extremely weak. To overcome this limitation, a Michelson interferometer is used, and the interferograms generated are Fourier transformed to generate the spectra. FT-Raman spectrometers have proved to be especially useful for exploring biological samples where background fluorescence poses a major obstacle in collecting meaningful Raman data.

11.3.2 Advanced Raman Techniques

With advances in laser technology and optics, at least 25 different variants of Raman spectroscopy have been developed. The impetus for such developments stemmed from a number of motivations, especially to enhance the sensitivity (e.g. surface-enhanced Raman), to acquire specific information of selective sites of macromolecules or molecular complexes (Resonance Raman), and to obtain spatial resolution (Raman microscopy) [5–7]. Here, we describe two important techniques which have proved to be particularly useful in biological applications. These are surface-enhanced Raman Spectroscopy (SERS) and coherent anti-Stokes Raman Spectroscopy (CARS). Brief descriptions of these two advanced techniques are presented here:

11.3.2.1 Surface Enhanced Raman Spectroscopy (SERS)

SERS is a surface-sensitive technique involving enhancement of Raman scattering for molecules adsorbed on rough metal surfaces or by nanostructures like plasmonic-magnetic silica nanotubes. The first observation of SERS was serendipitous and was made in 1974 by Fleischman et al. [15] in course of their attempts to distinguish two types of a heterocyclic molecule on the surface of a silver electrode which was used to reduce absorption effects. The increase in Raman signal by SERS has been attributed to two principal mechanisms [6]: (a) Electromagnetic enhancement

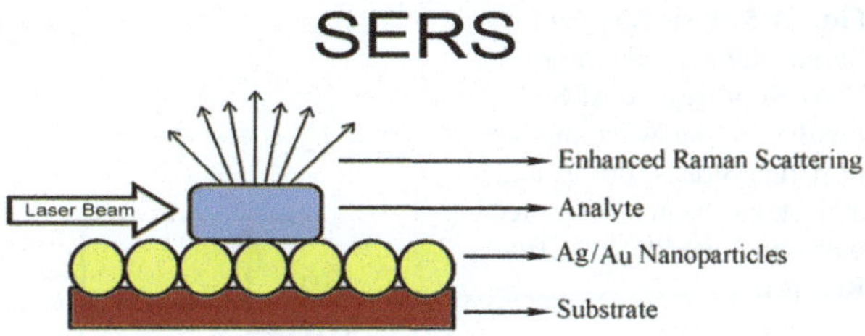

Fig. 11.4 Surface-enhanced Raman scattering [SERS] setup. (Adapted with slight modification from Refs. [7, 14])

where local surface plasmons concentrate the local field proximal to the surface of the metal in 'hot spots' situated on the sharp edges of the nanostructures. (b) Chemical enhancement via charge transfer (CT) between the metal surface and the analyte, which enhances the Raman signal by 10^2–10^4 times. In this context, it is noteworthy that while the CT mechanism applies to specific molecules, the electromagnetic mechanism applies for all analytes. The most common materials used to fabricate SERS are silver and gold, because of their superior plasmonic response. The basic optical set-up for an SERS experiment is presented in Fig. 11.4.

11.3.2.2 Coherent Anti-Stokes Raman Spectroscopy (CARS)

CARS involves a third-order nonlinear four-wave optical mixing process. A pump beam and probe beam of frequencies, $\acute{\omega}_p$ and $\acute{\omega}_{prb}$, respectively, are mixed with a third beam of frequency $\acute{\omega}_s$ (Stokes beam) and are allowed to impinge on the sample. The frequency difference ($\acute{\omega}_p - \acute{\omega}_s$) must match the frequency associated with the Raman-active vibrational mode, $\acute{\omega}$. The frequency of the Stokes beam is normally tuned to fulfill this condition, that is, $\acute{\omega} = \acute{\omega}_p - \acute{\omega}_s$. Next, a probe photon of frequency $\acute{\omega}_{prb}$ acts as a perturbation for the anti-Stokes scattering process to occur at the frequency $\acute{\omega}_{AS} = \acute{\omega}_p - \acute{\omega}_s + \acute{\omega}_{prb}$. The basic optical set-up used for a CARS experiment [6, 7] is depicted in Fig. 11.5.

11.4 Applications to Biomolecules: Overview

Biomolecules are produced by living systems and are associated with various physiologically relevant functions. The four major classes of biomolecules are proteins, nucleic acids, lipids, and carbohydrates. Since Raman spectroscopy is based on light scattering, it is of noninvasive nature and is therefore suitable for exploring biomolecular systems under in vitro, in vivo, and ex vivo situations, leading to its wide applications as an analytical tool in biology. Each molecule has its characteristic Raman spectroscopic signature, which is determined by the three-dimensional structure of the molecule. An analysis of this spectrum can therefore yield a signature of the overall molecular structure as well as on intramolecular and intermolecular interactions. Since lasers were invented in the 1960s, followed by other technological breakthroughs, there has been a continuing outburst on Raman studies on different types of biomolecules. The initial developments were reviewed in comprehensive texts by A.T. Tu and P.R. Carey, published in 1982 [4, 11], and

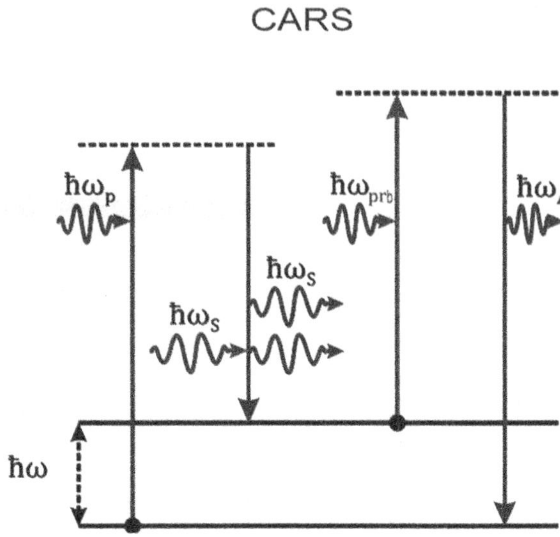

Fig. 11.5 Coherent Anti-Stokes Raman scattering (CARS) process. CARS involves a four-wave mixing of pump, Stokes, probe, and anti-Stokes beams. (Adapted with some modification from Ref. [6])

later additional books and monographs have appeared on specific areas of applications including SERS and confocal microscopy, detailing subsequent progress in biological and biomedical applications [14, 16–18]. In this chapter, we shall present perspectives from an historical context, highlighting the applications of Raman spectroscopy in the conformational analyses of the two most important classes of biomacromolecules, namely, proteins and DNA.

11.4.1 Proteins: Conformational and Related Studies

Proteins, together with amino acids and peptides, were probably the first group of biorelevant molecules to which Raman spectroscopy was applied, beginning with the pioneering efforts by J.T. Edsall's group at Harvard University, USA, which started within a decade of the discovery of the Raman effect in 1928. Even in the pre-laser era, Edsall recorded a Raman spectrum of the protein (an enzyme) lysozyme as far back as 1958, using Hg line excitation. However, the feebleness of the spectrum permitted assignment of only six lines [19]. Following the invention of lasers in the 1960s, laser-excited Raman spectra were published during the 1970s onward by various workers, including Lord et al., Yu et al., Frushour and Koenig, Tu et al., Fasman, Peticolas, and others for a wide range of peptides, polypeptides, and proteins. For reviews of these early ventures, the reader is referred to excellent texts by Tu and by Carey [4, 11]. Systematic analyses of Raman (together with IR) spectra of synthetic polypeptides facilitated spectra-secondary structure correlations in critical detail [4, 11, 20].

What is the protocol followed for the use of Raman spectroscopy to analyze the secondary structure of proteins? This is essentially based on the use of the vibrational

Table 11.1 Raman bands for amide I and amide III in secondary structures of protein [4, 22]

Conformation type	Amide I	Amide III
α-helix	1645–1660	1265–1300[a]
β-sheet	1665–1680	1230–1240
Disordered	1660–1670	1240–1260
[b]β-turns	1665–1677	1265–1286

[a]Weak; may be confused with side chain bands
[b]Type III

modes of the peptide bond –CONH–, which is associated with nine normal modes of vibration, called amide bands [4, 11, 20, 21]. These are designated as amide A and B, and amide I–VII modes, in decreasing order of frequency. The typical frequencies and vibrational assignments of these modes are as follows: amides A and B (3300 and 3100 cm^{-1}, respectively), the doublet arising due to Fermi resonance interaction between NH stretching and the first overtone of amide II), amide I (1600–1690 cm^{-1} mainly due to C=O stretching), amide II (1480–1575 cm^{-1}, C–N stretch, NH in plane bending), amide III (1230–1300 cm^{-1}, C–N stretching, NH in plane bending, involving a different phase relationship compared to amide II), amide IV (625–767 cm^{-1}, OCN bending), amide V (640–800 cm^{-1}, mainly NH out of plane bending), amide VI (540–600 cm^{-1}, mainly C=O out of plane bending), amide VII (200 cm^{-1}, mainly CN torsion). Assignments of the amide bands and spectra-structure correlations were originally established by detailed spectroscopic measurements (via Raman and IR), together with normal mode calculations on the peptide model, *N*-methylacetamide, and various homo-polypeptide models. Pioneers in this field of research were Miyazawa, Elliott and Ambrose, Krimm, and others [20]. Among the amide bands, amide I and amide III are most prominent in the Raman spectra of proteins, and their frequencies are useful for detecting the presence of different secondary structures (alpha helix, beta sheet, beta turns, disordered structure). The approximate frequency (cm^{-1}) ranges for the most intense amide I and amide III Raman bands are listed in Table 11.1.

In addition to the identification of different secondary structures from frequency data, the intensities of these bands (amide I and amide III) have also proved to be useful in quantitative analyses of secondary structure [23], which is reminiscent of protocols developed earlier for applications of far UV CD spectroscopy for estimating secondary structure compositions of proteins via multicomponent analysis.

Apart from the importance of amide bands, which provide information on protein conformation, the side chain Raman frequencies are a rich source of information for the microenvironments of the side chains of proteins. These include the Fermi resonance doublet of tyrosine (850/830 cm^{-1} intensity ratio) reflecting buried vs. exposed Tyr residues, the 1360/1340 cm^{-1} intensity ratio of tryptophan which can be correlated with the environment of the indole group and is useful as a hydrophobicity indicator, that is, for diagnosing buried versus exposed tryptophan residues, and vibrational modes of sulfur containing residues, especially changes in the S–S stretching band around 510 cm^{-1}, which provides details on protein folding, aggregation, and other aspects of relevance to protein structure and function [4, 11, 21].

Use of advanced Raman methodologies has opened up new dimensions in structural studies on proteins. For example, using chiroptical measurements based on Raman optical activity (ROA) motifs and folds in proteins can be explored [24, 25]. High analytical sensitivity can be achieved with surface-enhanced resonance Raman spectroscopy (SERRS). This was demonstrated, for example, for the enzyme glucose oxidase at a silver electrode, where release of free flavin adenine dinucleotide (FAD) was detected from enhanced Raman signals for a solution with extremely low concentrations (10^{-10} M). Interesting studies have been reported on the use of 2D Raman spectroscopy to explore pH-induced changes in protein conformation and dynamics [39]. Baranska and coworkers published an exhaustive vibrational spectra library, incorporating Raman data on as many as 26 proteins of different structure and function, using different laser lines (488, 532, and 1064 nm). The proteins included different structural classes, encompassing alpha-helical, beta-sheet, and mixed structures [21]. Another important direction which has been pursued by S. Krimm and coworkers, and others, concerns detailed vibrational analysis on polypeptides using Raman (along with infrared) spectroscopic measurements combined with theoretical analysis involving normal mode calculations [20]. Studies on conformational polymorphism in calcium and magnesium salts of polyglutamic acid revealed an interesting correlation of the Raman-active CC skeletal sketching band (900–1000 cm^{-1} region) with the Ramachandran φ angle. Also, dependence of the conformationally sensitive amide III frequency on side chain composition was demonstrated [26–29]. Most particularly, based on Raman and Normal mode analysis data, strong support was obtained for an earlier proposal (stemming from far UV CD studies) for the existence of a left-handed three-fold 'extended helical' conformation in polypeptides with charged side chains [29]. UV resonance Raman studies by Asher's group (published in 2001) have been of special interest. Asher's work revealed an interesting dependence of amide III on the dihedral angle ψ, thus providing a unique and sensitive UV resonance Raman structural probe [30].

11.4.2 Nucleic Acids: Conformational and Drug–DNA Interaction Studies

Raman spectroscopic studies on nucleic acids and its constituents were initiated by Lord and Thomas in the late 1960s, soon after the availability of lasers. A great deal of attention has been devoted to the study of the interactions of DNA and RNA with polypeptides, proteins, metal ions, and drugs, for which Raman spectroscopy turned out to be a highly sensitive technique, providing critical insights regarding the binding sites of the ligands in the nucleic acid matrix, as well as conformational changes involved. Since Raman bands provide insights into local structural details, in contrast to electronic CD, where gross structural changes can only be probed (owing to the delocalized nature of electronic transitions), the superiority of Raman spectroscopy as an analytical tool in Structural Biology and Biophysics became immediately apparent.

Raman bands of nucleic acids arise from the in-plane vibrations of the bases (adenine, guanine, thymine, cytosine, and uracil) and from the phosphodiester backbone. The phosphodiester stretching vibration (at ca. 815 cm^{-1}) is sensitive to conformational changes, while the phosphoionic bond stretching vibration (at ca. 1100 cm^{-1}) is relatively insensitive to structural changes and serves as an intensity standard. Raman spectra of DNA or RNA provide valuable insights regarding base stacking and interbase H-bonding interactions. B–Z transitions of synthetic polynucleotides due to drug binding have been an active area of exploration. Resonance Raman spectroscopy (with resonance enhancement of nucleic acid base vibrational modes by using far UV excitation, where the bases selectively absorb and undergo resonance enhancement) provides a useful means of selectively probing the DNA and drug vibrations, especially if the electronic absorption of the drug occurs in the near-UV or visible range. While duplex DNA structures (right-handed B and A forms, and left-handed Z form) were the exclusive focus of attention in earlier studies, much recent attention centers on 4-stranded, noncanonical G-quadruplex DNA structures, in view of their role in downregulating high telomerase activity in cancer cells, as well as their relevance to neurodegenerative and other human diseases. In this context, a detailed Raman study by Palacky et al. [31] on the polymorphism of human telomeric quadruplexes is noteworthy. Very recently, Vorlickova et al. performed a critical assessment to establish the potential of Raman spectroscopy in recognizing different G-quadruplex topologies [32]. For their exhaustive research, they chose as model structures a 22-mer human telomeric fragment $AG_3(TTAG_3)$ and its two variants with adenine a basic lesions at positions 19 and 7, respectively, characterized by three contrasting G4 folds: a basket form, a (3 + 1) hybrid form, and a parallel G4. This study revealed that the reliability of Raman spectroscopy in this context is comparable to that of electronic CD, and moreover, Raman studies have the potential to provide additional useful structural information of significant interest.

Table 11.2 lists Raman marker band frequencies for different DNA structures, including duplex DNA (B, A, and Z forms) as well as quadruplex DNA (for parallel as well as antiparallel quadruplex).

11.4.3 Drug–DNA Interaction Probed via Raman Spectroscopy

Interaction of DNA with therapeutically important small molecules has been a topic of continuing attention in relation to the biophysical applications of Raman spectroscopy. Recent interest in quadruplex DNA as drug design targets for the treatment of a variety of human disorders, and the application of Raman spectroscopy in the quest for drugs targeting quadruplex DNA of different topologies represents a promising field of investigation. Recent years have witnessed considerable interest on plant flavonoids as novel therapeutics having high potency and low cytotoxicity, making them viable alternatives to conventional drugs. Various spectroscopic

Table 11.2 Nonresonance Raman marker bands for DNA

Frequency (cm^{-1})	Description
805–815	Phosphodiester symmetric stretching (A DNA)
835 (weak)	Phosphodiester symmetric stretching (B DNA)
1265 (strong)	Cytosine band: Z-DNA
ca. 499	PO^{2-}, dT, dG, stronger in APQD
ca. 581	dG, stronger in PQD
ca. 611	C2′ endo/syn dT in diagonal or lateral loops in APQD or hairpin
ca. 671	C2′ endo/syn dG, stronger shoulder in APQD
ca. 686	C2′ endo/anti-dG, stronger in PQD
ca. 781	O–P–O, dG, dT, dA, stronger in PQD
ca. 1322	C2′ endo/syn dG in non-QD structures
ca. 1326	C2′ endo/syn dG, stronger in APQD
ca. 1338	C2′ endo/anti-dG, stronger in PQD
ca. 1371	dT, stronger in APQD
ca. 1483	dG-N7 strong Hoogsteen H-bond in PQD or APQD, downshift or hypochromism in PQD
ca. 1581	dG-N2H interbase H-bond in PQD or APQD
ca. 1719	dG-O6 interbase H-bond in PQD or APQD

Data compiled from Refs. [31–33]. *pQD* parallel quadruplex, *APQD* antiparallel quadruplex

methods are useful for exploring the interactions of such compounds with their biorelevant target molecules. With this scenario in mind, we embarked on the exploration of the prospect of the dietary flavonoid fisetin as a quadruplex DNA ligand, with its potential utility as an anticancer drug [34]. Various spectroscopic and other biophysical techniques were used in this work. We used Raman spectroscopy as an important part of the repertoire of our experimental tools. Salient aspects of this research are presented here since it exemplifies the application of Raman spectroscopy in a novel research direction, namely drug discovery involving natural product-based drugs targeting quadruplex DNA, significantly complementing insights from studies via other common spectroscopic techniques such as electronic absorption, fluorescence, and circular dichroism (CD).

Raman spectroscopy (using 632.8 nm excitation) was used to examine the binding of fisetin (3, 4, 3′, 4′-OH flavone, structure shown in Fig. 11.6) with the quadruplex forming sequence d $(T_2AG_4)_4$. The data are presented in Fig. 11.7. The intense Raman bands at 1084, 1478, and 1582 cm^{-1} are consistent with the quadruplex structure, in general, while the antiparallel topology is evidenced by the presence of the shoulders around 684 and 1330 cm^{-1}. Interestingly, the Raman spectrum of fisetin–DNA complex shows distinct changes indicating fisetin–G4 DNA interaction. In particular, the line at 1455 cm^{-1} for the free ligand (due to C–O and C–C stretching of the B ring of fisetin) is totally absent in the presence of DNA, suggesting the absence of free movement of the B ring, which is indicative of the interaction of the B ring with DNA via H-bonding. Most particularly, since the

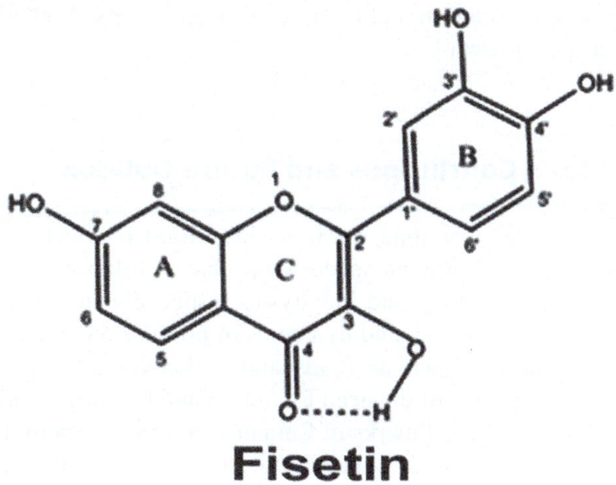

Fig. 11.6 Chemical structure of fisetin, a therapeutically important plant flavonoid

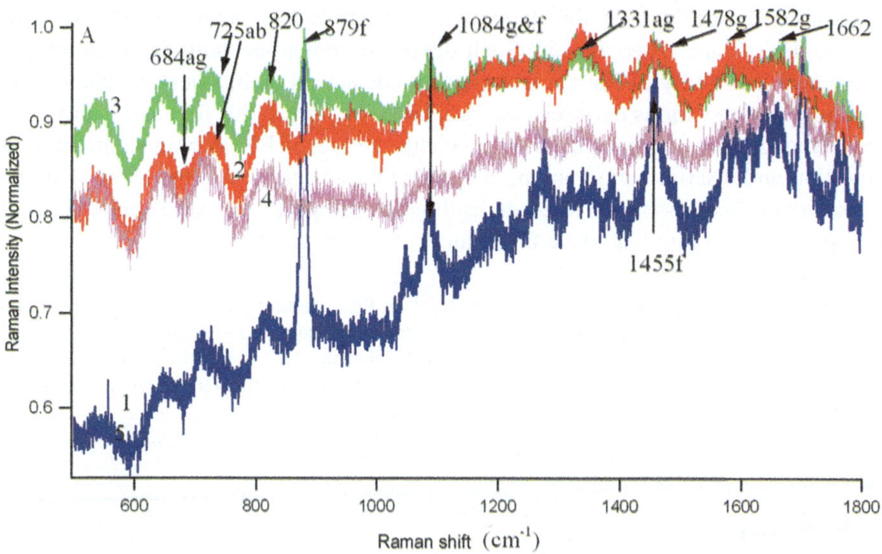

Fig. 11.7 Raman spectral profiles showing studies on quadruplex DNA-fisetin interaction. Blue: 1 mM fisetin; red: 1 mM (T2AG4)4 DNA; green: (T2AG4)4 DNA interacting with 1 mM fisetin; pink: 10 mM Tris buffer (pH 7) with 26 mM NaCl, in which the solutions were prepared. (Figure adapted from Ref. [34] in modified form)

major spectral changes are found to occur for the thymine and adenine marker lines (at 1660, 725, and 825 cm^{-1}), it appears that fisetin binds the DNA at the loop region which is made up of adenine and thymine bases. Thus, this particular study exemplifies the strength of Raman spectroscopy about the drug-binding site in DNA in specific detail.

11.5 Conclusions and Future Outlook

In a little over nine decades since Raman effect was discovered by Professor C.V. Raman, Raman spectroscopy has developed into a powerful research tool in structural biology and biophysics, especially in relation to the elucidation of the structure, function, and dynamics of proteins and nucleic acids and other important biomolecules such as lipids and carbohydrates. The noninvasive nature of the technique, lack of the need for labels, and the intrinsically high chemical specificity, which are the hallmarks of Raman spectroscopy, permit its widespread applications in a broad range of situations in vitro, ex vivo, and in vivo. Raman spectroscopy coupled with microscopy has enabled enormously powerful prospects of live cell imaging and associated studies. Latest developments in Raman spectroscopy and confocal Raman microscopy are opening up vast possibilities in biomedical applications, especially in the diagnosis of cancers, and other diseases such as atherosclerosis and osteoarthritis. Such diagnostic potential stems from the exquisitely high sensitivity of Raman techniques to detect subtle biochemical changes in cells underlying various diseases. The reliability and thresholds for cancer diagnosis are being greatly augmented by refinements in the SERS technique, permitting detection and discrimination among different types of cancers at the level of single cells. Great advances can be foreseen leading to routine and versatile applications of Raman techniques in diagnostic medicine.

Acknowledgments The author is deeply grateful to Prof. Sandip K. Sengupta for insightful discussions and many valuable suggestions. He wishes to express his heartfelt thanks to the Department of Biophysics, Molecular Biology and Bioinformatics for infrastructural support. He also wishes to thank Dr. Snehasish Bhattacharjee and Mr. Niladri Bhattacharyya for their kind assistance.

References

1. Raman CV, Krishnan KS (1928) A new type of secondary radiation. Nature 121:501
2. Raman CV, Krishnan KS (1928) A new class of spectra due to secondary radiation. Part I. Indian J Phys 2:399–419
3. Compton AH (1923) A quantum theory of the scattering of X-rays by light elements. Phys Rev 21:483
4. Tu AT (1982) Raman spectroscopy in biology: principles and applications. Wiley, New York
5. Adar F, Delhaye M, Da Silva E (2007) Evolution of the instrumentation for detection of the Raman effect as driven by available technologies and by developing applications. J Chem Educ 84:50–60
6. Jones RR, Hooper DC, Zhang L, Wolverson D, Valev VK (2019) Raman techniques: fundamentals and frontiers. Nanoscale Res Lett 14(1):231

7. Das RS, Agrawal YK (2011) Raman spectroscopy: recent advancements, techniques and applications. Vib Spectrosc 57:163–176
8. Paudel A, Raijada D, Rantanen J (2015) Raman spectroscopy in pharmaceutical product design. Adv Drug Deliv Rev 89:3–20
9. Smith R, Wright KL, Ashton L (2016) Raman spectroscopy: an evolving technique for live cell studies. Analyst 141:3500–3600
10. Jia M, LiS ZL, Lu X, Zhang H (2018) Analysis of biomolecules based on the surface enhanced Raman spectroscopy. Nano 8(9):730
11. Carey PR (1982) Biochemical applications of Raman and resonance Raman spectroscopies. Academic Press, Boston
12. Smekal A (1923) Zur Quantentheorie der Dispersion. Naturwissenschaften 11:873–875
13. Long DA (2002) The Raman effect. Wiley, New York
14. Rostron P, Gaber S, Gaber D (2016) Raman spectroscopy, review. Int J Eng Tech Res 6:50–64
15. Fleischmann M, Hendra PJ, McQuillan AJ (1974) Raman spectra of pyridine adsorbed at a silver electrode. Chem Phys Lett 26:163–166
16. Spiro TG (1987) Biological applications of Raman spectroscopy. Wiley, New York
17. Peticolas WL (2004) Raman spectroscopy of DNA and proteins. Methods Enzymol 246:189–436
18. Thomas GJ Jr (1999) Raman spectroscopy of proteins and nucleic acid assemblies. Annu Rev Biophys Biomol Struct 28:1027
19. Garfinkel D, Edsall JT (1958) Raman spectra of amino acids and related compounds X. the Raman spectra of certain peptides and of lysozyme. J Am Chem Soc 80(15):3818–3823
20. Krimm S, Bandekar J (1986) Vibrational spectroscopy and conformation of peptides, polypeptides and proteins. Adv Protein Chem 38:181–364
21. Rygula A, Majzner, Marzec KM, Kaczor A, Pilarczyk M, Baranska M (2008) Raman spectroscopy of proteins. J Raman Spectrosc 44:1061–1076
22. Ishizaki H, Balaram P, Nagaraj R, Venkatachalapathi YV, Tu AT (1981) Determination of beta-turn conformation by laser Raman spectroscopy. Biophys J 36:509–517
23. Williams RW (1986) Protein secondary structure analysis using Raman amide I and amide III spectra. Methods Enzymol 130:311–331
24. Zhu F, Isaacs NW, Hecht L, Barron LD (2005) Raman optical activity: a tool for protein structure analysis. Structure 13:1409–1419
25. Blanch EW, Hecht L, Barron LD (2006) Raman optical activity. In: Busch KW, Busch MA (eds) Chiral analysis. Elsevier, Amsterdam, pp 545–594
26. Sengupta PK, Krimm S (1985) Vibrational analysis of peptides, polypeptides and proteins. XXXII. α-poly (L-glutamic acid). Biopolymers 24:1479–1491
27. Sengupta PK, Krimm S (1984) Vibrational analysis of peptides, polypeptides and proteins. XXI. β-calcium-poly (L-glutamate). Biopolymers 23:1565–1594
28. Sengupta PK, Krimm S (1987) Raman and normal-mode studies of the extended-helix conformation in polypeptide chains. Biopolymers 26(S0):S99–S107
29. Sengupta PK, Krimm S (1985) Raman and normal mode studies of the polypeptide chain conformations in crystalline magnesium and calcium poly(glutamate)s. Spectrochim Acta Part A Mol Spectrosc 41(1–2):205–207
30. Asher SA, Ianoul A, Mix G, Boyden MN, Karnoup A, Diem M, Schweitzer-Stenner R (2001) Dihedral psi angle dependence of the amide III vibration: a uniquely sensitive UV resonance Raman structural probe. J Am Chem Soc 123:11775–11781

31. Palacky J, Vorlickova M, Kejnovska I, Mojzes P (2013) Polymerphism of human telomere quadruplex structure controlled by DNA concentration: a Raman study. Nucleic Acid Res 41: 1005–1016
32. Palacky J, Mojzes P, Kejnovska I, Vorlickova M (2020) Does Raman spectroscopy recognize different G-quadruplex arrangements. J Raman Spectrosc 51:301–312
33. Hildebrandt P, Lecomte S (1999) Encyclopedia of spectroscopy and spectrometry, 2nd edn. Elsevier, Amsterdam
34. Sengupta B, Pahari B, Blackmon L, Sengupta PK (2013) Prospect of bioflavonoid Fisetin as a quadruplex DNA ligand. PLoS One 8(6):65383

The manufacturer's authorised representative in the EU is Springer Nature Customer Service Centre GmbH, Europaplatz 3, 69115 Heidelberg, Germany. If you have any concerns regarding our products, please contact ProductSafety@springernature.com

Printed and bound by CPI Group (UK) Ltd, Croydon, CR0 4YY
30/09/2026
02225239-0006